Applications of
Chinese
Herbal
Compatibility

Project Editor: **Li Rui**

Book Designer: **Guo Miao**

Cover Designer: **Guo Miao**

Typesetter: **Wei Hong-bo**

Applications of
Chinese
Herbal
Compatibility

He Xiu-chuan

Chief Physician, Cangzhou City Hospital of TCM

Translated by **Ijuen Frieda Mah**, L.Ac.

Edited by **Lara Deasy**, B.Sc.TCM, M.B.TCM

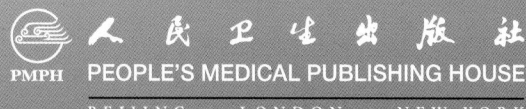
PMPH **PEOPLE'S MEDICAL PUBLISHING HOUSE**

BEIJING · LONDON · NEW YORK

PMPH PEOPLE'S MEDICAL PUBLISHING HOUSE

Website: http://www.pmph.com

Book Title: Applications of Chinese Herbal Compatibility
中药配伍应用心得

Contact address: Bldg 3, 3 Qu, Fangqunyuan, Fangzhuang, Beijing 100078, P.R. China, phone/fax: 8610 6769 1034, E-mail: pmph@pmph.com

For text and trade sales, as well as review copy enquiries, please contact PMPH at pmphsales@gmail.com

First published: 2008
ISBN: 978-7-117-09208-1/R · 9209

Cataloguing in Publication Data:
A catalogue record for this book is available from the CIP-Database China.

Printed in The People's Republic of China

ISBN 978-7-117-09208-1

9 787117 092081 >

About the Author

He Xiu-chuan, was born in Hejian city, Hebei province in 1939. He graduated from Tianjin University of Chinese Medicine in 1962. He is the instructor for the second class of the Reknowned Doctors' Clinical Experiences Course. He was the vice-president of medical services of Cangzhou City Hospital of Chinese Medicine, the chairman of the board at Cangzhou Chinese Medicine Academic Society, a board member of the Hebei Province Chinese Medicine Academic Society, a senior member of the Hebei Province Chinese Medicine Professional Appraisal Committee, a member of the Hebei Chinese Medicine Magazine Editorial Committee, and the seventh and eighth term Hebei province representative for the National People's Congress. He has practiced Chinese Medicine for more than forty years and specializes in internal medicine, gynaecology, paediatrics, and complex disorders. He has published more than forty papers.

Foreword

Chinese herbal compatibility is an important part of Chinese medicine. It is a key part to all formulae. Herbal compatibility combines medicinals to induce a new complex transformation. It changes the primary herb's function and expands its main treatment scope, hence, it can be said that Chinese doctors treat disease through correct herbal combination. For this reason, doctors of Chinese medicine throughout the generations have paid a great deal of attention to herbal compatibility. In addition, throughout their long clinical experience, they have left us with their precious records. This inheritance, study, and continuous clinical practice are important for spreading Chinese medicine and improving Chinese medicine's clinical effectiveness.

He Xiu-chuan, Director of Chinese medicine, graduated from Tianjin University of Traditional Chinese Medicine in 1962. He was the instructor for the second class of Reknowned Doctors' Clinical Experiences Course. He is a diligent and rigorous scholar and instead of remaining limited to one school of thought, he can extract and use the best parts from many schools. He pursues accurate diagnosis, refined medicinal compatibility, and innovations in clinical practice. Based on the inheritance of traditional Chinese medicine and through decades of clinical practice and meticulous research, he has complied the "Applications of Chinese Herbal Compatibility". There are more than 120 compatible herb groups. For each group, he has analyzed its function, application, compatibility, and provided an example from a real case. It is a beneficial reference for any doctor's practice. At the moment of this book's publication, I have written this to express my congratulations.

Yu Sheng-long
former Deputy Director General of
State Administration of TCM
Beijing, China
January 2007

Preface

Combining two or more herbs in clinical practice is called compatibility. The majority of Chinese doctors use herbal compatibility to make a formula and treat disease. The purpose of using compatibility is to improve the herbs treatment effectiveness, prevent side effects of herbal interactions, and give comprehensive treatment to complicated cases. Compatibility is the essence of Chinese medicine and the key to a Chinese formula.

Our ancestors used herbs to treat illness. They first started by using a single herb. Later, they realised that a single herb could not completely relieve a patient's sickness. Then, they gradually tried to use two or more herbs to treat disease. In doing this they found their treatments were more effective than by just using a single herb. This was the start of herbal compatibility. Through long term practice they concluded that different herbs in combination can generate complicated changes, expand the scope of application, and enhance effectiveness, therefore, it is evident that it is a big leap from the application of single herbs to herbal combinations and compatibility. More than two thousand years ago, the classic book of Chinese herbs *The Divine Husbandman's Classic of the Materia Medica (Shén Nóng Běn Cǎo Jīng, 神农本草经)* named seven kinds of herbal combinations: mutual accentuation, mutual enhancement, mutual counteraction, mutual antagonism, mutual suppression, mutual incompatibility, and the single effect. This is also called the "seven affects" of herbal application. Aside from the single effect, they are all rules for herbal combination. As society developed and people increased their medical knowledge of diseases, doctors found that simply combining herbs did not fit clinical needs. This led to combining many herbal combinations into complex formulae. Formulae, also called "*tāng tóu, 汤头*", are a more complicated form of herbal combination. They consist of a chief, deputy, assistant, and envoy. The development of formulae was the consummate step in Chinese medical science. Using herbal combination in Chinese medicine's development process was a significant jump from using single herbs, to joining them together to treat disease. Study of herbal combination and a grasp of medicinal compatibility will raise clinical effectiveness, expand the scope of herbal medicine, and lead to the composition of new and effective formulae that have vital significance.

In 1991, I followed the famous Chinese medicine doctor Fu Xiang-fang, professor of clinical practice at Tianjin University of Traditional Chinese Medicine. Professor Fu also learned from the current medical expert Mr. Zhang Xi-chuan, who is an expert in herbal compatibility, his exquisite combinations could improve severe illnesses in no time. These studies were of great benefit to me and influenced my interest in herbal compatibility. After years of having paid attention to the famous experts and through clinical verification, I summarized my conclusions on herbal compatibility. Meanwhile, I engaged in continuous research to find out the most effective herbal compatibilities. Now I have compiled many years of clinical experience into a book. There are a total of 120 herbal combinations. For each combination, there is an analysis of function, application, compatibility, and a case history. Detailed explanations of each herbal combination give readers a convenient reference that will be an aid in clinic practice.

It should be noted that the dosages used in this book relate to those used in China, and therefore dosages administered in practice should be relevant to the individual patient, and their condition.

Despite our best efforts, it is recognised that human error may be reflected in the text. We welcome all suggestions from our readers to improve future editions of this book.

I would like to acknowledge Mr. Yu Sheng-long's (from the State Administration of Traditional Chinese Medicine) concern and support, in the compilation of this book, and for contributing the preface.

He Xiu-chuan
Chief Physician
Cangzhou City Hospital of TCM
Hebei Province, China
February 2008

Contents

Rén shēn's (人参 , Radix et Rhizoma Ginseng) nature is sweet and bland, it enters the spleen, lung, and heart meridians. It is able to strongly tonify the original qi, boost qi, and calm the spirit. The *Divine Husbandman's Classic of the Materia Medica* (神农本草经 , *Shén Nóng Běn Cǎo Jīng*) (hereafter abbreviated as *Běn Jīng* -- 本经) classifies it as a first class herb. It states *"the taste is sweet, it is slightly cold, it mainly tonifies the five viscera, calms the spirit, anchors the ethereal and corporeal soul, stops fright palpitations, expels pathogenic qi, brightens the eyes, opens the heart, and benefits intelligence"*. *Rén shēn* (Radix et Rhizoma Ginseng), since the *Běn Jīng*, has often been confused with *dǎng shēn* (党参 , Radix Codonopsis) as being the same product up until the *Qing* Dynasty physician Wu Yi-luo's *Thoroughly Revised Materia Medica* (本草从新 , *Běn Cǎo Cóng Xīn*) started to differentiate them. Hence, in classical formulae, most *rén shēn* (Radix et Rhizoma Ginseng) are *dǎng shēn* (Radix Codonopsis). In clinical practice, the tonifying effect of *rén shēn* (Radix et Rhizoma Ginseng) is stronger than that of *dǎng shēn* (Radix Codonopsis). It is used more often for severe patients. *Dǎng shēn* (Radix Codonopsis) is weaker and is used for patients with spleen and/or lung qi deficiency.

1. *Rén Shēn* (Radix et Rhizoma Ginseng) ——
Gān Jiāng (干姜 , Rhizoma Zingiberis)

Function

To supplement qi and assist yang.

Application

This is often used for heart and spleen yang deficiency syndromes such as heart yang deficiency causing chest pain, or spleen yang deficiency causing middle burner deficiency cold with abdominal pain and diarrhoea. Modern clinical applications include coronary heart disease with heart yang weakness and cold stagnation blocking the vessels, and middle burner deficiency cold causing chronic enteritis, gastritis, reduced digestive function, or gastric or duodenal ulcer.

Compatibility Analysis

Rén shēn (Radix et Rhizoma Ginseng) strongly tonifies the original qi. *Gān jiāng* (干 姜 , Rhizoma Zingiberis) is acrid and very hot, and can warm yang and scatter cold. Of these two herbs, one warms and the other tonifies, together they can arouse the heart and spleen yang to dispel stagnated cold pathogens. You Zai-jing said: *"tonifying deficient yang can dispel yin"*. This two-herb combination can promote heart yang, as in *Rén Shēn Tāng* (Ginseng Decoction, 人参汤), which consists of *rén shēn* (Radix et Rhizoma Ginseng*), gān jiāng* (Rhizoma Zingiberis*), bái zhú* (白术 , Rhizoma Atractylodis Macrocephalae), and *gān cǎo* (甘草 , Radix Glycyrrhizae) from the *Essentials From the Golden Cabinet* (金匮要略 , *Jīn Guì Yào Lüè*), to treat heart yang deficiency with yin cold stagnation causing chest pain. In addition, they can enhance spleen yang to treat middle *jiao* deficiency cold causing vomiting, diarrhoea and abdominal pain, as in *Lǐ Zhōng Wán* (Centre Rectifying Pill, 理中丸), which consists of *rén shēn* (Radix et Rhizoma Ginseng*), gān jiāng* (Rhizoma Zingiberis), *bái zhú* (Rhizoma Atractylodis Macrocephalae), and *gān cǎo* (Radix Glycyrrhizae). This formula has the same ingredients as *Rén Shēn Tāng* (Ginseng Decoction) but is in pill form rather than as a decoction, which allows it to treat the heart yang first and the middle *jiao*'s spleen yang later.

Case Study

Initial Visit

Male patient, 64 years old, November 9th, 1992.

The patient was physically weak and suffered from panting. Each time the patient encountered cold he experienced cough and panting. Recently when the patient was exposed to cold, he also felt oppression of the chest, with copious phlegm and distressing pain in the area in front of the heart. The EKG showed a low and flat V5T wave. His pulse was wiry and slippery and showed a white but slightly greasy tongue coating. This was weak chest yang, cold phlegm stagnating, and a stagnated qi mechanism. The treatment plan was to enhance heart yang, disperse cold, and transform phlegm.

Prescription

白参(先煎)	*bái shēn*	8g	Radix et Rhizoma Ginseng (decocted first)
干姜	*gān jiāng*	9g	Rhizoma Zingiberis
丹参	*dān shēn*	24g	Radix et Rhizoma Salviae Miltiorrhizae
葛根	*gé gēn*	20g	Radix Puerariae Lobatae
川芎	*chuān xiōng*	12g	Rhizoma Chuanxiong
檀香	*tán xiāng*	10g	Lignum Santali Albi
茯苓	*fú líng*	20g	Poria
半夏	*bàn xià*	9g	Rhizoma Pinelliae
陈皮	*chén pí*	9g	Pericarpium Citri Reticulatae
甘草	*gān cǎo*	6g	Radix Glycyrrhizae

This was decocted in water, divided into two equal doses, one pack per day.

Second Visit

November 12th: The patient took four packs of the prescription and the symptoms of cough, panting, chest oppression, and obstructed breathing were reduced. The pain in the area in front of the heart also disappeared. To the above prescription, *chuān bèi* (川贝 , Bulbus Fritillariae Cirrhosae) 10g, and *xìng rén* (杏仁 , Semen Armeniacae Dulcis) 10g, were added, and the patient continued to take the prescription.

Third Visit

November 17th: The patient had taken another four packs. There was no more coughing or panting, the chest oppression had also disappeared, and the EKG was normal. The patient was advised to take three more packs of the prescription to consolidate the effect.

2. *Rén Shēn* (Radix et Rhizoma Ginseng) ——
Fù Zǐ (附子, Radix Aconiti Lateralis Praeparata)

Function

To boost qi, secure yang to restrain collapse.

Application

This combination treats qi and blood sudden desertion, rapid panting, spontaneous sweating, and coldness of the hands and feet. In the modern clinic it is used for severe syndromes of heart failure presenting with heart and kidney yang deficiency, desertion of yang and qi, and a minute pulse that is almost expired. Clinically, it can also be used for kidney yang deficiency and weakness, or cold.

Compatibility Analysis

Rén shēn (Radix et Rhizoma Ginseng) strongly tonifies the original qi. The original qi is our body's most primary qi. If original qi is deficient and deserts, life may be in danger. *Rén shēn* (Radix et Rhizoma Ginseng) can tonify the basal qi and restrain collapse. It is a noble herb among the qi tonics. *Fù zǐ* (附子, Radix Aconiti Lateralis Praeparata) is acrid and hot, in the upper body it can help the heart yang to unblock the vessels, in the lower body, it can supplement kidney yang and boost fire. It is an important herb for returning yang and restraining collapse. When the two herbs are combined together, *rén shēn* (Radix et Rhizoma Ginseng) helps *fù zǐ* (Radix Aconiti Lateralis Praeparata) to return yang and boost qi. *Fù zǐ* (Radix Aconiti Lateralis Praeparata) helps *rén shēn* (Radix et Rhizoma Ginseng) to tonify qi and secure yang. Together, they boost qi and secure yang, to restrain collapse. This simple combination is very effective as both herbs benefit each other. It is truly a great combination to restrain collapse. *Rén Shēn Fù Zǐ Tāng* (Ginseng and Aconite Decoction, 人参附子汤) in the *Effective Formulae from Generations of Physicians* (世医得效方, *Shì Yī Dé Xiào Fāng*) and Zhang Zhong-jing's *Sì Nì Jiā Rén Shēn Tāng* (Counterflow Cold Decoction Plus Ginseng, 四逆加人参汤) are all based on this combination.

Case Study

Initial Visit

Female patient, 45 years old, April 10th, 1973.

The patient was usually in good health. However, three months ago she became pregnant. She felt she was too old to have a baby, so she aborted it herself, at home. It is not known what method she used to abort the foetus, but it caused severe bleeding. Initially, she did not want to tell anyone about the bleeding, but when it did not stop and symptoms of heart palpitations and severe sweating occurred, she let a family member invite me to visit her. Upon examination, the patient was found to have cold limbs, sweating, and a thin minute pulse on the verge of expiring. Her blood pressure was 50/20mm Hg. The pattern was identified as sudden desertion of qi and blood with severe damage to the original qi. In order to save her it was necessary to boost qi and return yang, immediately, in order to stem desertion.

Prescription

| 高丽参 | *gāo lì shēn* | 30g | Radix Ginseng Coreensis |
| 熟附子 | *shú fù zǐ* | 10g | Radix Aconiti Lateralis Praeparata |

This was decocted in water, divided into three equal doses, and one dose was taken every two hours.

After one pack of the medicine, her limbs warmed, the sweating stopped, and the pulse changed to light, but with some force. Her blood pressure rose to 75/40mm Hg. Treatment methods were then changed and she was prescribed a formula to nourish blood and stop bleeding, tonify qi and strengthen the spleen to complete her recovery.

3. *Rén Shēn* (Radix et Rhizoma Ginseng) ── *Mài Mén Dōng* (麦门冬, Radix Ophiopogonis)

Function

To tonify the lung and boost the qi, nourish yin and downbear counterflow.

Application

This is mainly used for lung and stomach qi and yin deficiency causing cough, panting due to qi counterflow, and weakness of the body and limbs.

Compatibility Analysis

Rén shēn (Radix et Rhizoma Ginseng) or *dǎng shēn* (Radix Codonopsis) supplements the middle and boosts the qi, harmonises the spleen and stomach, nourishes yin and generates fluid. It is an important herb to treat qi and blood deficiency. *Mài mén dōng* (麦门冬, Radix Ophiopogonis) enters the lung and stomach channels. It can nourish yin and moisten the lung, boost the stomach and generate fluids. These herbs combined together are best for treating lung and stomach deficiency of both qi and yin. The stomach is the root of the acquired constitution. Stomach channel qi and yin deficiency can cause the lung qi to be insufficient. If the lung qi is deficient there will be shortness of breath, cough and panting, and tiredness of the body and limbs. *Shēn* combined with *mài* cultivates earth in order to engender metal. This combination is called *Shēn Mài Yǐn* (Ginseng and Ophiopogonis Beverage, 参麦饮). It was originally recorded in the book of *Symptom, Cause, Pulse, and Treament* (症因脉治, *Zhèng Yīn Mài Zhì*). If one adds *wǔ wèi zǐ* (五味子, Fructus Schisandrae Chinensis) the formula is called *Shēng Mài Yǐn* (Pulse Engendering Beverage, 生脉饮). It is a formula from the *Discussion on the Distinguishing of External and Internal Damage* (内外伤辨感论, *Nèi Wài Shāng Biàn Gǎn Lùn*) that mainly treats "*heat injuring the original qi causing tiredness of the limbs and body, shortness of breath, thirst, and incessant sweating, also known as metal restraining fire. When water has lost its support, this will cause cough, hasty panting, shortness of breath, and weakness of the body and limbs*". Recent research has shown that this formula is able to strengthen the heart and raise blood pressure. It can be used to treat cardiogenic shock. If one adds *bàn xià* (半夏, Rhizoma Pinelliae), *jīng mǐ* (粳米, Oryza sativa L.), and *gān căo* (Radix Glycyrrhizae) this forms *Mài Mén Dōng Tāng* (Ophiopogonis Decoction, 麦门冬汤), a formula from the *Essentials From the Golden Cabinet,* and is indicated for lung wilting, where lung heat has caused cough with spitting of saliva, shortness of breath with panting, dry throat and mouth, and emaciation. Modern clinical applications include dry cough and chronic nasopharyngitis. The author has used this formula to treat PMS quite effectively.

Case Study

Initial Visit

Male patient, 46 years old. May 12th, 1989.

The patient caught a cold 10 days ago. At the time he showed symptoms of fever, aversion to cold, and a stuffy and runny nose. The patient self medicated with two capsules of a common cold remedy but they had no effect. Relying on his strong constitution, he took another 3 capsules that still had no effect. He took a further 3 capsules, and then another 8 capsules. After this he experienced a great sweat that would not stop. However, the sweating did not reduce the fever. The body temperature had reached above 38.5℃. He visited a doctor, who treated him with *Xiăo Chái Hú Tāng* (Minor Bupleurum Decoction, 小柴胡汤), but the fever did not abate. By then it had already been four or five days, and he invited me to visit him. Upon examination, he presented with a hot body, sweating, dry mouth, cough, shortness of breath, tiredness and weakness of the four limbs, a floating, thin, fast pulse without strength, and a thin white tongue coating. This was a case where an external disorder was mistakenly treated with sweating and had injured qi and yin. Excessive sweating had injured yin, and yin had consumed qi, and gradually the disease had become more and more severe. The damaged qi could not control the superficial pores, and thus the sweat did not stop. Injured qi also resulted in shortness of breath and tiredness, and damaged yin caused dry mouth. With both qi and yin injured, the lung lost its moistness and nourishment and resulted in cough. The treatment principle was to nourish yin and boost qi, clear heat and downbear counterflow.

Prescription

麦冬	*mài dōng*	60g	Radix Ophiopogonis
党参	*dăng shēn*	20g	Radix Codonopsis
山药	*shān yào*	30g	Rhizoma Dioscoreae
半夏	*bàn xià*	9g	Rhizoma Pinelliae
生石膏	*shēng shí gāo*	20g	Gypsum Fibrosum
知母	*zhī mŭ*	15g	Rhizoma Anemarrhenae
甘草	*gān căo*	6g	Radix Glycyrrhizae

This was decocted in water and divided into three equal doses. One pack per day.

Second Visit

May 14th: After taking one pack of the above prescription, the body heat and sweating were both reduced. After two packs the sweating stopped and the panting was calmed. Only the body weakness and tiredness were still not relieved. The original prescription was modified accordingly.

Prescription

麦冬	*mài dōng*	30g	Radix Ophiopogonis
党参	*dǎng shēn*	15g	Radix Codonopsis
山药	*shān yào*	15g	Rhizoma Dioscoreae
半夏	*bàn xià*	15g	Rhizoma Pinelliae
仙鹤草	*xiān hè cǎo*	30g	Herba Agrimoniae
知母	*zhī mǔ*	9g	Rhizoma Anemarrhenae
甘草	*gān cǎo*	10g	Radix Glycyrrhizae
大枣	*dà zǎo*	10pcs	Fructus Jujubae

This was decocted in water and divided into two equal doses. One pack per day.

Third Visit

May 17th: The patient had basically recovered. He was advised to rest and to take two more packs of the prescription to strengthen the result.

4. *Rén Shēn* (Radix et Rhizoma Ginseng) ——
Dài Zhě Shí (代赭石, Haematitum)

Function

To supplement the qi and downbear counterflow, harmonise the stomach and stop vomiting.

Application

This is often used for stomach deficiency causing nausea and vomiting, kidney deficiency that is not astringing causing qi to counterflow upwards along the penetrating vessel causing stomach fullness, or yin and yang deficiency causing hasty panting. The author has applied it in the clinic in cases of stomach and intestinal dysfunction causing nausea and vomiting, diaphragm spasm, and pre-menstrual vomiting or epistaxis in women.

Compatibility Analysis

Rén shēn (Radix et Rhizoma Ginseng) (*dǎng shēn* - Radix Codonopsis) supplements the centre and boosts the qi, and harmonises the spleen and stomach. *Dài zhě shí* (代 赭石, Haematitum) is red in colour and cold in nature, enters the spleen, stomach, and pericardium channels. It is heavy and good at downbearing counterflow qi. The stomach is located in the middle *jiao*, its nature is to descend. If the stomach is deficient and cannot descend properly, the stomach qi will counterflow upwards and cause nausea, vomiting, and incessant hiccough. *Shēn* (Radix et Rhizoma Ginseng) combined with *zhě* (Haematitum) can supplement the middle qi, and descend stomach qi. It is useful in supplementing deficiency, descending counterflow, and harmonising the stomach. In addition, the penetrating vessel belongs to the *yangming*. If stomach qi counterflows upwards, this will cause the penetrating vessel to do likewise. Hence, *shēn* (Radix et Rhizoma Ginseng) and *zhě* (Radix et Rhizoma Ginseng) are also useful in treating counterflowing penetrating vessel qi.

The earliest record for the combination of *shēn* (Radix et Rhizoma Ginseng) and *zhě* (Haematitum) is in *Xuán Fù Huā Dài Zhě Shí Tāng* (Inula and Haematite Decoction, 旋覆花代赭石汤), which includes *rén shēn* (Radix et Rhizoma Ginseng), *dài zhě shí* (Haematitum), *xuán fù huā* (旋覆花 , Flos Inulae), *bàn xià* (Rhizoma Pinelliae), *gān cǎo* (Radix Glycyrrhizae), *shēng jiāng* (Rhizoma Zingiberis Recens), *dà zǎo* (大枣 , Fructus Jujubae), and appeared in Zhang Zhong-jing's *Discussion on Cold Damage* (伤寒论 , *Shāng Hán Lùn*). It is indicated for deficient stomach channel qi counterflowing upward causing "glomus and fullness below the heart, and incessant belching". The famous doctor of recent times Dr. Zhang Xi-chun especially likes the *shēn* (Radix et Rhizoma Ginseng) and *zhě* (Haematitum) combination. For example, his *Shēn Zhě Péi Qì Tāng* (Ginseng and Haematite Bank Up Qi Decoction, 参赭培气汤) treats stomach disharmony, and the *Shēn Zhě Zhèn Qì Tāng* (Ginseng and Haematite Suppress Qi Decoction, 参赭镇气汤) treats upward counterflow of qi. Both formulae have good clinical results.

Case Study

Initial Visit

A female patient, 41 years old, May 22 nd, 1985.

The patient had moved location, with her husband, as he had changed from a military to a civilian career. She had arrived in Cangzhou from Ningxia twenty days ago. Due to poor planning they were temporarily living in guest housing. She was exhausted from moving and travelling, and nervous due to the new environment and this has caused stomach discomfort, nausea and poor appetite. Her condition gradually worsened and she presented at the clinic with fullness and distention in the chest and abdomen, discomfort from qi counterflow, frequent nausea and vomiting, and inability to eat food. Her pulse was weak and wiry. The tongue coating was thin and white. The pattern was diagnosed as stomach deficiency and qi counterflow, and loss of regulation of proper descending. The treatment principle was to harmonise the stomach and descend counterflow.

Prescription

代赭石	*dài zhě shí*	30g	Haematitum
党参	*dǎng shēn*	20g	Radix Codonopsis
半夏	*bàn xià*	15g	Rhizoma Pinelliae
苏梗	*sū gěng*	10g	Caulis Perillae
藿香	*huò xiāng*	10g	Herba Agastachis
麦冬	*mài dōng*	30g	Radix Ophiopogonis
甘草	*gān cǎo*	10g	Radix Glycyrrhizae

This was decocted twice, in water. The resulting liquid was combined and taken in many frequent doses. If vomiting occurred after taking the decoction, water was used to rinse the mouth, and then it was taken again. One pack per day.

Second Visit

May 25th: After taking one pack of the above formula the nausea and vomiting stopped. After three packs she was back to her normal diet and all the symptoms had disappeared.

Dà zăo's (大枣 , Fructus Jujubae) taste and property are sweet and neutral. It enters the spleen channel of the foot *taiyin* and the stomach channel of the foot *yangming* meridians. It can harmonise the stomach and strengthen the spleen, tonify qi and generate fluid, and harmonise nutritive and defensive qi. The *Divine Husbandman's Classic of the Materia Medica* stated: *"It can eliminate pathogenic qi in the heart and abdomen; and it can stabilise the middle jiao and nourish the spleen qi. It can harmonise the stomach qi, open the nine orifices, benefit the twelve meridians, tonify deficient qi and fluid, supplement poor constitution, can act as a tranquilliser, alleviate heaviness of the four limbs, and it can harmonise a hundred types of herbs."* Although *dà zăo* (Fructus Jujubae) is a food, it should not be neglected. If it is combined correctly, it can get very good results.

1. *Dà Zăo* (Fructus Jujubae) ——
Xiān Hè Căo (仙鹤草, Herba Agrimoniae)

Function

To tonify qi and tonify deficiency.

Application

This is used for shortness of breath and lack of strength, whole body lassitude, or both spirit and body fatigue during remission of severe diseases.

Compatibility Analysis

This combination is a formula from experience, originating in the south of China. It can treat loss of strength and injury by over-exertion. *Xiān hè căo* (Herba Agrimoniae) is also called *lóng yá căo* (龙芽草 , Herba Agrimoniae), but is usually called *tuō lì căo* (脱力草 , Herba Agrimoniae) and is said to have a tonifying function. This function does not correlate with the records in the herbal books. However, in practice, it really has the function to relieve fatigue and treat injuries due to over-exertion. Modern research has used the alcohol solute of *xiān hè căo* (Herba Agrimoniae) to test on animals. It has the function to contract internal organs' blood vessels, raise blood pressure, promote breathing, increase blood clotting, and to energise lassitude of skeletal muscle. Modern reports have indicated that *dà zăo* (Fructus Jujubae) has the function to increase laboratory animals' muscle strength and body weight. The two herbs combined together can tonify qi and deficiency, and relieve fatigue. The author has used them effectively to treat chronic fatigue syndrome.

Case 1

Initial Visit

A female patient, 39 years old, August 15th, 1989.

The patient had shortness of breath, tiredness, weakness, a stifling chest sensation, anorexia, a poor appetite, and palpitations after exertion. It persisted for a couple of days, without recovery. Her pulse was deep, thready and forceless. Her tongue was pale with a thin white coating. The diagnosis was middle *jiao* qi deficiency with spleen qi not transporting nutrition to the four limbs and the whole body.

Prescription

仙鹤草	*xiān hè cǎo*	24g	Herba Agrimoniae
生黄芪	*shēng huáng qí*	24g	Radix Astragali (raw)
百合	*bǎi hé*	20g	Bulbus Lilii
檀香	*tán xiāng*	6g	Lignum Santali Albi
砂仁	*shā rén*	12g	Fructus Amomi
甘草	*gān cǎo*	6g	Radix Glycyrrhizae
大枣	*dà zǎo*	10 pcs	Fructus Jujubae

This was immersed in clean water for half an hour, then decocted three times and divided in two, to be taken twice. One pack per day.

Second Visit

August 22nd: After she took six packs of the above formula, all symptoms had gone. She was told to take three to five packs monthly for three months to strengthen the results. Now, after 10 years, it has never reoccurred again.

Case 2

Initial Visit

A male patient, 58 years old, March 23rd, 1995.

Two months previously, the patient caught the flu with a fever. After a local doctor's treatment, he basically recovered except for body lassitude and lower limb heaviness, the lassitude was aggravated especially after activities. After many different examinations and treatments, he did not get a satisfactory diagnosis. All routine laboratory tests showed no abnormality. His pulse was wiry and thready. His tongue coating was thin and white. His throat was slightly red. The syndrome differentiation indicated that after external pathogens had invaded, they injured both qi and yin, and also the pathogens had not been totally cleared. The treatment principle was to tonify qi and nourish yin, plus clear out the residual pathogens.

Prescription

仙鹤草	*xiān hè cǎo*	30g	Herba Agrimoniae
生黄芪	*shēng huáng qí*	24g	Radix Astragali (raw)
太子参	*tài zǐ shēn*	15g	Radix Pseudostellariae
百合	*bǎi hé*	20g	Bulbus Lilii
茯苓	*fú líng*	15g	Poria
桔梗	*jié gěng*	9g	Radix Platycodonis
柴胡	*chái hú*	12g	Radix Bupleuri
甘草	*gān cǎo*	6g	Radix Glycyrrhizae
大枣	*dà zǎo*	10 pcs	Fructus Jujubae

This was decocted with water, divided in two, to be taken once in the morning and once at night, one pack each day.

Second Visit

April 5ᵗʰ: After he took seven packs of the above formula, his lassitude was significantly reduced. After activities, he did not feel any lassitude. However, due to personal reasons, he could not come for a visit. After he stopped the prescription, he felt the lower limbs' lassitude so he visited again. His tongue and pulse were slightly better than before, so, following the original prescription *chuān niú xī* (川牛膝 , Radix Cyathulae) 15g was added. He took another seven packs and recovered.

Note

In recent years, the author used this compatibility to prescribe *Xiān Qí Dà Zǎo Tāng* (Hairyvein Agrimonia, Astragalus and Jujube Decoction, 仙芪大枣汤) to treat chronic fatigue syndrome (CFS) with good results. The ingredients are listed below for reference.

Ingredients

仙鹤草	*xiān hè cǎo*	30g	Herba Agrimoniae
生黄芪	*shēng huáng qí*	24g	Radix Astragali (raw)
太子参	*tài zǐ shēn*	15~20g	Radix Pseudostellariae
百合	*bǎi hé*	20g	Bulbus Lilii
砂仁	*shā rén*	6g	Fructus Amomi
柴胡	*chái hú*	12g	Radix Bupleuri
甘草	*gān cǎo*	6g	Radix Glycyrrhizae
大枣	*dà zǎo*	10 pcs	Fructus Jujubae

This was decocted with water, divided in two to drink twice, one pack per day.

Modification

1. With low grade fever, add *dì gǔ pí* (地骨皮 , Cortex Lycii), *qīng hāo* (青蒿 , Herba Artemisiae Annuae) and *chuān xiōng* (Rhizoma Chuanxiong).

2. For dry, sore, itching or red swollen throat, add *jié gěng* (Radix Platycodonis), *bǎn lán gēn* (板蓝根 , Radix Isatidis) and *jīn guǒ lán* (金果榄 , Radix Tinosporae).

3. For insomnia, add *yè jiāo téng* (夜交藤 , Caulis Polygoni Multiflori) and *zǎo rén* (枣仁 , Semen Ziziphi Spinosae).

4. For anorexia and abdominal distention, add *jiāo sān xiān* (焦三仙 , Scorch-fried Fructus Hordei Germinatus et Crataegi et Massa Fermentata Medicinalis), *fó shǒu* (佛手 , Fructus Citri Sarcodactylis) and *sū gěng* (苏梗 , Caulis Perillae).

5. For upper limbs with significant lassitude, add *sāng zhī* (桑枝 , Ramulus Mori).

6. For lower limbs with significant lassitude, add *chuān niú xī* (Radix Cyathulae).

Compatibility Analysis

Although the aetiology of chronic fatigue syndrome is still unknown, many medical experts think that it is related to a viral infection or auto-immunization. Most patients have a history of an external pathogen invasion. Currently, there is no effective treatment. There is no record of chronic fatigue syndrome in traditional Chinese medicine; however, based on our clinical observation, the majority of patients have had a history of external pathogen invasion with fever. If the external pathogen invasion and fever is prolonged, or if it is not treated properly by over sweating, then it injures the upright qi, consuming and injuring the yin fluid, which induces both qi and yin deficiency. When there is upright qi deficiency, the lung and spleen qi cannot transport and go downwards; hence, it induces lassitude, shortness of breath, and weakness of the four limbs. Injured yin fluid cannot transport and induces lassitude and deterioration after activities. Therefore, the treatment principle is mainly to tonify qi and nourish yin, while strengthening the spleen and soothing the liver, in order to nourish earth to generate metal. This formula is used a lot: *Xiān hè cǎo* (Herba Agrimoniae), *huáng qí* (Radix Astragali), *tài zǐ shēn* (Radix Pseudostellariae) and *gān cǎo* (Radix Glycyrrhizae) are used to tonify the middle *jiao*'s qi. *Bǎi hé* (Bulbus Lilii) and *dà zǎo* (Fructus Jujubae) nourish yin, generate fluid and strengthen the spleen. *Chái hú* (Radix Bupleuri) soothes liver stagnation to ensure the smooth flow of qi. A small amount of *shā rén* (Fructus Amomi) can arouse the spleen and assist the above mentioned herbs in tonifying without stagnating. This formula uses a lot of tonifying middle *jiao* qi and yin herbs, based on the theory that the spleen is the foundation of the post heaven qi and governs the four limbs, while also nourishing earth to generate metal. If the spleen is strong, then the lung qi is abundant. Although it does not tonify the lung, lung qi will generate by itself.

Dà huáng's (大黄 , Radix et Rhizoma Rhei) property and taste are cold and bitter. It enters the spleen channel of the foot *taiyin*, the stomach channel of the foot *yangming*, the liver channel of the foot *jueyin*, the pericardium channel of the hand *jueyin* and the large intestine channel of the hand *yangming* meridians. It functions to sedate the heat and unblock the intestines, break stagnation and invigorate stasis. The *Divine Husbandman's Classic of the Materia Medica* classified it as a third class medicinal. It states: *"It treats lower body blood stasis, cold or heat induced blood obstruction. It breaks abdominal abscesses and accumulations, food and fluid stagnation; it cleans the stomach and intestines, pushes the old out in order to generate the new; frees water and food transportation; it harmonises the middle jiao to transform food; and it calms and harmonises the five zang organs."* From the modern clinical view point, the *Divine Husbandman's Classic of the Materia Medica* listed *dà huáng's* (Radix et Rhizoma Rhei) function and main treatments as broad. This proves that two thousand years ago, our ancestors were already very familiar with *dà huáng* (Radix et Rhizoma Rhei) and its application was popular. In Zhang Zhong-jing's *Discussion on Cold Damage* and the *Essentials from the Golden Cabinet*, there are 49 formulae using *dà huáng* (Radix et Rhizoma Rhei). This covers almost 25% of the formulae in the whole book. This indicates how widely used it was.

1. *Dà Huáng* (Radix et Rhizoma Rhei) ——
Gān Cǎo (甘草, Radix Glycyrrhizae)

Function

To clear and sedate heat stagnation, descend rebellious qi and harmonise the middle *jiao*.

Application

Dà Huáng Gān Cǎo Tāng (Rhubarb and Liquorice Decoction, 大黄甘草汤) in the *Discussion on Cold Damage* treats *"vomiting after eating"*. The vomiting after eating is due to the *yangming* meridian's heat stagnation, stomach qi rebels upward inducing vomiting after eating. It can also be used for chronic food stagnation transformed to heat, inducing nausea, acute gastritis, chronic constipation, etc.

Compatibility Analysis

Dà huáng (Radix et Rhizoma Rhei) is bitter, cold and good at purging. It can purge fire and unblock the *yangming* meridians. *Gān cǎo* (Radix Glycyrrhizae) harmonises the middle *jiao*. It can reduce the harshness of *dà huáng* (Radix et Rhizoma Rhei). It makes *dà huáng* (Radix et Rhizoma Rhei) purge but without injuring the middle *jiao*. These two herbs combined together can purge heat stagnation in the *yangming* meridian, descend stomach qi, and harmonise the stomach to stop nausea. This is the method of unblocking the inferior to stop the superior qi rebelling, as in: *"If you want southern scent, open the northern window first"*.

Case Study

Initial Visit
A female patient, 40 years old, August 10th, 1986.

The patient had a history of habitual constipation. In recent years, it had become severe. Every three to five days, she had one bowel movement. She needed *Kāi Sài Lù* (开塞露 , a glycerine enema) to help move her bowels. There were accompanying symptoms of dry mouth, a stifling chest sensation, dizziness, anorexia, irritability, etc. Her pulse was wiry and thready. Her tongue was red with a scanty coating. These are symptoms of stomach yin deficiency and *yangming* heat stagnation. With the stomach yin deficiency, the stomach has lost its harmony to move qi downwards, the rebellious stomach qi induced irritability, dizziness and stifling chest sensation. As the *yangming* heat stagnation could not diffuse, it induced the dry mouth, anorexia and constipation. Stagnated *yangming* heat really injured stomach yin. The injured stomach yin induced increased *yangming* heat stagnation. This induced habitual constipation becoming worse as days passed by, thus preventing recovery for a couple of years. Therefore a formula to clear heat and purge stagnation, harmonise stomach and nourish yin was prescribed. The treatment plan was to use modified *Dà Huáng Gān Cǎo Tāng* (Rhubarb and Liquorice Decoction).

Prescription

大黄	dà huáng	6g	Radix et Rhizoma Rhei
甘草	gān cǎo	6g	Radix Glycyrrhizae
麦冬	mài dōng	20g	Radix Ophiopogonis
石斛	shí hú	20g	Caulis Dendrobii
代赭石	dài zhě shí	20g	Haematitum
生山药	shēng shān yào	24g	Rhizoma Dioscoreae (raw)
桃仁	táo rén	9g	Semen Persicae

This was decocted with water, divided in two, and taken twice, one pack per day.

Second Visit

August 15th: After she took three packs of the above formula, her bowel movement was smoother than before, and regular once a day. However, if she stopped the formula, she was still constipated. The symptoms of dry mouth, a stifling chest sensation and anorexia were reduced, and there was no dizziness. She continued to follow the same formula.

Third Visit

August 22nd: She continued for another seven packs, and her symptoms basically disappeared. Her bowels continued to move once a day. To reinforce the treatment effectiveness, the decoction was changed to a powder, to be taken constantly to prevent reoccurrence.

Prescription

大黄	dà huáng	30g	Radix et Rhizoma Rhei
甘草	gān cǎo	30g	Radix Glycyrrhizae
当归	dāng guī	50g	Radix Angelicae Sinensis
生首乌	shēng shǒu wū	60g	Radix Polygoni Multiflori (raw)
郁李仁	yù lǐ rén	30g	Semen Pruni
代赭石	dài zhě shí	60g	Haematitum

All of the above herbs were ground to powder and put into a dark coloured bottle to prepare for later use. A 3g dose was taken, with warm water, once in the morning and another one at night.

Later, at a follow up visit: When taking the powder, her bowel movements occurred once a day. All her symptoms were gone except that she was occasionally constipated but did not require the help of *Kāi Sài Lù* (a glycerine enema). The patient was instructed to take the same formula constantly.

Note

The author used the above prescription to treat many patients with habitual constipation due to yin deficiency and all had good results. This formula uses *dà huáng* (Radix et Rhizoma Rhei) as the king herb to clear *yangming* heat and purge stagnation. It uses *hé shǒu wū* (Radix Polygoni Multiflori), *dāng guī* (Radix Angelicae Sinensis) and *yù lǐ rén* (Semen Pruni) as assistants to tonify blood, nourish yin and lubricate the intestines. *Dài zhě shí* (Haematitum) is the deputy, to descend stomach qi and break the stagnation, in order to help the bowels move. This formula also can be used for postpartum women who have constipation due to blood deficiency, and elderly people who have habitual constipation. If it is qi deficiency induced constipation, add *huáng qí* (黄芪 , Radix Astragali) and *bái zhú* (Rhizoma Atractylodis Macrocephalae). *Bái zhú* (Rhizoma Atractylodis Macrocephalae) has the function to increase the intestines' secretions and can be used in large doses. However, this formula is not suitable for yang deficiency induced constipation.

2. *Dà Huáng* (Radix et Rhizoma Rhei) ——
Máng Xiāo (芒硝, Natrii Sulfas)

Function

To purge heat, and free bowel movement.

Application

This is used for *yangming* heat stagnation, dry and hard stools syndrome such as stool stagnation, abdominal distention, thirst, a yellow tongue coating, a slippery and fast pulse, etc. Modern clinics use it for acute abdominal syndromes such as intestinal obstruction, pancreatitis, cholecystitis, appendicitis, etc. It can also be used for the acute jaundice type hepatitis and the closed type of stroke that has *yangming* heat stagnation.

Compatibility Analysis

Dà huáng (Radix et Rhizoma Rhei) is bitter and cold. It can flush excess. *Máng xiāo* (Natrii Sulfas) is salty and cold. It can moisten dryness and soften the hardness. *Dà huáng* (Radix et Rhizoma Rhei) is good at flushing and purging but does not soften the hardness, therefore, if only *dà huáng* (Radix et Rhizoma Rhei) is used to free bowel movement, usually, the stool would not evacuate quickly. If combined with *máng xiāo* (Natrii Sulfas), to make the dry stool soft, then, *dà huáng* (Radix et Rhizoma Rhei) can function better to purge heat, flush stagnation and push the old stools out to generate the new. These two herbs, used together, increase the power to purge downward and unblock the stagnation; hence, many purging heat stagnation formulae use this compatibility. For example in the *Discussion on Cold Damage*, *Dà Chéng Qì Tāng* (Major Qi-Coordinating Decoction, 大承气汤) is composed of *dà huáng* (Radix et Rhizoma Rhei, *máng xiāo* (Natrii Sulfas), *zhǐ shí* (枳实 , Fructus Aurantii) and *hòu pò* (厚朴 , Cortex Magnoliae Officinalis) and treats *yangming fu* excess, fullness, dryness, excess and hardness; *Tiáo Wèi Chéng Qì Tāng* (Stomach-Regulating and Qi-Coordinating Decoction, 调胃承气汤) is composed of *dà huáng* (Radix et Rhizoma Rhei), *máng xiāo* (Natrii Sulfas) and *gān cǎo* (Radix Glycyrrhizae) and treats *yangming fu* fullness, dryness, firmness and hardness; *Dà Xiàn Xiōng Tāng* (Major Chest Bind Decoction, 大陷胸汤) treats the combined dampness and heat pathogens inducing clumping in the chest). In the *Revised Popular Guide to the Discussion on Cold Damage* (通俗伤寒论 , *Tōng Sú Shāng Hán Lùn*), *Xiàn Xiōng Chéng Qì Tāng* (Chest Bind Qi-Coordinating Decoction, 陷胸承气汤) is composed of *dà huáng* (Radix et Rhizoma Rhei), *máng xiāo* (Natrii Sulfas), *guā lóu* (瓜蒌 , Fructus Trichosanthis), *bàn xià* (Rhizoma Pinelliae) and *huáng lián* (黄连 , Rhizoma Coptidis) and treats heat and phlegm clumped in the chest. In the *Essentials from the Golden Cabinet*, the *Dà Huáng Mǔ Dān Pí Tāng* (Rhubarb and Moutan Decoction, 大黄牡丹皮汤) (*dà huáng*-Radix et Rhizoma Rhei, *máng xiāo* - Natrii Sulfas, *táo rén* - Semen Persicae, *dōng guā zǐ* - 冬瓜子 , Semen Benincasae) treats intestinal abscesses.

Case Study

Initial Visit

A female patient, 30 years old, June 8th, 1986.

The patient had an abdominal pain for three days, stomach distention and fullness, no appetite, occasional nausea, and no bowel movement for three days. On examination, when pressing on the abdomen, it was full and tender in the right lower quadrant, and there was rebound tenderness at the McBurney's point. Her pulse was deep, wiry and forceful. Her tongue was yellow and greasy. This was the initial stage of an intestinal abscess; the symptoms were induced by toxic heat stagnating in the intestines. It required urgent treatment with a heat clearing, detoxifying, bowel freeing and stagnation dispersing method.

Prescription

丹皮	*dān pí*	10g	Cortex Moutan
大黄	*dà huáng*	30g	Radix et Rhizoma Rhei
桃仁	*táo rén*	10g	Semen Persicae
冬瓜仁	*dōng guā rén*	15g	Semen Benincasae
生苡米	*shēng yì mǐ*	30g	Semen Coicis (raw)
赤芍	*chì sháo*	15g	Radix Paeoniae Rubra
败酱草	*bài jiàng cǎo*	20g	Herba Patriniae
木香	*mù xiāng*	6g	Radix Aucklandiae
代赭石	*dài zhě shí*	60g	Haematitum
芒硝	*máng xiāo*	12g	Natrii Sulfas
甘草	*gān cǎo*	6g	Radix Glycyrrhizae

Water was used to decoct twice and then the decoctions were combined together. The mix was divided into three servings and served warm, one pack per day.

Second Visit

June 9th: After taking one pack of the above formula, her bowel movement was freed. She had diarrhoea twice with thick and sticky stools, and reduced abdominal pain. After two packs, the abdominal pain stopped and she could eat. Her tongue coating was also reduced. After taking two more packs without the *máng xiāo* (Natrii Sulfas), the patient recovered.

3. *Dà Huáng* (Radix et Rhizoma Rhei) —— *Yīn Chén* (茵陈, Herba Artemisiae Scopariae)

Function

To clear heat and leach out dampness, to free bowel movement and eliminate jaundice.

Application

This is used for damp-heat type jaundice with symptoms of a yellowish colouring all over the body, and eyes, bright yellowish in colour, as well as fever, abdominal distention, thirst, a yellow and greasy tongue coating, difficulty urinating or brownish urine, a deep excessive or slippery fast pulse, etc. Current clinical applications use it for the acute jaundice type hepatitis, gallbladder stone syndrome and cholecystitis that have the above symptoms.

Compatibility Analysis

Dà huáng (Radix et Rhizoma Rhei) is bitter and cold. It can clear heat and benefit the gallbladder, free the *fu* organs and expel nodules, invigorate blood and transform blood stasis. *Yīn chén hāo* (Herba Artemisiae Scopariae) can clear heat and leach out dampness; it is an important herb to treat jaundice. *Yīn chén hāo* (Herba Artemisiae Scopariae) helps *dà huáng* (Radix et Rhizoma Rhei) to clear and purge to make the *fu* organs qi flow freely, eliminating dampness and clearing heat, as well as eliminating jaundice quicker. They can be used for any yang type jaundice due to damp-heat. Traditional Chinese medicine believes that yang jaundice is due to damp-heat stagnation obstructing the middle *jiao*. It prevents the liver and gallbladder from flowing freely, thus, causing the liver to lose its free flowing function and then the bile to overflow inducing jaundice, hence, to purge the heat and free bowel movement, leach out dampness and eliminate jaundice is the best treatment principle for this combination. Only *dà huáng* (Radix et Rhizoma Rhei) and *yīn chén hāo* (Herba Artemisiae Scopariae) can do this. Modern clinical practice has proven that *dà huáng* (Radix et Rhizoma Rhei) and *yīn chén hāo* (Herba Artemisiae Scopariae) can unblock bile ducts and intestinal flow, to make bile flow to the intestine normally, and help to eliminate jaundice. The commonly used *Yīn Chén Hāo Tāng* uses the compatibility of *dà huáng* (Radix et Rhizoma Rhei) and *yīn chén hāo* (Herba Artemisiae Scopariae).

Initial Visit

A male patient, 17 years old, January 1st, 1999.

The patient came to visit due to recent anorexia, stomach and abdominal distention and fullness, whole body lassitude, with brownish urine. His liver function was investigated: Total bilirubin 78μmol/L, direct bilirubin 44μmol/L, indirect bilirubin 34μmol/L, TTT 9U, ALT 200U/L, HBs Ag (-). His pulse was wiry and forceful. His tongue coating was turbid and greasy. His sclera were remarkably yellowish. Diagnosis: Yang jaundice (damp-heat type with more damp than heat), jaundice type hepatitis. The treatment plan was to clear heat, leach out dampness and eliminate jaundice.

Prescription

茵陈蒿	*yīn chén hāo*	20g	Herba Artemisiae Scopariae
大黄	*dà huáng*	9g	Radix et Rhizoma Rhei
白花蛇舌草	*bái huā shé shé cǎo*	30g	Herba Hedyotis Diffusae
猪苓	*zhū líng*	15g	Polyporus
柴胡	*chái hú*	15g	Radix Bupleuri
蒲公英	*pú gōng yīng*	15g	Herba Taraxaci
泽兰	*zé lán*	15g	Herba Lycopi
虎杖	*hǔ zhàng*	15g	Rhizoma et Radix Polygoni Cuspidati
白术	*bái zhú*	12g	Rhizoma Atractylodis Macrocephalae
枳壳	*zhǐ qiào*	12g	Fructus Aurantii
板蓝根	*bǎn lán gēn*	15g	Radix Isatidis
甘草	*gān cǎo*	6g	Radix Glycyrrhizae

Water was used to decoct, one pack per day, two servings, taken once in the morning and once at night.

Second Visit

February 1st: After taking seven packs of the above prescription, the brownish urine became lighter, the stomach and abdominal distention reduced, appetite increased but the tongue coating was still turbid. Treatment continued with the original formula.

Third Visit

February 8th: The patient felt all of his symptoms had gone. The liver function test showed: Total bilirubin 15 μmol/L, direct bilirubin 9 μmol/L, TTT 7.6 U, ALT 34 U/L. Treatment was again continued with the original formula, to be taken every other day for 15 more packs.

Fourth Visit

March 10th: All liver functions tested normal.

4. *Dà Huáng* (Radix et Rhizoma Rhei) ——
Táo Rén (桃仁, Semen Persicae)

Function

To invigorate blood and dispel blood stasis.

Application

This is used for all kinds of blood stasis syndrome. It is irrevelant whether it is a new or chronic blood stasis syndrome, or if the blood stasis is in the superior or inferior part of the body, it can be used for all.

Compatibility Analysis

Dà huáng (Radix et Rhizoma Rhei) moves instead of adhering, clears heat and purges downward. The *Divine Husbandman's Classic of the Materia Medica* said: "*It descends blood stasis and blood blockage induced cold or heat, it breaks abdominal abscesses and accumulations.*" It can be seen that *dà huáng* (Radix et Rhizoma Rhei) is not only for clearing heat and freeing bowel movement, but is also a common herb for invigorating blood and transforming blood stasis. *Táo rén* (Semen Persicae) is acrid, bitter and neutral. It enters the blood. Acridity can disperse and break stasis in the blood, as well as moisten the intestines and free bowel movement. With these two herbs combined together, the power for breaking blood and eliminating stasis is even stronger. Zhang Zhong-jing always treated blood stasis using this compatibility, as in *Táo Rén Chéng Qì Tāng* (Peach Kernel Qi-Coordinating Decoction, 桃仁承气汤) for treating heat and blood stasis in the lower abdomen and gallbladder; *Dǐ Dǎng Tāng (Wán)* (Dead-On Decoction (Pill), 抵当汤 (丸)) for treating chronic, manic depression, lower abdomen hardness and fullness, with incontinence of urine; *Dà Huáng Zhè Chóng Wán* (Rhubarb and Ground Beetle Pill, 大黄蟅虫丸) for treating dried blood tuberculosis; *Dà Huáng Mǔ Dān Pí Tāng* (Rhubarb and Moutan Decoction) for treating blood and heat stagnation in the intestines, etc. Later, the *Fù Yuán Huó Xuè Tāng* (Origin-Restorative Blood-Quickening Decoction, 复元活血汤) in *Elucidation of Medicine* (医学发明 , *Yī Xué Fā Míng*) also used this compatibility of *dà huáng* (Radix et Rhizoma Rhei) and *táo rén* (Semen Persicae) for treating blood stasis inside the chest.

Case Study

Initial Visit

A male patient, 20 years old, May 20th, 1972.

The patient fell down a ladder accidently and his left ribs were injured by a water container. The pain was hard to bear and he could not move or turn his body laterally. Talking or coughing aggravated the pain. He called the author to visit urgently. His pulse was wiry, tight and forceful. This pulse indicates blood stasis induced pain. Immediately, he was prescribed modified *Fù Yuán Huó Xuè Tāng* (Origin-Restorative Blood-Quickening Decoction).

Prescription

大黄	dà huáng	10g	Radix et Rhizoma Rhei
桃仁	táo rén	12g	Semen Persicae
红花	hóng huā	9g	Flos Carthami
延胡索	yán hú suǒ	15g	Rhizoma Corydalis
柴胡	chái hú	15g	Radix Bupleuri
当归	dāng guī	15g	Radix Angelicae Sinensis
瓜蒌	guā lóu	15g	Fructus Trichosanthis
杭芍	háng sháo	15g	Radix Paeoniae Lactiflorae
川楝子	chuān liàn zǐ	10g	Fructus Toosendan
甘草	gān cǎo	9g	Radix Glycyrrhizae
黄酒	huáng jiǔ	1cup	Yellow Rice Wine (as a guide)

Water was used to decoct and it was taken warm. He took one pack a day once in the morning and once at night. After taking three packs the pain stopped.

5. *Dà Huáng* (Radix et Rhizoma Rhei) ——
Huáng Lián (黄连, Rhizoma Coptidis)

Function
To clear heat and sedate fire, to eliminate clumping and detoxify toxins.

Application
This combination treats all kinds of excess heat and fire syndromes such as haemoptysis, or epistaxis induced by searing fire causing restless movement of blood or high fever, irritability, headache, red eyes, mouth and tongue ulcers, constipation or dysentery with pus and blood due to excess heat of the *san jiao*. Nowadays, clinics use it for acute gastritis, gastrointestinal tract bleeding, angioneurotic headache, suppurative inflammation of the head, etc.

Compatibility Analysis
Dà huáng (Radix et Rhizoma Rhei) is bitter and cold. It can clear *yangming* stomach excess fire, free the *fu* organs to purge heat and descend fire downward. *Huáng lián*'s (Rhizoma Coptidis) property is bitter and cold and good at clearing invisible excess fire of the heart channel of the hand *shaoyin* and the stomach channel of the foot *yangming* meridians. These two herbs combined together have a stronger power to clear and purge. They can clear the upper, middle and lower *jiao*'s excess fire and purge the *san jiao*'s (three burners') stagnated heat. It can be used to treat any syndrome induced by excess heat or searing fire.

Case Study

Initial Visit
A male patient, 38 years old, June 13th, 1988.

The patient had left side migraine for more than 10 days. Everyday, before or after noon, the headache became more severe. He had treated it with acupuncture and medicine but they did not work. His tongue coating was thin and yellow. He complained of stomach distention and fullness, and constipation with bowel movements taking a couple of days. This was a *yangming* excess heat syndrome. Heat stagnation induced the stomach distention and constipation. Due to the bowel constipation and stagnation inducing the *yangming* heat pathogen, the heat could not go downward but rebelled upward, to cause the headache. Heat stagnated headache belongs to fire, hence, the headache got severe around noon time. The treatment plan was to clear heat, descend fire and free the bowels.

Prescription

大黄	*dà huáng*	8g	Radix et Rhizoma Rhei
黄连	*huáng lián*	9g	Rhizoma Coptidis
黄芩	*huáng qín*	10g	Radix Scutellariae
川芎	*chuān xiōng*	15g	Rhizoma Chuanxiong
菊花	*jú huā*	15g	Flos Chrysanthemi
柴胡	*chái hú*	12g	Radix Bupleuri
丹参	*dān shēn*	20g	Radix et Rhizoma Salviae Miltiorrhizae
甘草	*gān cǎo*	6g	Radix Glycyrrhizae

Water was used to decoct, it was divided into two servings, one pack per day.

Second Visit

August 16th: After taking three packs of the above prescription, the bowels moved, the headache disappeared, the tongue coating changed to white. The author was concerned that the headache could return, and so asked him to take more packs. After two more packs of the same formula, the patient recovered.

6. *Dà Huáng* (Radix et Rhizoma Rhei) ——
Dài Zhě Shí (代赭石, Haematitum)

Function

To purge heat, descend fire, and stop bleeding.

Application

This mainly treats lung and stomach exuberant heat, pushing blood to move restlessly, inducing upper gastrointestinal tract bleeding and nasal cavity bleeding, etc.

Compatibility Analysis

Dà huáng (Radix et Rhizoma Rhei) is bitter and cold. It is good at freeing and descending, it can guide fire downward. It is a frequently used herb to treat heat in the blood and rebellious qi induced haematemesis and epistaxis. *Dài zhě shí* (Haematitum) is bitter and cold. It enters the meridians of the heart channel of the hand *shaoyin* and the liver channel of the foot *jueyin*. It is heavy and has the property to drain downward.

It enters the blood and can tranquilise the liver to descend fire, cool blood and stop bleeding. *Dà huáng* (Radix et Rhizoma Rhei) with *dài zhě shí* (Haematitum) combined, their coldness and bitterness can have more power to unblock, descend, purge heat and stop bleeding, as both herbs can eliminate stasis in order to generate the new. Hence, they have the function to stop bleeding without encouraging stasis. They are a good combination to treat blood heat induced haematemesis or epistaxis.

Dà huáng (Radix et Rhizoma Rhei) is known to stop haematemesis and epistaxis. Its clinical applications are broad. Zhang Xi-chun used *dà huáng* (Radix et Rhizoma Rhei) with *ròu guì* (Cortex Cinnamomi) called *Mì Hóng Dān* (Confidential Red Pill, 秘 红 丹) especially to treat haematemesis. In recent years, there have been reports that *dà huáng* (Radix et Rhizoma Rhei) and *bái jí* (白及 , Rhizoma Bletillae) have been used to treat liver cirrhosis complicated with upper gastrointestinal tract bleeding, with good results. These are all herbal pairs that use *dà huáng* (Radix et Rhizoma Rhei) as the king herb to stop bleeding, and are a useful reference.

Case 1

Epistaxis.

Initial Visit

A male patient, 40 years old, May 15th, 1985.

The patient had a history of epistaxis for more than 20 years. It occurred every spring and autumn. This time, he had epistaxis relentlessly for more than a month. Everyday, after midday, there was bleeding with a small amount of blood. If he used cold water to clean it, then it stopped. He checked into an hospital's Otolaryngology Department. There was mucous membrane hyperaemia in his left nasal cavity. It was more obvious in the inferior part of the nose bridge. He took adrenobazon, *Yún Nán Bái Yào* (Yunnan White Medicine, 云 南 白 药), vitamin C and externally used *Fù Fāng Bò Hé Yóu* (Compound Mint Oil, 复 方 薄 荷 油), and chloromycetin nasal drops, but nothing worked. He also had migraines with it. A routine blood test and the platelet count were normal. His pulse was wiry and fast. His tongue was red with a white coating. He was physically well developed. His diagnosis was lung and stomach heat stagnation pushing blood to flow upward and induce the symptoms. The treatment principle was to descend the rebellious, to purge heat, nourish blood and stop bleeding.

Prescription

代赭石	*dài zhě shí*	30g	Haematitum
大黄	*dà huáng*	9g	Radix et Rhizoma Rhei
半夏	*bàn xià*	9g	Rhizoma Pinelliae
麦冬	*mài dōng*	30g	Radix Ophiopogonis
白茅根	*bái máo gēn*	30g	Rhizoma Imperatae
旱莲草	*hàn lián cǎo*	20g	Herba Ecliptae
仙鹤草	*xiān hè cǎo*	30g	Herba Agrimoniae
三七粉	*sān qī fěn*	3g	Radix et Rhizoma Notoginseng Powder (dissolved)
甘草	*gān cǎo*	6g	Radix Glycyrrhizae

Water was used to decoct, and then divided into two servings, taken in the morning and at night, one pack per day.

Second Visit

May 23rd: After taking three packs, the bleeding stopped but the migraine was not relieved. To the above formula, *chuān xiōng* (Rhizoma Chuanxiong) 12g and *gé gēn* (葛根 , Radix Puerariae Lobatae) 20g, were added. He continued to take it.

Third Visit

August 9 th: He visited for treatment for his headache and he was told to take six packs of May 23 rd's prescription. There was no more epistaxis after that.

Case 2

Gastric Haemorrhage.

Initial Visit

A female patient, 40 years old, July 8th.

The patient had a history of chronic stomachache. One week ago, due to anger with a family member, she suddenly had stomachache with haemoptysis. She had an x-ray, and a barium test from a Radiography Department and was diagnosed with a gastric ulcer haemorrhage, but the treatments failed, so she visited our outpatients' clinic. It was obvious from her appearance that she had chronic disease. Her face was pale without lustre, she was short of breath, had anorexia, and her stomachache was aggravated on pressure, her stool tested positive for blood, and she had haematemesis with dark red blood. Her pulse was deficient, big but forceless. Her tongue was pale and the coating was thin and white. The treatment plan was to tonify qi, harmonise the stomach, descend rebellion and stop bleeding.

Prescription

党参	*dǎng shēn*	20g	Radix Codonopsis
代赭石	*dài zhě shí*	15g	Haematitum
大黄	*dà huáng*	6g	Radix et Rhizoma Rhei
藿香	*huò xiāng*	9g	Herba Agastachis
百合	*bǎi hé*	15g	Bulbus Lilii
海螵蛸	*hǎi piāo xiāo*	15g	Os Sepiae seu Sepiellae
半夏	*bàn xià*	9g	Rhizoma Pinelliae
白芍	*bái sháo*	20g	Radix Paeoniae Alba
甘草	*gān cǎo*	6g	Radix Glycyrrhizae

Water was used to decoct, and then divided into two servings, served warm, one pack per day.

Second Visit

September 12th: After she took two packs of the above formula, her haemoptysis stopped, and her stomachache also reduced, but she was still short of breath and had anorexia. To the above formula, *shēng huáng qí* (raw Radix Astragali) 20g and *shā rén* (Fructus Amomi) 6g were added. She continued to take the formula.

Third Visit

September 21ˢᵗ: After taking ten packs of the above modification, all symptoms had gone and the bloody stool tested negative.

There had been no reoccurrence by the follow up visit, six months after stopping the treatment.

7. *Dà Huáng* (Radix et Rhizoma Rhei) ——
Zhū Líng (猪苓, Polyporus)

Function

To promote urination, free bowel movement and purge the turbid.

Application

This is used for chronic nephritis with anorexia, constipation, abdominal distention, proteinuria, and elevated creatine and urea nitrogen in patients with chronic nephrosis.

Compatibility Analysis

With chronic nephritis, if the kidney is injured, and the glomerular filtration rate is lower than 1/3 of the normal rate, then the blood serum creatine and the blood urea nitrogen will increase. *Dà huáng* (Radix et Rhizoma Rhei) has the function to reduce the symptoms, extend life, and lower the blood serum creatine and the blood urea nitrogen. It works mainly through reducing the amino acid reabsorption in the intestinal tract, inhibiting urea synthesis in the liver and kidney, increasing the free amino acid concentration in the blood, using body urea nitrogen to synthesise protein and inhibiting muscle protein decomposition. *Zhū líng* (Polyporus) is sweet, bland and neutral with weak qi and a light taste, but has stronger urine promoting functions. It can increase the glomerular filtration rate and can help *dà huáng* (Radix et Rhizoma Rhei) to reduce blood creatine and urea nitrogen. Moreover, in recent years immunological research indicated that *dà huáng* (Radix et Rhizoma Rhei) and *zhū líng* (Polyporus) have the function to regulate immunity and may delay a prognosis of kidney function failure.

Case Study

Initial Visit

A female patient, 46 years old, May 8ᵗʰ, 1992.

The patient had nephritis for a couple of years. She had been admitted into Beijing and Tianjin hospitals and had been diagnosed as having chronic nephritis. It recently manifested in whole body lassitude, anorexia, poor appetite, etc. A routine urine test showed: urine protein ++, occult blood +, blood urea nitrogen 15mmol/L, blood pressure 150/85mm Hg. Her pulse was deep and thready. Her tongue coating was thin and white. Her lower limbs were slightly swollen. The diagnosis was deficiency of both the spleen and kidney, and abnormal water and dampness transformation disorder. The spleen deficiency induced the lassitude, anorexia and poor appetite. The kidney deficiency, impaired water and dampness transformation and induced swollen limbs. The kidney governs two yins and the spleen astringes essences. If both kidney and spleen are deficient urine protein and occult blood will be seen. The treatment principle was to strengthen the spleen, tonify the kidney, raise the clear and descend the turbid.

Prescription

生黄芪	*shēng huáng qí*	30g	Radix Astragali (raw)
白术	*bái zhú*	15g	Rhizoma Atractylodis Macrocephalae
猪苓	*zhū líng*	15g	Polyporus
大黄	*dà huáng*	6g	Radix et Rhizoma Rhei
红花	*hóng huā*	9g	Flos Carthami
蝉蜕	*chán tuì*	10g	Periostracum Cicadae
益母草	*yì mǔ cǎo*	15g	Herba Leonuri
白茅根	*bái máo gēn*	20g	Rhizoma Imperatae
熟地	*shú dì*	15g	Radix Rehmanniae Praeparata
山茱萸	*shān zhū yú*	12g	Fructus Corni
山药	*shān yào*	20g	Rhizoma Dioscoreae
甘草	*gān cǎo*	6g	Radix Glycyrrhizae

This was decocted with water, one pack per day, divided into two servings.

Second Visit

May 15th: After she took seven packs of the above prescription, her appetite increased and her lassitude symptoms reduced; her urine protein +, blood urea nitrogen 13 mmol/L. It was considered to be effective so, the same formula was kept for treatment.

Third Visit

June 10th: She used the previous formula with modifications and took more than 20 packs, all the clinical symptoms disappearaed, urine protein ±, blood urea nitrogen reduced to 8.5 mmol/L. She was recommended to take it every other day continuously.

Dà fù pí's (大腹皮 , Pericarpium Arecae) property and taste are acrid and warm. It enters the spleen channel of the foot *taiyin* and the stomach channel of the foot *yangming* meridians. *Dà fù pí* (Pericarpium Arecae) is the fruit skin of *bīng láng* (槟榔 , Semen Arecae). Its quality is light and its taste is acrid. It is good at promoting qi movement and soothing stagnation, relaxing the middle *jiao* and eliminating fullness. It also has functions to promote urination and reduce swelling. Therefore, it is effectively used for qi stagnation and water accumulation induced stomach and abdomen distention and fullness, oedema, and scanty urine symptoms.

1. *Dà Fù Pí* (Pericarpium Arecae) ——
Dà Fù Zǐ (大腹子, Semen Arecae)

Function

To promote qi movement and urination, to expel nodules and reduce distention.

Application

This is used for food and water stagnation, stomach and abdominal distention and fullness, or oedema due to athlete's foot, or ascites in the early stages of liver cirrhosis patients who have a strong constitution.

Compatibility Analysis

Dà fù pí (Pericarpium Arecae) and *dà fù zǐ* (大腹子 , Semen Arecae) (榔片 , sliced Semen Arecae) are from the same plant. They are the fruit of *bīng láng* (Semen Arecae). The outside fibre shell is the *dà fù pí* (Pericarpium Arecae). The inside is the flat, big rounded seed, *dà fù zǐ* (Semen Arecae). It is also called *láng yù* (榔玉 , jade of Semen Arecae), or *bīng láng* (Semen Arecae). *Dà fù pí* (Pericarpium Arecae) is acrid and slightly warm. It is good at moving qi downward and easing the middle *jiao*, promoting urination and eliminating oedema. *Dà fù zǐ* (Semen Arecae) is acrid, bitter and warm. Its functions are similar to *dà fù pí* (Pericarpium Arecae) but more powerful, and can kill warmth to eliminate accumulation. Huang Gong-xiu said: *"Due to its bitter taste, it governs descending, therefore, there is no unbreakable firmness, no uneliminated distention, no undigested food, no unmoving phlegm, all qi can descend, no worm that cannot be killed, and no stools that cannot be freed."* *Dà fù pí* (Pericarpium Arecae) mainly promotes qi movement while promoting urination. *Dà fù zǐ* (Semen Arecae) mainly promotes urination and breaks accumulations while descending qi. The two herbs used together can increase the functions of promoting qi movement and urination, reducing stagnation and eliminating distention. *Dà Fù Pí Sǎn* (Areca Husk Powder, 大腹皮散) and *Dà Fù Pí Tāng* (Areca Husk Decoction, 大腹皮汤) in *Indispensable Tools for Pattern Treatment* (证治准绳 , *Zhèng Zhì Zhǔn Shéng*) are all formulae that use these two herbs together.

Case Study

Initial Visit

A female patient, 56 years old, June 30th, 1987.

For the previous 10 days, the patient felt her stomach had abdominal distention which was becoming worse, she had anorexia, no appetite with hiccough, was spitting clear water, and had nausea after eating; her stomach and abdomen were more distended, and she could not flex or extend her body. Her pulse was deep, slippery and forceful. Her tongue coating was white and turbid. This all pointed to food stagnation and water retention. The food was stagnated in the middle *jiao*; the transforming and transporting function had been disturbed and induced anorexia, lack of appetite, with symptoms worse after eating. Water and dampness stagnation induced spitting clear water, hiccoughs, and the white and turbid tongue coating. The treatment plan was to promote qi movement to eliminate accumulation, and to promote urination to reduce distention.

Prescription

槟榔片	*bīng láng piàn*	10g	Semen Arecae (sliced)
大腹皮	*dà fù pí*	10g	Pericarpium Arecae
藿香	*huò xiāng*	9g	Herba Agastachis
山楂	*shān zhā*	20g	Fructus Crataegi
陈皮	*chén pí*	9g	Pericarpium Citri Reticulatae
清半夏	*qīng bàn xià*	9g	Rhizoma Pinelliae
香附	*xiāng fù*	10g	Rhizoma Cyperi
枳壳	*zhǐ qiào*	10g	Fructus Aurantii
乌药	*wū yào*	10g	Radix Linderae
柴胡	*chái hú*	12g	Radix Bupleuri
甘草	*gān cǎo*	6g	Radix Glycyrrhizae

Water was used to decoct, and then divided into two servings, to be taken warm, in the morning and night, one pack per day.

Second Visit

July 3rd: After she took four packs of the above formula, the abdominal distention and nausea reduced a lot, and appetite increased. She followed the original formula for another four packs.

Third Visit

July 7th: All symptoms had disappeared; she recovered and so stopped taking the formula.

Gān jiāng's (干姜 , Rhizoma Zingiberis) property and taste are acrid and warm. It enters six meridians: the heart channel of the hand *shaoyin*, the lung channel of the hand *taiyin*, the spleen channel of the foot *taiyin*, the stomach channel of the foot *yangming*, the kidney channel of the foot *shaoyin* and the large intestine channel of the hand *yangming*. Its functions can warm the middle *jiao* and expel cold. It is the main herb to treat internal cold. The *Divine Husbandman's Classic of the Materia Medica* said: *"It governs chest fullness, cough and qi rebelling upward, it warms the middle jiao and stops bleeding, it induces sweating, and it expels wind-damp painful obstruction, and bleeding from the bowel with dysentery."* The clinical combinations and applications are broad and can treat many diseases, such as with *Xiǎo Qīng Lóng Tāng* (Minor Green-Blue Dragon Decoction, 小青龙汤) in the *Discussion on Cold Damage* which uses *gān jiāng* (干姜 , Rhizoma Zingiberis) and *wǔ wèi zǐ* (五味子 , Fructus Schisandrae Chinensis) together to treat cold induced cough. On the one hand, it warms the lung and on the other hand, it astringes the lung. In *Lǐ Zhōng Wán* (Centre-Rectifying Pill) it combines *gān jiāng* (Rhizoma Zingiberis) and *bái zhú* (Rhizoma Atractylodis Macrocephalae) to treat vomiting and diarrhoea, with one warming the spleen and the other strengthening the spleen, providing mutual benefit for each other's treatment, making it more effective. In *Sì Nì Tāng* (Counterflow Cold Decoction, 四逆汤), it uses *gān jiāng* (Rhizoma Zingiberis) and *fù zǐ* (Radix Aconiti Lateralis Praeparata) together to resuscitate yang and save the collapse syndrome. In *Bàn Xià Gān Jiāng Sǎn* (Pinellia and Dried Ginger Powder, 半夏干姜散) it uses *gān jiāng* (Rhizoma Zingiberis) and *bàn xià* (Rhizoma Pinelliae) together to warm the stomach and stop vomiting. *Èr Jiāng Wán* (Lesser Galangal and Ginger Pill, 二姜丸) in the *Formulary of the Bureau of Medicines of the Taiping Era* (和剂局方 , *Hé Jì Jú Fāng*) uses *gān jiāng* (Rhizoma Zingiberis) and *gāo liáng jiāng* (Rhizoma Alpiniae Officinarum) together, to treat the heart and abdominal cold pain.

1. *Gān Jiāng* (Rhizoma Zingiberis) ——
Gān Cǎo (甘草, Radix Glycyrrhizae)

Function

To warm yang and tonify qi.

Application

This can be used for all kinds of deficiency cold syndromes such as stomach cold induced pain, acid regurgitation, stomach and abdominal distention and fullness, borborygmus and diarrhoea, cough and asthma, chest pain, dizziness and vomiting, abdominal pain due to menstruation, a slow pulse, a pale tongue, no thirst.

Compatibility Analysis

Gān jiāng (Rhizoma Zingiberis) is acrid and warm, to warm the middle *jiao* and expel cold. *Gān cǎo* (Radix Glycyrrhizae) is sweet and neutral to tonify the middle *jiao* and qi. The two herbs together make the combination of the acrid and sweet. This combination can warm up and tonify yang qi, hence, all kinds of yang qi deficiency, as well as yin cold pathogen congealed and stagnated syndromes, can apply it with good effect. Based on modern medical research, a slow pulse indicates the heart rate has slowed down. Acid regurgitation and stomach pain indicate hyperfunction of the stomach and intestinal peristalsis and their secretions. Cough and asthma are due to hypercontraction of the trachea smooth muscle. Dizziness and blood pressure drops are due to peripheral blood vessel expansion and induced temporary shortness of blood. Menstrual abdominal pain is more often due to uterus spasm. All of the above syndromes are due to hyperactivity of the parasympathetic nerves. *Gān Cǎo Gān Jiāng Tāng* (Liquorice and Dried Ginger Decoction, 甘草干姜汤) can cause reflux sympathetic nerve activity to balance the parasympathetic nerves and soothe the stomach's smooth muscle spasms, therefore, it is effective for the above stated symptoms. Clinically, there are more formulae using the *gān jiāng* (Rhizoma Zingiberis*)* and *gān cǎo* (Radix Glycyrrhizae) compatibility such as: *Gān Cǎo Gān Jiāng Tāng* (Liquorice and Dried Ginger Decoction), *Sì Nì Tāng* (Counterflow Cold Decoction), *Lǐ Zhōng Wán* (Centre Rectifying Pill), *Dà Jiàn Zhōng Tāng* (Major Centre-Fortifying Decoction, 大建中汤), *Gān Jiāng Líng Zhú Tāng* (Liquorice, Dried Ginger Decoction, Poria, and Atractylodes Macrocephala Decoction, 甘姜苓术汤), etc., in the *Discussion on Cold Damage.*

Case Study

Initial Visit

A female patient, 46 years old, March 15th, 1986.

The patient generally had a poor constitution. She recently suffered from overexertion, due to family reasons. She also had irregular meals with raw and cold food, which injured the stomach and induced stomachache, diarrhoea or loose stool, nausea and no appetite for a week. On pressing on her abdomen, it was soft. The patient liked pressure and warmth, every night, she used a hot water bottle to warm her stomach and felt comfortable. Her tongue was pale and the coating was white, she had no thirst. Her pulse was deep and slow. This was the spleen and stomach deficiency cold syndrome.

Prescription

干姜	*gān jiāng*	9g	Rhizoma Zingiberis
甘草	*gān cǎo*	8g	Radix Glycyrrhizae
藿香	*huò xiāng*	9g	Herba Agastachis
苏梗	*sū gěng*	9g	Caulis Perillae
陈皮	*chén pí*	9g	Pericarpium Citri Reticulatae
沉香	*chén xiāng*	6g	Lignum Aquilariae Resinatum
木香	*mù xiāng*	6g	Radix Aucklandiae
枳壳	*zhǐ qiào*	10g	Fructus Aurantii
白术	*bái zhú*	12g	Rhizoma Atractylodis Macrocephalae
砂仁	*shā rén*	6g	Fructus Amomi

Water was used to decoct, then it was divided into two servings, one pack per day.

Second Visit

March 18th: After she took three packs of the above formula, her stomachache stopped, and the loose stool and nausea also reduced, but she still had anorexia and no appetite. *Jiāo sān xiān* (焦三仙 , Scorch-fried Fructus Hordei Germinatus et Crataegi et Massa Fermentata Medicinalis) 30g was added for three more packs.

Third Visit

March 21ˢᵗ: All her symptoms basically disappeared and her appetite had increased. She followed the same formula for three more packs.

2. *Gān Jiāng* (Rhizoma Zingiberis) ─── *Fù Zǐ* (附子, Radix Aconiti Lateralis Praeparata)

Function

To resuscitate yang and save collapse, to warm the middle *jiao* and expel cold.

Application

This is most often used for yang qi deficiency or weakness, internal excess yin and cold. Symptoms include cold in the four limbs, aversion to cold and spirit lassitude, somnolence, diarrhoea with indigested food, abdominal cold and pain, no taste inside the mouth, no thirst, a pale tongue with a white coating, and a deep, thready and slow pulse.

Compatibility Analysis

Fù zǐ (Radix Aconiti Lateralis Praeparata) is acrid, warm and very hot. Its properties are harsh and dry. It moves yang without holding it. In the upper *jiao*, it can help the heart yang to unblock blood vessels. In the middle *jiao*, it can warm the spleen and strengthen its transforming and transporting functions. In the lower *jiao*, it can tonify the kidney yang and benefit fire. It is an important herb to warm the interior and tonify yang. *Gān jiāng* (Rhizoma Zingiberis) warms the middle *jiao* and expels cold to help *fù zǐ's* (Radix Aconiti Lateralis Praeparata) power of resuscitation. It moves without holding and can stabilise *fù zǐ's* (Radix Aconiti Lateralis Praeparata) property of strengthening the power to resuscitate yang and warm the middle *jiao*, hence, our ancestors had the saying: "*Without jiāng (Rhizoma Zingiberis), fù zǐ (Radix Aconiti Lateralis Praeparata) would not be hot.*" The *Seeking Accuracy in the Materia Medica* (本 草 求 真 , *Běn Cǎo Qiú Zhēn*) said: "*Gān jiāng (Rhizoma Zingiberis) is very hot but not toxic. It holds but does not disperse. For any stomach deficiency cold, with basal yang separating, using it with fù zǐ (Radix Aconiti Lateralis Praeparata) can have an immediate effect on resuscitating yang.*" *Gān Jiāng Fù Zǐ Tāng* (Dried Ginger and Aconite Decoction, 干姜附子汤) in the *Discussion on Cold Damage* uses the compatibility of *gān jiāng* (Rhizoma Zingiberis) and *fù zǐ* (Radix Aconiti Lateralis Praeparata) to treat sweating injuring yang inducing irritability. If *zhì gān cǎo* (Radix et Rhizoma Glycyrrhizae Praeparata cum Melle) is added it becomes *Sì Nì Tāng* (Counterflow Cold Decoction). It functions to resuscitate yang and save the rebellious. Use *zhì gān cǎo* (Radix et Rhizoma Glycyrrhizae Praeparata cum Melle), for its sweet and relaxing properties, to prevent injury from *gān jiāng* (Rhizoma Zingiberis) and *fù zǐ's* (Radix Aconiti Lateralis Praeparata) dry and hot properties, to warm the middle *jiao* while tonifying it. If *rén shēn* (Radix et Rhizoma Ginseng) is added it is called *Sì Nì Jiā Rén Shēn Tāng* (Counterflow Cold Decoction Plus Ginseng, 四逆加人参汤). It can tonify both yin and yang and is useful for all kinds of yang deficiency diseases and cold excess coma.

Add *fú líng* (Poria) to the above formula to become *Fú Líng Sì Nì Tāng* (Poria Counterflow Cold Decoction, 茯苓四逆汤), which functions to warm the yang and promote urination. It can be used for heart failure and chronic nephritis oedema induced by deficiency cold. The application for the *gān jiāng* (Rhizoma Zingiberis) and *fù zǐ* (Radix Aconiti Lateralis Praeparata) compatibility is broad. If the diagnosis is accurate, it can more often than not, immediately reduce the severity, however, when applying *fù zǐ* (Radix Aconiti Lateralis Praeparata), *shú fù zǐ* (熟附子 , Radix Aconiti Lateralis Praeparata) should be used. If the dosage is large, it should be decocted for longer, in order to reduce its toxicity.

Case Study

Initial Visit

A female patient, 45 years old, June 15th, 1991.

The patient had chronic nephritis for more than three years. Recently, she had whole body oedema more prevalent in the lower limbs. On pressing, the finger sank in. The facial oedema was of the deficiency type and was lustreless, she had scanty urination, turbid urine with bubbles, accompanied by low back pain, anorexia, nausea and vomiting often, with a deep thready pulse, a pale tongue, with a white and slippery coating. Her urine protein +++, occasionally red blood cells were visible. The kidney picture showed that the kidneys did not have full function. This was the acute outbreak of chronic nephritis. It was diagnosed as oedema of the spleen and kidney yang deficiency type, in Chinese medicine. The *Fú Líng Sì Nì Tāng* (Poria Counterflow Cold Decoction) was immediately modified for treatment.

Prescription

熟附子	*shú fù zǐ*	10g	Radix Aconiti Lateralis Praeparata
干姜	*gān jiāng*	6g	Rhizoma Zingiberis
茯苓	*fú líng*	20g	Poria
党参	*dǎng shēn*	20g	Radix Codonopsis
半夏	*bàn xià*	9g	Rhizoma Pinelliae
益母草	*yì mǔ cǎo*	20g	Herba Leonuri
大黄	*dà huáng*	5g	Radix et Rhizoma Rhei
猪苓	*zhū líng*	15g	Polyporus
白术	*bái zhú*	15g	Rhizoma Atractylodis Macrocephalae
泽泻	*zé xiè*	15g	Rhizoma Alismatis

Water was used to decoct, it was divided into two servings, one pack per day.

Second Visit

June 18th: After she had taken three packs of the above formula, her nausea and vomiting stopped, and her appetite increased. *Chán yī* (蝉衣 , Periostracum Cicadae) 10g was added to the above formula and she continued to take it.

Third Visit

June 22nd: After she took another four packs, the oedema basically disappeared, her urine protein ++, her appetite was normal, her pulse became floating, her tongue coating was thin and white. The formula was changed to be a strengthening the spleen and tonifying the kidney formula, she used it for treatment for more than a month, then, her urine protein was between ± and +, and she had no oedema any more.

3. *Gān Jiāng* (Rhizoma Zingiberis) —— *Huáng Lián* (黄连, Rhizoma Coptidis)

Function

To naturally adjust cold and heat, to harmonise the middle *jiao* and expel nodules.

Application

This is used for cold and heat stagnated abdominal pain, vomiting, diarrhoea or dysentery, and vomiting due to cold and heat repulsion. Nowadays, it is applied in clinic for acute gastroenteritis, gastritis, pancreatitis, as well as stomach and intestinal neurosis syndrome, etc.

Compatibility Analysis

This compatibility uses cold and heat simultaneously. Ke Qin said: "*Cold and heat mutually obstructing is the ge syndrome. Cold and heat clumped together is the stuffiness syndrome. Even though both ge and stuffiness have the same aetiology but the progression is different.*" Stuffiness and *ge* have different clinical expressions but their pathologies are all due to a cold and heat complex, disharmony in the raising and descending of the qi mechanism, inability to separate the clear from the turbid, and unsmooth movement of the qi. The less severe one is *ge* and the more severe one is stuffiness. The initial stage is *ge*, the more chronic then becomes stuffiness. The treatment principle is to neutrally regulate cold and heat, break the stuffiness and expel nodules. *Gān jiāng* (Rhizoma Zingiberis) is acrid and warm, revives yang and expels cold. *Huáng lián* (Rhizoma Coptidis) is bitter and cold; it clears heat, and descends to expel heat nodules. With *ge* expelled and nodules descended, the symptoms will resolve by themself. Li Shi-zhen said: "*One cold and one hot, yin and yang are mutually enhanced. It's the most beautiful part of the formula as it does not cause injury due to any one excess.*" Zhang Zhong-jing's *Gān Jiāng Huáng Qín Huáng Lián Rén Shēn Tāng* (Dried Ginger, Scutellaria, Astragalus and Ginseng Decoction, 干姜黄芩黄连人参汤) treats vomiting after eating due to cold *ge* syndrome. All of *Bàn Xià Xiè Xīn Tāng* (Pinellia Heart-Draining Decoction, 半夏泻心汤), *Gān Cǎo Xiè Xīn Tāng* (Liquorice Heart-Draining Decoction, 甘草泻心汤) and *Shēng Jiāng Xiè Xīn Tāng* (Fresh Ginger Heart-Draining Decoction, 生姜泻心汤) treat stuffiness inferior to the Heart. They all use the acrid, breaking, bitter and descending compatibility of *gān jiāng* (Rhizoma Zingiberis) and *huáng lián* (Rhizoma Coptidis).

Case 1

Initial Visit

A female patient, 33 years old, August 10th, 1987.

The patient had stomach distention, fullness and pain for many years. Recently, due to eating too much raw and cold food, she had stomachache with a burning sensation, acid regurgitation accompanied by stomach distention and fullness, discomfort due to qi rebellion, hiccoughs, a desire to vomit, and lassitude, etc. Her pulse was deep, thready and wiry. Her tongue coating was thin and white. The diagnosis was a repelling syndrome. cold and heat clumped together. The treatment plan was to neutrally regulate cold and heat, eliminate stuffiness and expel the *ge*.

Prescription

黄连	*huáng lián*	8g	Rhizoma Coptidis
黄芩	*huáng qín*	10g	Radix Scutellariae
干姜	*gān jiāng*	9g	Rhizoma Zingiberis
党参	*dǎng shēn*	12g	Radix Codonopsis
檀香	*tán xiāng*	9g	Lignum Santali Albi
砂仁	*shā rén*	6g	Fructus Amomi
枳壳	*zhǐ qiào*	12g	Fructus Aurantii
白术	*bái zhú*	12g	Rhizoma Atractylodis Macrocephalae
藿香	*huò xiāng*	9g	Herba Agastachis
甘草	*gān cǎo*	6g	Radix Glycyrrhizae

Water was used to decoct, divided into two servings, to be taken in the morning and night, one pack per day.

Second Visit

August 13th: After taking three packs, the stomach distention, fullness, hiccough and qi rebellion reduced significantly. There was no vomiting after eating. Acid regurgitation was also reduced. To the original prescription *hǎi piāo xiāo* (Endoconcha Sepiae) 15g and *dà bèi mǔ* (大贝母 , Bulbus Frittillariae Thunbergii) 9g were added for treatment.

Third Visit

August 17th: After another four packs, all symptoms reduced significantly. Her appetite was back to normal. She was told to take four more packs to strengthen the effects.

Case 2

Initial Visit

A female patient, 44 years old, April 12th, 1987.

The patient had a history of diarrhoea and abdominal pain. Every time when angry or after overexertion, it became worse. In recent months, the diarrhoea and abdominal pain became more severe. Everyday she had diarrhoea two to three times. Her stool was loose with sticky stuff and she had a tenesmus sensation. A regular stool check did not show any abnormality. She also had accompanying irritability and chest distention, nausea with desire to vomit, and anorexia, etc. Her tongue was red, the coating was white but slightly greasy. Her pulse was wiry. She was diagnosed with repelling of cold and heat syndrome with fire in the superior, and cold in the inferior. Fire in the superior and stomach heat not going downward induced the irritability and nausea. The inferior cold was spleen yang not rising inducing abdominal pain and diarrhoea. The treatment plan was to neutrally regulate cold and heat, and to regulate qi and harmonise the middle *jiao*.

Prescription

黄连	*huáng lián*	10g	Rhizoma Coptidis
干姜	*gān jiāng*	9g	Rhizoma Zingiberis
杭白芍	*háng bái sháo*	15g	Radix Paeoniae Lactiflorae
白术	*bái zhú*	12g	Rhizoma Atractylodis Macrocephalae
陈皮	*chén pí*	9g	Pericarpium Citri Reticulatae
防风	*fáng fēng*	10g	Radix Saposhnikoviae
香附	*xiāng fù*	12g	Rhizoma Cyperi
木香	*mù xiāng*	9g	Radix Aucklandiae
当归	*dāng guī*	12g	Radix Angelicae Sinensis
半夏	*bàn xià*	9g	Rhizoma Pinelliae
甘草	*gān cǎo*	6g	Radix Glycyrrhizae

This was decocted with water, served warm, one pack a day.

Second Visit

April 25th: After three packs, the abdominal pain and diarrhoea had reduced significantly, the nausea and vomiting symptoms were relieved, and the appetite had increased. Three more packs of the same formula were used for treatment.

Third Visit

April 30th: Basically, all the symptoms disappeared. She was told to pay attention to her diet, and keep her spirits up. She was treated with a strengthening the spleen and harmonising the stomach formula to complete the treatment.

Tǔ fú líng's (土茯苓 , Rhizoma Smilacis Glabrae) property and taste are neutral, sweet and bland. It enters the liver channel of the foot *jueyin* and the stomach channel of the foot *yangming* meridians and functions to clear heat and eliminate dampness, cool blood and detoxify toxins. It is an important herb to treat syphilis. Recently, *tǔ fú líng* (Rhizoma Smilacis Glabrae) has been used to treat hepatitis B. It is believed it has the function to eliminate dampness and detoxify toxins, and to reduce transaminase. The author also uses it often to treat viral hepatitis B and it is very effective.

1. *Tǔ Fú Líng* (Rhizoma Smilacis Glabrae) —— *Wǔ Wèi Zǐ* (五味子, Fructus Schisandrae Chinensis)

Function

To eliminate dampness and detoxify, to protect the liver and reduce transaminase secretions.

Application

It is used for viral hepatitis B and all kinds of hepatitis that have an elevated transaminase combined with dampness.

Compatibility Analysis

Tǔ fú líng (Rhizoma Smilacis Glabrae) is sweet, bland and can leach out dampness, clear blood and detoxify toxins. *Wǔ wèi zǐ* (Fructus Schisandrae Chinensis) is sweet, sour and warm. It functions to tonify qi and generate fluid. Recent research has shown *wǔ wèi zǐ* (Fructus Schisandrae Chinensis) has the function to protect the liver and reduce transaminase secretion. It is a common herb to treat hepatitis. However, *wǔ wèi zǐ*'s (Fructus Schisandrae Chinensis) tastes are sour, warm and astringent. For some hepatitis patients who have damp-heat, it is not appropriate, especially for hepatitis with more dampness, where the dampness pathogen is entangled, persistant and hard to eliminate. Hence, the viral hepatitis is prolonged without recovery. If only *wǔ wèi zǐ* (Fructus Schisandrae Chinensis) is used it will aggravate it and trap dampness, and keep the pathogen in. If combined with *tǔ fú líng* (Rhizoma Smilacis Glabrae) it can reduce or eliminate the astringent property of *wǔ wèi zǐ* (Fructus Schisandrae Chinensis). Laboratory research has proven that *tǔ fú líng* (Rhizoma Smilacis Glabrae) has a certain function to control the hepatitis B virus and is effective for protecting the liver. The two herbs combined together can enhance the function to reduce transaminase, meanwhile reducing the resistance build up of *wǔ wèi zǐ* (Fructus Schisandrae Chinensis) in reducing transaminase.

Case 1

Initial Visit

A male patient, 35 years old, October 19th, 1991.

The patient was hospitalised in July 1991 for a burn injury. After one month of hospitalization, he developed jaundice. The results of a liver function test: Jaundice index 70U, serum thymol turbidity test (TTT) 5.9U, ALT 390U/L, HBs Ag positive. He was then transferred to an infection hospital to be treated for one more month, after which his liver function test showed: Jaundice index and TTT were all normal. Only the ALT 290U/L was still elevated. Later he was referred by someone to our hospital for treatment. The patient had good physical development, no yellowish signs, a wiry and forceful pulse, a turbid and greasy tongue coating, with a dark tongue with stasis spots. His chief complaint was stomach and abdominal distention and fullness, with a poor appetite. The treatment plan was to clear heat and detoxify, to harmonise the stomach and eliminate dampness. A formula was chosen accordingly.

Prescription

白花蛇舌草	bái huā shé shé cǎo	30g	Herba Hedyotis Diffusae
蒲公英	pú gōng yīng	20g	Herba Taraxaci
丹参	dān shēn	9g	Radix et Rhizoma Salviae Miltiorrhizae
柴胡	chái hú	15g	Radix Bupleuri
泽兰	zé lán	15g	Herba Lycopi
虎杖	hǔ zhàng	15g	Rhizoma et Radix Polygoni Cuspidati
藿香	huò xiāng	9g	Herba Agastachis
茵陈	yīn chén	10g	Herba Artemisiae Scopariae
枳壳	zhǐ qiào	10g	Fructus Aurantii
苏梗	sū gěng	9g	Caulis Perillae
赤芍	chì sháo	15g	Radix Paeoniae Rubra
甘草	gān cǎo	6g	Radix Glycyrrhizae
土茯苓	tǔ fú líng	20g	Rhizoma Smilacis Glabrae
五味子	wǔ wèi zǐ	10g	Fructus Schisandrae Chinensis

Water was used to decoct, it was divided into two servings, one pack per day.

Second Visit

October 25th: After taking seven packs of the above formula, his appetite increased, his stomach and abdominal distention also reduced. *Yīn chén hāo* (Herba Artemisiae Scopariae) was removed from the above formula to continue treatment.

Third Visit

November 8th: He took another fourteen packs, and the clinical symptoms basically disappeared. His liver function test showed: ALT reduced to 72U/L. He continued with the last formula.

Fourth Visit

December 18th: The above formula was modified and he took more than forty packs more. His liver function test showed: All indexes were within the normal range. The Hbs Ag had become negative. He used the last formula but reduced the dosage of *wǔ wèi zǐ*

(Fructus Schisandrae Chinensis) to 6g, and *tǔ fú líng* (Rhizoma Smilacis Glabrae) 15g and he took it every other day for a month. Then returned for a check up.

Fifth Visit

On January 20th, 1992 his liver function was retested and showed up normal, HBs Ag (−).

Case 2

Initial Visit

A male patient, 44 years old, July 28th, 1995.

The patient had an history as a carrier of the hepatitis B virus. For the previous ten days, he had lassitude in his four limbs, with soreness and weakness, a poor appetite, stomach and abdominal distention and fullness, hypochondriac pain, etc. After testing his liver function it showed: TTT 12.2U, ALT 246U/L, the rest (−), of the five items indicating hepatitis: HBs Ag (+), HbeAg (+), Anti HB c (+). His pulse was deep and wiry. His tongue coating was thin and his tongue was slightly dark. The treatment plan was to nourish the liver, strengthen the spleen, and detoxify and reduce transaminase secretion.

Prescription

白花蛇舌草	*bái huā shé shé cǎo*	30g	Herba Hedyotis Diffusae
蒲公英	*pú gōng yīng*	15g	Herba Taraxaci
丹参	*dān shēn*	20g	Radix et Rhizoma Salviae Miltiorrhizae
柴胡	*chái hú*	15g	Radix Bupleuri
五味子	*wǔ wèi zǐ*	9g	Fructus Schisandrae Chinensis
生黄芪	*shēng huáng qí*	20g	Radix Astragali (raw)
砂仁	*shā rén*	9g	Fructus Amomi
枳壳	*zhǐ qiào*	12g	Fructus Aurantii
土茯苓	*tǔ fú líng*	20g	Rhizoma Smilacis Glabrae
白术	*bái zhú*	12g	Rhizoma Atractylodis Macrocephalae
泽兰	*zé lán*	15g	Herba Lycopi
虎杖	*hǔ zhàng*	15g	Rhizoma et Radix Polygoni Cuspidati
沙苑子	*shā yuàn zǐ*	15g	Semen Astragali Complanati
丹皮	*dān pí*	12g	Cortex Moutan
甘草	*gān cǎo*	6g	Radix Glycyrrhizae

The herbs were soaked in clear water for 30 minutes, then decocted twice and mixed together, and divided into two servings, to be taken in the morning and night, one pack per day.

Second Visit

August 16th: The above formula was modified and twenty packs were taken. The clinical symptoms reduced significantly. The liver function test showed: TTT 10.8U, ALT 67U/L. The original formula was modifed and he continued the treatment.

Third Visit

September 13th: All his symptoms were gone. The liver function test showed: TTT 6.4U, ALT 42U/L. He still used the original formula's modification and continued with treatment.

Fourth Visit

October 4th: All liver function indexes were within the normal range, the five indicators for hepatitis B became three small positive values. The patient was told to take the same formula every other day for one more month.

After one year follow up visit, there was no reoccurrence.

Sān qī (三七 , Radix et Rhizoma Notoginseng) is sweet, bitter, and slightly warm. It enters the liver channel of the foot *jueyin* and the stomach channel of the foot *yangming* meridians. It functions to stop bleeding, transform stasis, eliminate swelling and stop pain. To stop bleeding is a special characteristic of *sān qī* (Radix et Rhizoma Notoginseng). *Sān qī* (Radix et Rhizoma Notoginseng) can effectively treat all kinds of bleeding, regardless of whether the bleeding is inside or outside the body. Meanwhile, it also has the function of invigorating blood and transforming stasis; hence, it has the characteristic of "stopping bleeding without leaving stasis". *Sān qī* (Radix et Rhizoma Notoginseng) also can reduce swellings and relieve pain. If its qi stagnation, blood stasis, or wind-damp pain, all can be treated with *sān qī* (Radix et Rhizoma Notoginseng) and have effective quick results. For treating injury, swelling and pain, abscesses, and toxic swellings, its expelling effect is also quick. Huang Gong-xiu said it "*can transform blood into water*". Zhang Xi-chun said: "*Only one single herb, sān qī* (Radix et Rhizoma Notoginseng), *can replace Xià Yū Xuè Tāng (Stasis-Precipitating Decoction, 下瘀血汤) in the Essentials From the Golden Cabinet, but it is more stable than Xià Yū Xuè Tāng (Stasis-Precipitating Decoction).*" Recently, *sān qī* has been used to treat angina pectoris in coronary disease, using its function to expel the stasis and alleviate the pain.

1. *Sān Qī* (Radix et Rhizoma Notoginseng) ── *Xī Yáng Shēn* (西洋参, Radix Panacis Quinquefolii)

Function

To tonify qi and invigorate blood, to transform blood stasis and alleviate the pain.

Application

It is used for angina pectoris in coronary heart disease, with insufficient blood supply from the cardiac muscle, due to heart qi deficiency and blood stasis.

Compatibility Analysis

Sān qī (Radix et Rhizoma Notoginseng) can transform blood stasis and alleviate pain. *Xī yáng shēn* (西洋参 , Radix Panacis Quinquefolii) is sweet, cold, and slightly bitter, it can tonify qi, generate fluid and nourish yin. The two herbs combined together can transform blood stasis without injury to the upright qi, they can tonify qi but not leave stasis. It can be used for any case of both qi and yin deficiency inducing palpitations, a stifling chest sensation, a suffocating sensation, and pain anterior to the heart. They are gentle and effective. Clinically, it is used alone. It can be ground into powder and put into capsules to take, or according to diagnosis added into a formula. All of them are effective treatments.

Case Study

Initial Visit

A male patient, 54 years old, February 15th, 1990.

The patient had palpitations, shortness of breath, a suffocating sensation in the area anterior to the heart. He had had his symptoms for a couple of months already. He was investigated by his factory's hospital. The EKG showed the cardiac muscle did not supply sufficient blood. He used *Fù Fāng Dān Shēn Piàn* (Compound Salvia Tablet, 复方

丹参片), *Sù Xiào Jiù Xīn Wán* (Fast Acting Heart Saving Pill, 速效救心丸), etc. to treat it, but symptoms were sometimes better and sometimes worse. There was no big change, after many EKG tests, over time. Hence, he tried to use Chinese medicinals to treat it. His pulse was wiry and deep, with no strength at the two *cun* pulses. His tongue coating was thin and white and his tongue was slightly dark. He felt suffocated and distention in the pectoral area, and sometimes shortness of breath, irritability, and insomnia. The diagnosis was the syndrome of heart qi deficiency with heart blood vessels stagnated and obstructed. The heart qi deficiency was indicated by the pulse being deep and the forcelessness of the two *cun* pulses. The heart blood vessel obstruction was indicated by the tongue being dark and the suffocating sensation and distention in the pectoral area. The treatment plan was to tonify qi, invigorate blood and the blood vessels.

Prescription

西洋参	*xī yáng shēn*	9g	Radix Panacis Quinquefolii
三七粉	*sān qī fěn*	6g	Radix et Rhizoma Notoginseng Powder (dissolved)
麦冬	*mài dōng*	12g	Radix Ophiopogonis
丹参	*dān shēn*	24g	Radix et Rhizoma Salviae Miltiorrhizae
五味子	*wǔ wèi zǐ*	9g	Fructus Schisandrae Chinensis
炙甘草	*zhì gān cǎo*	9g	Radix et Rhizoma Glycyrrhizae Praeparata cum Melle
檀香	*tán xiāng*	10g	Lignum Santali Albi
葛根	*gé gēn*	24g	Radix Puerariae Lobatae

Water was used to decoct, and was divided into two servings, one pack per day.

Second Visit

February 20th: After taking five packs of the above formula, his chest distention, and shortness of breath symptoms had reduced. He still had irritability and insomnia sometimes. *Bǎi hé* (Bulbus Lilii) 24g and *zǎo rén* (枣仁 , Semen Ziziphi Spinosae) 15g, were added, and the patient continued taking the formula.

Third Visit

March 5th: After taking more than ten packs of the last formula, the clinical symptoms basically disappeared. He then used an equal amount of *xī yáng shēn* (Radix Panacis Quinquefolii) and *sān qī* (Radix et Rhizoma Notoginseng) to grind into powder and put into 0.5g capsules. Each time, he took two to four capsules, three times a day.

Third Visit

April 1st: After taking the formula, he never had the symptoms again. The EKG reported that blood supply to the cardiac muscle had become better than before.

Sān léng's (三棱 , Rhizoma Sparganii) property and taste are neutral and bitter. It enters the liver channel of the foot *jueyin* and the spleen channel of the foot *taiyin* meridians. Its functions can promote the movement of qi and break up blood stasis; it can eliminate stasis and alleviate pain. This product has a stronger power to break up blood and eliminate stasis. Hence, it is more often used for blood stasis induced amenorrhoea, postpartum stasis blood clumping or accumulation, etc. In addition, it has the function to reduce accumulation and strengthen the spleen. It is used for food accumulation and stagnation as well as stomach and abdominal distention and fullness, etc. However, dosages need to be small, if in a large dosage, it invigorates the blood.

1. *Sān Léng* (Rhizoma Sparganii) ——
É Zhú (莪术, Rhizoma Curcumae)

Function

To invigorate the blood and break up stasis, to reduce accumulation and dissipate nodules, to promote qi movement and alleviate pain.

Application

It is used for abdominal abscesses and accumulations; hypochondriac and abdominal pain induced by qi stagnation or blood stasis; and to strengthen the stomach, reduce accumulation and eliminate distention.

Compatibility Analysis

The two herbs, *sān léng* (Rhizoma Sparganii) and *é zhú*'s (莪术 , Rhizoma Curcumae) functions are similar. Clinically they are used together. If we compare the two herbs, *é zhú* (Rhizoma Curcumae) has a slightly stronger force to promote qi movement and break up qi stagnation; it can open the stomach and reduce food accumulation. *Sān léng* (Rhizoma Sparganii) is slightly better for breaking up blood and eliminating stasis. However, to invigorate blood, qi needs to be regulated first. Qi is the leader of the blood, qi moves then blood follows. Both herbs can promote qi movement; hence, they are effective to treat qi stagnation and blood stasis. Zhang Xi-chuan thought that: "*Sān léng* (Rhizoma Sparganii) *and é zhú* (Rhizoma Curcumae) *have a similar property and they are gentle. This compatibility can slowly eliminate female stasis of blood, even if it is very severe. Harsh breaking up herbs cannot achieve this outstanding merit. It is sān léng* (Rhizoma Sparganii) *and é zhú's* (Rhizoma Curcumae) *inherent capability.*" Modern clinical medical research proved that the two herbs have a better function to promote the body to self absorb blood clots, hence, it can eliminate and absorb the clumping of an extrauterine pregnancy. For example, *Yì Wèi Rèn Shēn Qù Yū Tāng* (Eliminating Clumping of an Extrauterine Pregnancy Decoction, 异位妊娠祛瘀汤) includes *sān léng* (Rhizoma Sparganii) and *é zhú* (Rhizoma Curcumae). If they are removed, the effectiveness will be not so obvious. The author likes to use this compatibility to treat all kinds of qi stagnation and blood stasis in his clinical practice.

Initial Visit

A female patient, 28 years old, December 25th, 1985.

There was a dove egg sized clump, on the superior area, of the patient's left breast for a couple of months. It was painful if it was pressed on. She had some hospital checks and it was diagnosed as mammary gland hypertrophy. Her pulse was wiry, thready and fast. Her tongue coating was thin and white. The treatment plan was to invigorate blood, eliminate stasis and soften hardness.

Prescription

丹参	dān shēn	30g	Radix et Rhizoma Salviae Miltiorrhizae
当归	dāng guī	15g	Radix Angelicae Sinensis
三棱	sān léng	9g	Rhizoma Sparganii
莪术	é zhú	9g	Rhizoma Curcumae
生牡蛎	shēng mǔ lì	30g	Concha Ostreae (raw)
川芎	chuān xiōng	10g	Rhizoma Chuanxiong
赤芍	chì sháo	10g	Radix Paeoniae Rubra
橘络	jú luò	10g	Vascular Aurantii
瓜蒌	guā lóu	15g	Fructus Trichosanthis
柴胡	chái hú	15g	Radix Bupleuri
元参	yuán shēn	15g	Radix Scrophulariae
甘草	gān cǎo	6g	Radix Glycyrrhizae

Water was used to decoct, one pack per day.

Second Visit

December 31st: After taking six packs of the formula the clump visibly reduced. Sometimes, she had insomnia. She used the above prescription, added *zǎo rén* (Semen Ziziphi Spinosae) 15g, and continued to take it again.

Third Visit

January 6th, 1986: After taking another seven packs, her clump reduced by more than half. It was not painful when pressed upon. Her sleep was okay. She continued to take the last formula.

Fourth Visit

January 13th: The clump basically disappeared. She continued with five more packs and recovered.

Initial Visit

A male patient, 33 years old, August 5th, 1987.

The patient had a history of stomachache for a couple of years. This time, he also had stomach distention, fullness and pain, anorexia, and hypochondriac discomfort on both sides for more than two months. He had treatment but did not improve; instead, the pain became severe. He came to hospital for a check up. An upper gastrointestinal tract barium radiograph diagnosed a duodenal ulcer. His upper abdomen was distended and full, tenderness on pressing was obvious. His pulse was wiry, thready and slippery. His tongue coating was white, and his tongue was red but slightly dark. The diagnosis was the disharmony of stomach and spleen inducing anorexia, and stomach distention. Prolonged sickness invaded to collaterals with qi and blood stagnation, and the obstruction induced stomachache, the dark tongue and wiry pulse. The treatment plan was to strengthen the spleen and harmonise the stomach, to invigorate blood and regulate the qi.

Prescription

百合	*bǎi hé*	24g	Bulbus Lilii
乌药	*wū yào*	10g	Radix Linderae
檀香	*tán xiāng*	6g	Lignum Santali Albi
砂仁	*shā rén*	3g	Fructus Amomi
三棱	*sān léng*	6g	Rhizoma Sparganii
莪术	*é zhú*	6g	Rhizoma Curcumae
丹参	*dān shēn*	20g	Radix et Rhizoma Salviae Miltiorrhizae
海螵蛸	*hǎi piāo xiāo*	15g	Endoconcha Sepiae
枳壳	*zhǐ qiào*	10g	Fructus Aurantii
白术	*bái zhú*	12g	Rhizoma Atractylodis Macrocephalae
枸杞	*gǒu qǐ*	15g	Fructus Lycii
甘草	*gān cǎo*	6g	Radix Glycyrrhizae

Second Visit

August 15th: The above formula was decocted with water and after taking seven packs, the stomach distention and fullness reduced a lot and there was no more pain except that he was slightly constipated. He used the original formula, added *dà bèi mǔ* (Bulbus Fritillariae Thunbergii) 10g, and continued to take it.

Third Visit

September 1st: After taking more than 10 packs of the last formula, all the symptoms disappeared.

Shān zhū yú's (山茱萸 , Fructus Corni) taste is sour and its property is warm. It enters the liver channel of the foot *jueyin* and the kidney channel of the foot *shaoyin* meridians. This herb can tonify the liver and kidney yin, and also can warm and tonify kidney yang. It is an important herb to neutrally tonify the liver and kidney. Its tastes are sour and astringent; hence, it has the function to astringe, secure and bind. It can be used for yang qi deficiency induced spermatorrhoea, frequent urination, persistant deficiency sweating and menorrhagia. Its clinical compatibility is broad.

1. *Shān Zhū Yú* (Fructus Corni) ——
Hú Táo Rén (胡桃仁, Semen Juglandis Regiae)

Function

To warm the kidney and tonify the lung, to reduce urine and stop leakage.

Application

This is used for elderly enuresis, or frequent nocturia.

Compatibility Analysis

Shān zhū yú (Fructus Corni) warms up the kidney and constrains the leakage. *Miscellaneous Records of Famous Physicians* (*Bié Lù*, 别录) said: *"Strengthen yin and tonify essence.....stop copious urine."* Zheng Quan said it can *"stop elderly urination that is out of control"*. Wang Hou-gu said: *"If slippery then qi has deserted, an astringent formula can secure it. Shān zhū yú (Fructus Corni) stops copious urine and secures essence, because of its sour and astringent tastes."* *Hú táo rén* (胡桃仁 , Semen Persicae) is sweet and warm. It enters the liver channel of the foot *jueyin* and the kidney channel of the foot *shaoyin* meridians. It can warm the kidney; tonify the lung and the *mingmen*. The two herbs combined together then enhance the power to warm the kidney and constrain the leakage. Also the lung is the upper source of water, if lung qi is plenty, urine will be secured. *Hú táo rén* (Semen Persicae) tonifies lung as well as enhances the astringing effect. Hence, an elderly person can use it effectively for kidney qi deficiency disturbing urinary bladder qi transformation, or kidney deficiency that cannot secure and bind, or lung qi deficiency that cannot raise and bind inducing frequent urination.

Case Study

Initial Visit

A male patient, 75 years old, November 2nd, 1974.

The patient had frequent nocturia. It occurred more than ten times a night and disturbed his sleep. His daytime urination was normal. His pulse was wiry and big. The two *chi* pulses were slightly weak. The tongue coating was thin and white. His diet was normal and there was no other discomfort. This was both kidney and lung deficiency syndrome of the elderly. The kidney governs the two lower orifices, if the kidney is not secure, then there is frequent nocturia. If there is lung deficiency then there is no strength to raise and secure, hence profuse urination. He was told to buy *hú táo* (胡 桃 , Semen Persicae) 1kg and peel off the skin and *shān yú ròu* (山萸肉 , Fructus Corni) 250g.

Every night, he used *shān yú ròu* (Fructus Corni) 20g, and *hú táo* (Semen Persicae) 30g, to decoct and took it with every meal, continuously for more than ten days. The frequency of his nocturia obviously reduced to three to four times every night. He was told to eat *hú táo rén* (Semen Persicae) frequently to clear up the problem.

2. *Shān Zhū Yú* (Fructus Corni) ——
Wú Gōng (蜈蚣, Scolopendra)

Function

To warm the kidney and to invigorate the yang.

Application

To treat male sexual impotence, and premature ejaculation.

Compatibility Analysis

Shān zhū yú (Fructus Corni) tonifies the kidney's yin and yang, it tonifies essence and constrains leakage. Zheng Quan said *"tonify kidney qi, increase sexual competence, enhance essence and bone marrow."* *Wú gōng*'s (蜈 蚣 , Scolopendra) property is warm and tastes sweet and salty. It enters the liver channel of the foot *jueyin* meridian. It is the herb to extinguish wind and stop conversion. Recent clinical reports indicate that it is effective to treat impotence. The liver channel of the foot *jueyin* meridian travels around the genital area and enters the lower abdomen. *Wú gōng* (Scolopendra) enters the liver channel of the foot *jueyin* meridian. Its property is freeing moving. It enters the genital organs to increase sexual competence. However, if using *wú gōng* (Scolopendra) to tonify yang and treat impotence, it only works temporarily. It should be used with kidney tonifying herbs to maintain its long term effect. *Shān zhū yú* (Fructus Corni) can tonify the kidney's yin and yang, increase sexual competence, bind essence and constrain the leakage. The two herbs combined together are better than the application of a single herb.

Case Study

Initial Visit

A male patient, 25 years old, June 30th, 1985.

The patient got married half a year previously. He came to visit due to suffering from impotence and spermatorrhoea for three months. On observation the patient's spirit appeared listless and his physical development was satisfactory. He complained of low back soreness and lassitude, and frequent nocturnal emissions during sleep. It occurred more than twice a week. During sexual intercourse, he had impotence and could not achieve an erection. His pulse was wiry, thready and fast, his tongue was red, with a thin white coating. On enquiry, he had had a history of masturbation. He was married for half a year, initially when the sexual intercourse disharmony started, he was over worried, sad and stressed. Later, it gradually developed into impotence, premature ejaculation and spermatorrhoea. The diagnosis was kidney yin deficiency, with deficiency minster fire flaring inducing spermatorrhoea impotence syndrome. The treatment plan was to use a formula to tonify the kidney and nourish yin, fulfill essence and assist yang.

Prescription

天冬	tiān dōng	20g	Radix Asparagi
党参	dǎng shēn	20g	Radix Codonopsis
生地	shēng dì	20g	Lignum Santali Albi
菟丝子	tù sī zǐ	15g	Semen Cuscutae
川黄柏	chuān huáng bǎi	9g	Cortex Phellodendri Chinensis
芡实	qiàn shí	10g	Semen Euryales
枸杞	gǒu qǐ	12g	Fructus Lycii
山萸肉	shān yú ròu	15g	Fructus Corni
杭白芍	háng bái sháo	15g	Radix Paeoniae Lactiflorae
覆盆子	fù pén zǐ	10g	Fructus Rubi
蜈蚣	wú gōng	1 pcs	Scolopendra

Water was used to decoct, to be taken one pack per day.

Second Visit

July 7th: He had no spermatorrhoea during the night, after taking seven packs of the above formula. His impotence also improved. During sexual intercourse, he achieved some erection. The formula was not changed due to positive effects and the patient continued to take the same formula.

Half a year later after more than 20 packs, the patient recovered from his impotence.

3. *Shān Zhū Yú* (Fructus Corni) ——
Shēng Lóng Gǔ (生龙骨, raw Os Draconis), *Shēng Mǔ Lì* (生牡蛎, raw Concha Ostreae)

Function

To secure collapse and constrain sweats, to tonify collaterals and stop bleeding.

Application

This is used for qi deficiency, superficial yang not secure and persistent deficiency sweats, a flaccid pulse almost stopped, and also used for acute or chronic haemoptysis.

Compatibility Analysis

Zhang Xi-chun said: "*Shān zhū yú (Fructus Corni) tastes sour and has a warm property. It can astringe and bind basal qi, cheer up the spirit, secure and bind the slippery and abandonment disorder. It can strongly tonify the liver qi and within its astringent property, it also has a dispersing character, hence, it opens the nine orifices, and promotes blood flow inside the vessels. It treats liver deficiency spontaneous sweating, moreover, it retains upright qi instead of trapping pathogens inside."* *Shēng lóng gǔ* (生 龙 骨 , raw Os Draconis) is sweet, astringent, and slightly cold. It has the function to anchor the yang and calm palpitations, secure and bind to stop sweating. *Shēng mǔ lì* (raw Concha Ostreae) is salty, neutral and slightly cold. It has the same function as *lóng gǔ* (Os Draconis). They both can tonify yin and anchor the yang, secure essence and stop sweating, hence, these two herbs are always used together. All three herbs, *shān zhū yú* (Fructus Corni), *shēng lóng gǔ* (raw Os Draconis) and *shēng mǔ lì* (raw Concha Ostreae) have retaining, securing and binding functions. If used together, the power is more obvious. It is in the category of *"mutual accentuation."* Clinically, if persistent spontaneous sweating, qi deficiency and flaccid pulse are observed, regardless of whether it is yin collapse going downward (urinary incontinence, and loss of essence), yang collapse going upward (spontaneous sweating, and asthma), or both yin and yang deficiency, all conditions can use them. Zhang Xi-chun's *Jì Jì Tāng* (Complementary Decoction, 既 济 汤) adopts this compatibility. Meanwhile, the three herbs used together, have the characteristic of retaining and stopping bleeding but without leaving stasis, as in Zhang Xi-chun's *Bǔ Luò Bǔ Guǎn Tāng* (Collaterals-tonifying and Channel-reinforcing Decoction, 补络补管汤) for treating haematemesis or haemoptysis, it is a formula that uses three herbs as the king herb.

Case 1

Initial Visit

A male patient, 53 years old, May 20th, 1987.

The patient had external invasion and fever for a couple of days. After using the medicine that releases the exterior and induces sweating, his body temperature reduced, but for the previous two days he had persistent sweating, accompanied by shortness of breath, lassitude, spirit fatigue and his condition was deteriorating daily. His pulse was floating, big, without strength. The tongue coating was thin and white. These were symptoms of over sweating exhausting yang and causing collapse. The treatment plan was to immediately use a retaining sweat and securing collapse formula.

Prescription

山茱萸	*shān zhū yú*	30g	Fructus Corni
生龙骨	*shēng lóng gǔ*	30g	Os Draconis (raw)
生牡蛎	*shēng mǔ lì*	30g	Concha Ostreae (raw)
太子参	*tài zǐ shēn*	15g	Radix Pseudostellariae
麦门冬	*mài mén dōng*	15g	Radix Ophiopogonis
五味子	*wǔ wèi zǐ*	9g	Fructus Schisandrae Chinensi
甘草	*gān cǎo*	6g	Radix Glycyrrhizae

Water was used to decoct twice and mixed together, then divided into two servings, one pack per day.

Second Visit

May 23ʳᵈ: After taking one pack of the above formula, he stopped sweating and after three packs he recovered.

Case 2

Initial Visit

A male patient, 28 years old, September 24ᵗʰ, 1994.

After exertion, the patient had an external invasion four months ago. He had fever, cough with a little phlegm, and phlegm with blood. After he was treated by a local hospital, the fever reduced but he still had persistent haemoptysis. It was also treated as pulmonary tuberculosis without effect. The patient went to a Beijing hospital to check it out and was diagnosed with bronchiectasis. He had been treated with ethamsylate and aminomethylbenzoic acid etc. but the haemoptysis still did not stop. A friend referred him to visit our hospital's outpatient clinic. The patient had good physical development but his chronic disease was apparent from his face. His chief complaint was cough, with a little phlegm containing a small amount of bloody red colour, every day 3 ~ 5 times. His tongue was slightly pale, and his coating was thin and white. His pulse was wiry and thready. A chest x-ray reported: The lungs' interstitial markings were rough, but did not show tuberculosis spots. It was only the overexertion and external invasion that injured the lung's collaterals. Once the collaterals were injured, blood overflowed outside and haemoptysis was the result. The best method to treat it was to use retaining, securing and binding, with tonifying the lung and stopping bleeding. Zhang Xi-cuan's *Bǔ Luò Bǔ Guǎn Tāng* (Collaterals-tonifying and Channel-reinforcing Decoction) was combined with modified *Huà Xuè Dān* (Blood-transforming Pill, 化血丹) to treat him.

Prescription

山茱萸	shān zhū yú	30g	Fructus Corni
生龙骨	shēng lóng gǔ	24g	Os Draconis (raw)
生牡蛎	shēng mǔ lì	24g	Concha Ostreae (raw)
大贝母	dà bèi mǔ	10g	Bulbus Fritillariae Thunbergii
三七粉	sān qī fěn	6g	Radix et Rhizoma Notoginseng Powder (dissolved)
血余炭	xuè yú tàn	12g	Crinis Carbonisatus
当归	dāng guī	15g	Radix Angelicae Sinensis
紫菀	zǐ wǎn	12g	Crinis Carbonisatus
甘草	gān cǎo	6g	Radix Glycyrrhizae

Water was used to decoct, then divided into two servings, one pack per day.

Second Visit

September 29ᵗʰ: After taking five packs of the above formula, the haemoptysis reduced, and the cough also reduced. He continued to take the origingal formula with *bái jí* (白及 , Rhizoma Bletillae) 12g added.

Third Visit

October 7th: After he continuously took seven packs, the haemoptysis basically stopped, except that sometimes there was a pink colour of blood inside the phlegm. He continued treatment with the above formula.

Fourth Visit

October 24th: After the patient had taken another seven packs, the haemoptysis stopped. The patient stopped the formula by himself. Recently, he saw pink silk like blood in the phlegm. He was told to quit smoking, not overexert himself and continue taking the formula as treatment until completely recovered.

4. *Shān Zhū Yú* (Fructus Corni) —— *Jiāng Huáng* (姜黄, Rhizoma Curcumae Longae)

Function

To nourish and soften the tendons, to activate stasis and alleviate pain.

Application

This is used for frozen shoulder (shoulder bursitis).

Compatibility Analysis

Shān zhū yú (Fructus Corni) is warm in property and enters the liver channel of the foot *jueyin* meridian. It has the function to nourish essence and blood, to soften the liver and tendons. Meanwhile, it unblocks blood vessels to treat wind arthralgia induced shoulder pain. The *Divine Husbandman's Classic of the Materia Medica* said it "*expels cold painful obstruction*". In 1984, a Chinese medicine magazine reported that only using *shān zhū yú* (Fructus Corni) alone to treat shoulder bursitis is effective. The author tried it for this and discovered it really did have this effect. *Jiāng huáng* (姜黄 , Rhizoma Curcumae Longae) is bitter, acrid and warm and has the function to move blood and free the collaterals, disperse wind and alleviate pain. It is especially good at treating upper limbs, arm and shoulder pain induced by wind-damp. The *Grand Materia Medica* said: "*It treats wind arthralgia arm pain.*" Clinically, *jiāng huáng* (Rhizoma Curcumae Longae) is an important herb to treat upper limb, shoulder and arm pain. However, as *jiāng huáng's* (Rhizoma Curcumae Longae) properties are warm and dry, its power to break up blood and descend qi is strong. If over-dosed it can consume the qi and injure blood. Hence, Zhang Shi-wan said: "*If arm pain is due to blood deficiency, taking it will make the pain severe*". If combined with *shān zhū yú* (Fructus Corni), then its disadvantage of consuming qi and injuring blood can be prevented. *Jiāng huáng* (Rhizoma Curcumae Longae) also can increase *shān zhū yú's* (Fructus Corni) effectiveness for treating shoulder pain. It can be said that they complement each other and are better off used together than by either one individually.

Case Study

Initial Visit

A male patient, 54 years old, May 5th, 1986.

The patient had right shoulder pain for more than three months. Elevation and horizontal abduction had a limited range of motion and was also very painful. The muscle was obviously looser than the left shoulder's. He had treated it with massage but did not have any results. His pulse was wiry and thready. His tongue coating was thin and white. He was diagnosed with the frozen shoulder syndrome. The treatment formula was to nourish blood and soften the tendons, to activate blood and the collaterals.

Prescription

山茱萸	shān zhū yú	40g	Fructus Corni
姜黄	jiāng huáng	10g	Rhizoma Curcumae Longae
当归	dāng guī	15g	Radix Angelicae Sinensis
丹参	dān shēn	30g	Radix et Rhizoma Salviae Miltiorrhizae
桑枝	sāng zhī	20g	Ramulus Mori
生苡米	shēng yǐ mǐ	24g	Semen Coicis (raw)
生黄芪	shēng huáng qí	20g	Radix Astragali (raw)
杭白芍	háng bái sháo	20g	Radix Paeoniae Lactiflorae
甘草	gān cǎo	6g	Radix Glycyrrhizae

Water was used to decoct, then divided into two servings, taken in the morning and at night.

Second Visit

June 21st: The above formula was modified and he took a total of more than thirty packs. The pain disappeared and he could freely elevate and abduct his shoulder backward.

Shān zhā's (山楂 , Fructus Crataegi) tastes and property are sour, sweet and warm. It enters the spleen channel of the foot *taiyin*, the stomach channel of the foot *yangming* and the liver channel of the foot *jueyin* meridians. Its functions are to reduce food stagnation and strengthen the stomach, to invigorate blood and transform stasis. It is a food category herb. Based on modern clinical research, taking it internally can increase enzyme secretion in the stomach and help digestion. Hence, it has the function to reduce food stagnation and strengthen the stomach. Also as it contains lipase, it can promote fatty food digestion. Hence, it has the effect to reduce meat accumulation and cholesterol. Clinically, combined with zé *xiè* (Rhizoma Alismatis) it can reduce cholesterol and treat steatorrhoeic hepatosis, high blood pressure, etc.

1. *Shān Zhā* (Fructus Crataegi) ——
Zé Xiè (泽泻, Rhizoma Alismatis)

Function

To transform stasis and resolve dampness, to reduce cholesterol.

Application

This can be used for high cholesterol and arteriosclerosis induced hypertension and steatorrhoeic hepatosis.

Compatibility Analysis

Steatorrhoeic hepatosis, a metabolic disorder of cholesterol, accompanied by hyperlipidaemia, belongs to the category in Chinese medicine of phlegm and dampness stagnation and obstruction, qi and blood stasis and stagnation. Dampness arthralgia and blood stasis is the main aetiology. *Shān zhā* (Fructus Crataegi) can reduce food stagnation and transform stagnation, invigorate blood and transform stasis. It has the function to reduce cholesterol. *Zé xiè* (Rhizoma Alismatis) promotes urination and leaches out dampness. Based on modern clinical research, it has the function to reduce cholesterol and glucose, as well as resist steatorrhoeic hepatosis. The two herbs combined together can eliminate dampness, transform stasis and reduce cholesterol; hence, it can effectively treat hyperlipidaemia and steatorrhoeic hepatosis.

Case Study

Initial Visit

A male patient, 42 years old, May 27th, 1994.

The patient used to be obese. He recently felt lower limb heaviness, slight oedema, lassitude, anorexia, right hypochondriac distention and discomfort. He was afraid he had liver or kidney disease and so, came to visit our clinic. A regular urine test was normal. The liver function test revealed only a single item of ALT was elevated; six items of type B hepatitis were negative. A B-ultrasound test of the liver showed that the perimeter had increased, the echo was fine, the first section enhanced, the latter section sounded weak. The report indicated moderate steatorrhoeic hepatosis. Lipids: Cholesterol 9.0mmol/L, triglyceral 2.1mmol/L. His pulse was wiry, slippery and slightly deep. His tongue coating was white, turbid and slightly greasy. This was the syndrome of dampness arthralgia and blood stasis. Obese patients have more dampness. Excess dampness induces heavy limbs and oedema, anorexia and lassitude, a slippery pulse and a greasy tongue coating. If dampness obstructs the meridians and collaterals, then qi and blood stagnates and obstructs. When there is blood stasis without moving it induces the hypochondriacal distention and liver enlargement. The treatment plan was to transform stasis and activate blood, to eliminate dampness and strengthen the spleen.

Prescription

生山楂	shēng shān zhā	24g	Fructus Crataegi (raw)
泽泻	zé xiè	15g	Rhizoma Alismatis
丹参	dān shēn	20g	Radix et Rhizoma Salviae Miltiorrhizae
泽兰	zé lán	15g	Herba Lycopi
益母草	yì mǔ cǎo	20g	Herba Leonuri
蒲公英	pú gōng yīng	15g	Herba Taraxaci
三棱	sān léng	8g	Rhizoma Sparganii
莪术	é zhú	8g	Rhizoma Curcumae
茯苓	fú líng	15g	Poria
枳壳	zhǐ qiào	12g	Fructus Aurantii
白术	bái zhú	12g	Rhizoma Atractylodis Macrocephalae
甘草	gān cǎo	6g	Radix Glycyrrhizae

Water was used to decoct, then divided into two servings, to be taken in the morning and at night, one pack per day.

Second Visit

August 20th: The above formula was modified for more than three months treatment. All his symptoms disappeared. His liver function and cholesterol were all normal. His liver B-ultrasound test did not show any abnormality.

2. *Shān Zhā* (Fructus Crataegi) ——
Jī Nèi Jīn (鸡内金, Endothelium Corneum Gigeriae Galli),
Chái Hú (柴胡, Radix Bupleuri)

Function

To soothe the liver and transform stasis, to reduce lipids and melt stones.

Application

This is used for gallbladder stones and bile duct stones inside the liver.

Compatibility Analysis

The formation of gallbladder stones, bile stagnation and accumulations, bile duct infection or cholesterol metabolism disorders are related to elevated bilirubin calcium.

To treat the gallbladder usually requires surgery, or Chinese medicinals to eliminate stones or medicinals to dissolve the stones. Surgical treatment is more often used for the acute episode with severe symptoms. Chinese medicinals is more often used to eliminate stones fror patients who have a smaller stone or one that can move into the gallbladder, gallbladder inflammation is not obvious. If the stone is bigger or it's inside the liver bile duct, it is generally not appropriate to do surgery or eliminate stones, and most adopt the dissolving the stone method. When using medicine to dissolve the stone, the treatment time is longer and more expensive. The author used Chinese medicinals to dissolve the stone, and the treatment effects were good. They mainly used *shān zhā* (山 楂, Fructus Crataegi). It was used to invigorate blood and transform stasis, to reduce food stagnation and cholesterol. *Jī nèi jīn* (鸡内金 , Endothelium Corneum Gigeriae Galli) reduces accumulation and transforms stasis, transforms and dissolves stones. *Chái hú* (柴胡 , Radix Bupleuri) soothes the liver and resolves stagnation and guides herbs to the affected part in order to help dissolve the stone. This compatibility soothes the liver and transforms stasis, reduces cholesterol in order to reduce bile concentration so as to help dissolve the stone.

Case Study

Initial Visit

A female patient, 55 years old, January 19th, 2000.

The patient was hospitalised last August due to stomachache, nausea and vomiting, and found a gallbladder stone and liver bile duct stone. He took *Dǎn Shí Tōng* (Gallstone Removal Capsule, 胆石通) and *Xiāo Yán Lì Dǎn* tablets (Anti-Inflammatory and Gallbladder-Disinhibiting Tablet, 消炎利胆片) for half a year but did not get better, hence, he visited our clinic. The patient sometimes had abdominal distention, discomfort around the liver area, and obvious distention after eating. The B-ultrasound showed: The left liver bile duct was expanded about 1.0cm and many echo clumps were visible. The biggest one was about 0.9cm, accompanied by a sound echo. The right liver bile duct and common bile duct did not show expansion. The gallbladder size and shape were normal. The wall was not thick or smooth, and inside a strong echo clump with a diameter of 0.7cm was evident, and was accompanied by a sound image; sound penetration was good inside the bladder. Diagnosis: 1. Gallbladder stone; 2. Liver bile duct with multiple stones. The patient had good physical development, a wiry pulse, a thin white tongue coating and a red tongue. The stone blocked the liver collaterals. When prolonged it will induce blood stasis and collateral obstruction. The treatment plan was to slowly soothe the liver and transform stasis, to reduce cholesterol and dissolve the stones.

Prescription

鸡内金	*jī nèi jīn*	150g	Endothelium Corneum Gigeriae Galli
生山楂	*shēng shān zhā*	150g	Fructus Crataegi (raw)
柴胡	*chái hú*	100g	Radix Bupleuri
郁金	*yù jīn*	100g	Radix Curcumae
金钱草	*jīn qián cǎo*	100g	Herba Lysimachiae

威灵仙	*wēi líng xiān*	80g	Radix et Rhizoma Clematidis
半夏	*bàn xià*	60g	Rhizoma Pinelliae
杭白芍	*háng bái sháo*	80g	Radix Paeoniae Lactiflorae
黄芩	*huáng qín*	60g	Radix Scutellariae
大黄	*dà huáng*	60g	Radix et Rhizoma Rhei
枳实	*zhǐ shí*	60g	Fructus Aurantii Immaturus
甘草	*gān cǎo*	60g	Radix Glycyrrhizae
延胡索粉	*yán hú suǒ fěn*	60g	Rhizoma Corydalis Powder

All of the above herbs were ground together into a fine powder. Each time he took 6g, three times each day, in the morning, at noon and at night. Meanwhile he also took the medicine AST 0.4g, three times a day.

Second Visit

October 16th: He took the above formula continuously for nine months, and all symptoms disappeared. The patient was re-examined in a hospital. The B-ultrasound report showed: The liver membrane was smooth, the shape and size was normal, the echo was uniform, the liver bile ducts structure was clear, there was no bile duct expansion inside or outside of the liver, the gallbladder was normal, no stone was discovered.

Note

AST is a kind of gastrin and gallbladder contraction receptor's inhibitor. From past clinical practice, it was more often used for digestive ulcers and chronic gastritis. In recent years, there were reports that AST has a strong function to promote bile secretion and can be used to treat gallbladder stones. The total effective rate for treating common bile duct stones, liver bile duct stones and gallbladder stones, reached 80%. AST benefits the gallbladder function. It is mainly achieved through bile ducts secreting inorganic salt and water. In addition, AST can reduce the concentration of cholesterol, calcium, iron, and the dissociation of bilirubin in the bile, and can improve the stone elimination process. From hundreds of observations of clinical cases, oral AST is safe and reliable to use for treating gallbladder stones. So far, no side effects have been reported.

From the author's clinical practice, if there was anyone who had a bile duct stone that could not be treated by the eliminating stone method, they then used Chinese herbs plus AST to treat it with good results.

Shān Yào (山药 , Rhizoma Dioscoreae)

Shān yào's (山药 , Rhizoma Dioscoreae) taste and property are sweet and neutral. It enters four meridians: the spleen channel of the foot *taiyin*, the stomach channel of the foot *yangming*, the lung channel of the hand *taiyin* and the kidney channel of the foot *shaoyin*. It has functions to tonify the spleen and stomach, and to tonify qi and nourish yin. It is widely used in tonic formulae. It is an important product to treat deficiency. The *Divine Husbandman's Classic of the Materia Medica* said: *"It treats the injured middle jiao, tonifies deficiency and weakness, expels cold, heat and pathogenic qi, tonifies the middle jiao and strengthens it, it generates flesh, and strengthens yin."* Clinically, it is more often used to treat spleen and stomach deficiency, poor appetite and fatigue, childhood malnutrition and diarrhoea syndromes, etc. If used to treat *xiao ke* syndrome, the result is also good.

Based on research, *shān yào* (Rhizoma Dioscoreae) contains mucin and amylase. Mucin, after digestion becomes a nutritive protein and carbohydrate compound. The amylase has the function to convert starch into glucose.

1. *Shān Yào* (Rhizoma Dioscoreae) ——
Huáng Qí (黄芪, Radix Astragali)

Function

To tonify qi and strengthen the spleen, to retain and bind, and tonify deficiency.

Application

This is used for *xiao ke* syndrome.

Compatibility Analysis

This compatibility is the senior Chinese medicine doctor Shi Jin-mo's empirical herbal pair. The famous Chinese medicine expert Zhu Shen-yu stated: *"Huáng qí (Radix Astragali) tonifies the middle jiao and qi, lifts yang and tightens up the skin. Shān yào (Rhizoma Dioscoreae) tonifies qi and yin, and retains and binds kidney essence. The two herbs combined can tonify qi and generate fluid, strengthen the spleen and tonify the kidney, prevent food essence leakage, reduce urine glucose, and decrease blood sugar."*

Case Study

Initial Visit

A female patient, 60 years old, November 18th, 1987.

The patient had a history of *xiao ke* syndrome for more than ten years. Recently she felt dizzy, had blurred vision, slightly swollen lower limbs with numbness, a dry mouth and thirst, with a wiry, slippery and thready pulse, a thin white tongue coating, a red and dark tongue, and empty stomach urine glucose ++++. The diagnosis was both qi and yin deficiency *xiao ke* syndrome.

Prescription

山药	*shān yào*	30g	Rhizoma Dioscoreae
地骨皮	*dì gǔ pí*	30g	Cortex Lycii
黄芪	*huáng qí*	20g	Radix Astragali

葛根	*gé gēn*	30g	Radix Puerariae Lobatae
钩藤	*gōu téng*	30g	Ramulus Uncariae Cum Uncis
茯苓	*fú líng*	15g	Poria
泽泻	*zé xiè*	10g	Rhizoma Alismatis
川牛膝	*chuān niú xī*	15g	Radix Cyathulae
麦冬	*mài dōng*	20g	Radix Ophiopogonis
半夏	*bàn xià*	9g	Rhizoma Pinelliae
生地	*shēng dì*	15g	Radix Rehmanniae
太子参	*tài zǐ shēn*	15g	Radix Pseudostellariae
白术	*bái zhú*	9g	Rhizoma Atractylodis Macrocephalae

Water was used to decoct, then divided into two servings, one pack per day.

Second Visit

December 3rd: After taking fourteen packs of the above formula, her dizziness disappeared, her dry mouth and thirst also reduced, but she still had limb numbness, blurred vision, and her empty stomach urine glucose +++. The treatment plan was to tonify qi and nourish yin.

Prescription

山药	*shān yào*	30g	Rhizoma Dioscoreae
地骨皮	*dì gǔ pí*	30g	Cortex Lycii
太子参	*tài zǐ shēn*	20g	Radix Pseudostellariae
黄芪	*huáng qí*	20g	Radix Astragali
山萸肉	*shān yú ròu*	20g	Fructus Corni
鸡血藤	*jī xuè téng*	20g	Caulis Spatholobi
黄连	*huáng lián*	6g	Rhizoma Coptidis

Note

As it was not convenient for the patient to visit, her daughter came on her behalf. The above formula was modified for treatment for two months. Her limb numbness basically disappeared, and her urine glucose was around ± .

*Chuān niú xī*s (川牛膝 , Radix Cyathulae) tastes bitter and sour and is neutral in property. It enters the liver channel of the foot *jueyin* and the kidney channel of the foot *shaoyin* meridians. It functions to invigorate blood and promote menstruation, relax the sinews and relieve cold arthralgia syndrome, it can tonify the liver and kidney, and guide blood to move downward. The *Divine Husbandman's Classic of the Materia Medica* said: "*It treats cold, damp, atrophy and cold arthralgia syndromes, convulsion of the four limbs, knee pain preventing flexion or extension, depleted qi and blood, injured by heat, and can induce abortion through invigorating blood and qi*". Li Shi-zhen stated: "*Niú xī (Radix Cyathulae) treating sickness, with wine, can tonify the liver and kidney. Used raw it can eliminate bad blood. It has only these two uses.*" At present, our clinical applications for *niú xī* (Radix Cyathulae) are differentiated according to whether its *chuān* or *huái*. The one that is produced in Sichuan is called *chuān niú xī* (Radix Cyathulae), it has a greater strength to free the flow of, and invigorate, blood. The other is produced in Henan and is called *huái niú xī* (怀牛膝 , Radix Achyranthis Bidentatae). It has a strong power to tonify the liver and kidney.

1. *Chuān Niú Xī* (Radix Cyathulae) ——
Dì Yú (地榆, Radix Sanguisorbae)

Function

To restrain metrorrhagia, and stop metrostaxis.

Application

It is used for women's uterine bleeding syndromes. Clinically, it is irrelevant whether it is due to deficiency, excess, cold or heat, all can be differentiated and use it.

Compatibility Analysis

Uterine bleeding syndromes have many aetiologies:

1. Due to qi deficiency that cannot hold, or due to blood heat flowing restlessly;
2. Due to liver and kidney deficiency, and essence and blood not secured;
3. Due to blood stasis and the blood failing to stay in its vessels.

However, clinically, the majority of cases are due to qi deficiency and blood stasis, or deficiency mixed with excess. To secure the primary and clear the source, to block the flow and stop bleeding is the treatment principle. *Niú xī* (牛膝 , Radix Cyathulae) has functions to tonify the liver and kidney meridians and to invigorate blood, to secure the root and clear the source. *Dì yú* (地榆 , Radix Sanguisorbae) can cool the blood, restrain blood flow and stop bleeding. The two herbs combined together, one moving and one stopping can stop blood without leaving stasis with very good results. Clinically we can use them individually or combined, depending on the diagnosis.

Case 1

Initial Visit

A female patient, 38 years old, March 31st, 1989.

The patient was pregnant for 50 days and had a dilatation and curettage abortion.

Her menstruation came two months after the abortion. She had continuous spotting for two more months, accompanied by clots, low back pain, dull abdominal pain. A doctor diagnosed her with "endometrial hyperplasia". She took testosterone, oestrostilben, etc., but none of them worked. Her tongue was pale and the coating was thin and white. Her pulse was deep and thready. The syndrome was qi deficiency and blood stasis, with blood failing to remain in its vessels.

Prescription

川牛膝	chuān niú xī	24g	Radix Cyathulae
地榆	dì yú	20g	Radix Sanguisorbae
生黄芪	shēng huáng qí	30g	Radix Astragali (raw)
当归	dāng guī	15g	Radix Angelicae Sinensis
香附	xiāng fù	12g	Rhizoma Cyperi
海螵蛸	hǎi piāo xiāo	20g	Endoconcha Sepiae
赤芍	chì sháo	12g	Radix Paeoniae Rubra
甘草	gān cǎo	6g	Radix Glycyrrhizae

Water was used to decoct, then divided into two servings, taken in the morning and night, one pack per day.

Second Visit

April 7th: After taking three packs of the above formula, the bleeding basically stopped. She took three more packs and recovered.

Case 2

Initial Visit

A female patient, 41years old, January 25th, 1987.

The patient's menstrual flow had increased in recent years. Each time it took more than ten days to stop. Two months previously, she overexerted herself, and her menstruation continued until her visit, without ceasing. She checked into a hospital and was diagnosed with uterine functional bleeding. She was treated with Chinese medicine and biomedicine but they did not work. She complained that her lower abdomen felt heavy and uncomfortable, the colour of her menstruation was red but slightly pale, this was accompanied by a bitter mouth, a stifling chest sensation, anorexia and palpitations. On observation, her face was lustreless, her tongue was pale, and her tongue coating was thin and white. Her pulse was wiry and thready. The diagnosis was liver stagnation, with spleen deficiency and the qi could not hold the blood. The treatment plan was to soothe the liver, strengthen the spleen, tonfy qi and stop bleeding.

Prescription

生黄芪	shēng huáng qí	30g	Radix Astragali (raw)
柴胡	chái hú	15g	Radix Bupleuri
檀香	tán xiāng	6g	Lignum Santali Albi
砂仁	shā rén	3g	Fructus Amomi

川牛膝	*chuān niú xī*	30g	Radix Cyathulae
地榆	*dì yú*	30g	Radix Sanguisorbae
升麻	*shēng má*	5g	Rhizoma Cimicifugae
桔梗	*jié gěng*	5g	Radix Platycodonis
白参	*bái shēn*	6g	Radix Campanumoeae
海螵蛸	*hǎi piāo xiāo*	15g	Endoconcha Sepiae
茜草	*qiàn cǎo*	10g	Radix et Rhizoma Rubiae

Water was used to decoct, then divided into two servings, served warm, one pack per day.

Second Visit

January 28th: After taking three packs of the above formula she stopped bleeding, and her dragging abdominal sensation also recovered. She still had the stifling chest sensation, and felt like she had something caught in her throat; hence, the treatment was changed to Modified *Bàn Xià Hòu Pò Tāng* (Pinellia and Officinal Magnolia Bark Decoction, 半夏厚朴汤).

Note

Dysfunctional uterine bleeding is a common disease in gynaecology. The author used the *chuān niú xī* (Radix Cyathulae) and *dì yú* (Radix Sanguisorbae) combination to treat more than ten patients with uterine bleeding with good results. He also prescribed one of his personal formula *Ān Chōng Zhǐ Xuè Tāng* (Penetrating Vessel Harmonising and Stop Bleeding Decoction, 安冲止血汤). It is detailed below for reference.

Ingredients:

川牛膝	*chuān niú xī*	20~30g	Radix Cyathulae
地榆	*dì yú*	20~30g	Radix Sanguisorbae
当归	*dāng guī*	15g	Radix Angelicae Sinensis
杭白芍	*háng bái sháo*	15g	Radix Paeoniae Lactiflorae
生地	*shēng dì*	12g	Radix Rehmanniae
川芎	*chuān xiōng*	10g	Rhizoma Chuanxiong
三七粉	*sān qī fěn*	4g	Radix et Rhizoma Notoginseng Powder (dissolved)

Modifications

1. For qi deficiency that cannot bind, with dripping blood, pale in colour and thin, add *shēng huáng qí* (raw Radix Astragali), *dǎng shēn* (Radix Codonopsis), *shēng má* (Rhizoma Cimicifugae) and *jié gěng* (Radix Platycodonis), etc.

2. For blood heat pushing blood to flow restlessly, a large amount and red in colour, add *hǎi piāo xiāo* (Endoconcha Sepiae), *qiàn cǎo* (Radix et Rhizoma Rubiae), *shēng lóng gǔ* (raw Os Draconis) and *shēng mǔ lì* (raw Concha Ostreae), use a large dosage of *shēng dì* (Radix Rehmanniae).

3. For blood stasis that obstructs and stagnates, blood that cannot flow back to its

route, is a dark colour with clots, and abdominal pain, add *dān pí* (Cortex Moutan), *pú huáng* (蒲黄 , Pollen Typhae), *wǔ líng zhī* (五灵脂 , Faeces Trogopterori), use a large dosage of *niú xī* (Radix Cyathulae).

4. For menopause patients with continuous dripping, add *xiān máo* (仙茅 , Rhizoma Curculiginis), *xiān líng pí* (仙灵脾 , Radix Clematidis) and *tù sī zǐ* (Semen Cuscutae), etc.

Compatibility Analysis

In the formula, *chuān niú xī* (Radix Cyathulae) tonifies the liver and kidney, promotes menstruation, invigorates blood and transforms stasis. *Dì yú* (Radix Sanguisorbae) is astringent, cools the blood and stops bleeding. Of the two herbs, one can unblock and the other can stop. It can stop bleeding without leaving stasis, and can transform stasis without injury to blood. Combined with *Sì Wù Tāng* (Four Agents Decoction, 四物汤), *dāng guī* (Radix Angelicae Sinensis), *chuān xiōng* (Rhizoma Chuanxiong), *bái sháo* (Radix Paeoniae Alba), *shēng dì* (Radix Rehmanniae), it can nourish blood and regulate menstruation and *sān qī* (Radix et Rhizoma Notoginseng) can be added to increase its function of eliminating stasis and stopping bleeding. The characteristic of the formula is to secure the root and clear the source to treat its root. Constraining flow and stopping bleeding is to treat its manifestation. It is a formula that takes care of both root and the branch. To secure the root is to tonify the liver and kidney, and calm the penetrating vessel. *Niú xī* (Radix Cyathulae) and *Sì Wù Tāng* (Four Agents Decoction) can treat this. To clear the source is to eliminate stasis obstruction, unblock blood vessels and make blood flow back within its vessels. *Chuān niú xī* (Radix Cyathulae), *chuān xiōng* (Rhizoma Chuanxiong) and *dāng guī* (Radix Angelicae Sinensis) can do this. To constrain blood flow and stop bleeding is similar to blocking leakages when building up a dam. *Dì yú* (Radix Sanguisorbae) and *sān qī* (Radix et Rhizoma Notoginseng) can do this. Modifications based on this formula can treat all kinds of aetiologies inducing dysfunctional uterine bleeding, with effective results.

2. *Chuān Niú Xī* (Radix Cyathulae) ——
Dài Zhě Shí (代赭石, Haematitum)

Function

To regulate the penetrating vessel and descend rebellion.

Application

This is used for female vomiting, pathological bleeding during menstruation and premenstrual syndrome (PMS) inducing qi to rebel upward causing symptoms such as headache, nausea and vomiting, cough and asthma, irritability, insomnia as well as breast distention and pain, etc.

Compatibility Analysis

Divine Pivot - On the Seas (灵枢·海论 , *Líng Shū - Hǎi Lùn*) said: "*The penetrating vessel is the sea of blood.*" The *Classic of Difficulties - Twenty-Ninth Difficulty* (难经·二十九

难 , *Nàn Jīng - Èr Shí Jiǔ Nàn*) pointed out: "*If the penetrating vessel is sick, qi will rebel and contract internal organs.*" Hence, if the penetrating vessel has irregularities, the clinical symptoms will include, firstly a change in menstruation and then, qi rebelling upward during menstruation. *Chuān niú xī* (Radix Cyathulae) enters the blood level. Its property is good at moving downward and unblocking blood vessels. *Dài zhě shí* (Haematitum) is red in colour and heavy in quality. It enters the blood level and is good at descending rebellion. The two herbs combined together, can regulate the penetrating vessel, and descend the rebellion. They are herbs for treatment "*if the penetrating vessel is sick, and qi is rebellious and there is contraction of internal organs.*" During the author's practice, when female patients with PMS had upper *jiao* symptoms or rebellious qi, he always used these two herbs combined with *Mài Mén Dōng Tāng* (Ophiopogon Decoction, 麦门冬汤) to treat it and always got instant results.

Case 1

Wheezing and asthma before menstruation.

Initial Visit

A female patient, 29 years old, April 18th, 1985.

Since the previous January, the patient had flu, every time, 10 days before her menstruation; she had cough, dyspnoea with rapid respiration, with a wheezy phlegm sound, preventing her from lying down. After menstruation, all her symptoms were relieved. She was admitted into hospital and used antibiotics to stop the cough, and medicine to calm the asthma, but they were ineffective. Later, when she was discharged from the hospital, she visited our hospital clinic for treatment. The patient was short of breath, breathing through an open mouth, with shoulders raised, a phlegm sound came from her throat, and was very painful and accompanied with an uncomfortable cold or hot sensation in the afternoon. Her tongue was red and her tongue coating was thin and white. Her pulse was thready and fast. This was an external pathogenic invasion. Pathogenic heat enters the uterus during deficiency. Hence, she had the afternoon cold and heat sensations. The pathogen enters the uterus, hence blocking the penetrating vessel which induced rebellious qi upward. This was accompanied by lung qi failing to disperse and descend, hence the cough and asthma. The treatment plan was to regulate the penetrating vessel and descend the rebellious.

Prescription

麦冬	*mài dōng*	30g	Radix Ophiopogonis
半夏	*bàn xià*	15g	Rhizoma Pinelliae
代赭石	*dài zhě shí*	20g	Haematitum
党参	*dǎng shēn*	15g	Radix Codonopsis
川牛膝	*chuān niú xī*	15g	Radix Cyathulae
甘草	*gān cǎo*	10g	Radix Glycyrrhizae
粳米	*jīng mǐ*	4g	Oryza Sativa L.

Water was used to decoct twice, then they were mixed together and divided into two servings, one pack per day.

Second Visit

April 20th: After taking two packs of the above formula the cough, asthma, and phlegm sound stopped and she could lie down. She still had the symptoms of afternoon cold and heat sensations. *Chái hú* (Radix Bupleuri) 12g, *huáng qín* (Radix Scutellariae) 9g, *jié gěng* (Radix Platycodonis) 12g, *bǎi hé* (Bulbus Lilii) 20g was then added to the original formula and she took another three packs.

Note

At a later follow up, this patient said that when her menstruation came, she never had another cough and asthma episode.

Case 2

Dizziness before menstruation.

Initial Visit

A female patient, 25 years old, February 12th, 1984.

Every month, three or four days before menstruation, she suffered dizziness in her head and eyes, with nausea. After menstruation, all the symptoms disappeared. She had these symptoms for more than a year. Her menstruation was scanty and black and dark in colour, she had slight pain in her lower abdomen. Her pulse was deep and wiry. Her tongue coating was thin and white. Her tongue was red but slightly dark. If the penetrating vessel is stagnated and obstructed, then qi cannot move downward but flows rebelliously upward. Once the penetrating vessel rebels upward, it makes liver or stomach fire flare upward to the clear orifices, thus inducing dizziness. The pathogen was in the penetrating vessel, so a treatment for regulating the penetrating vessel and descending the rebellious was required. When the penetrating vessel qi descends, the head and eyes will be cleared.

Prescription

川牛膝	*chuān niú xī*	15g	Radix Cyathulae
代赭石	*dài zhě shí*	20g	Haematitum
麦冬	*mài dōng*	20g	Radix Ophiopogonis
半夏	*bàn xià*	10g	Rhizoma Pinelliae
党参	*dǎng shēn*	10g	Radix Codonopsis
荷叶	*hé yè*	10g	Folium Nelumbinis
甘草	*gān cǎo*	6g	Radix Glycyrrhizae

Water was used to decoct, then it was divided into two servings, served warm. She was instructed to take 3 ~ 4packs, four days before her menstruation.

Second Visit

March 5th: She continuously took three packs before menstruation, her dizziness reduced a lot, and her nausea disappeared, but there was still mild lower abdominal pain. *Yán hú suŏ* (Rhizoma Corydalis) 12g, *pú huáng* (Pollen Typhae) 9g and *wŭ líng zhī* (Faeces Trogopterori) 9g was added to the original formula to regulate the penetrating vessel's stasis. She took them, the same as she was instructed before.

Thereafter, when her menstruation came, all her symptoms had gone.

Case 3

Insomnia before menstruation.

Initial Visit

A female patient, 38 years old, February 24th, 1986.

A year previously, four or five days before menstruation, she had overexerted herself. This resulted in her feeling irritabile, having insomnia, then overnight not being able to sleep and was accompanied by dizziness, low back soreness and lassitude, etc. After menstruation, all symptoms disappeared gradually back to normal. It reoccurred and exacerbated as time passed by. She had taken the Chinese medicine formula *Ān Shén Wán* (Spirit-Quietening Pill, 安神丸) and biomedicine diazepam etc., but they did not work, and so, she came to visit. Her pulse was wiry, slow and deep. Her tongue was pale red, and her tongue coating was thin and white. This was the syndrome of kidney deficiency and rebellious qi in the penetrating vessel. The overexertion injured the kidney qi causing kidney qi deficiency with low back ache and lassitude. The kidney qi deficiency could not bind the penetrating vessel qi, so the penetrating vessel qi rebelled upward. It disturbed the heart and spirit, so irritability and insomnia ensued. The treatment plan was to tonify the kidney, regulate the penetrating vessel, descend rebellion and calm the spirit.

Prescription

山药	*shān yào*	30g	Rhizoma Dioscoreae
川牛膝	*chuān niú xī*	15g	Radix Cyathulae
代赭石	*dài zhĕ shí*	15g	Haematitum
麦冬	*mài dōng*	30g	Radix Ophiopogonis
党参	*dăng shēn*	15g	Radix Codonopsis
半夏	*bàn xià*	10g	Rhizoma Pinelliae
茯苓	*fú líng*	20g	Poria
枣仁	*zăo rén*	15g	Semen Ziziphi Spinosae
夜交藤	*yè jiāo téng*	30g	Caulis Polygoni Multiflori
甘草	*gān căo*	10g	Radix Glycyrrhizae

Water was used to decoct, then it was divided into two servings, one pack per day.

Second Visit

February 28th: After taking four packs of the above formula she could sleep and had no more irritability. She was instructed to take three packs, four days before each menstruation, continuously for three months. At a later follow up visit she said she had not had any reoccurrence of the symptoms.

Case 4

Epistaxis before menstruation (inverted menstruation).

Initial Visit

A female patient, 41 years old, May 18th, 1988.

The patient had breast distention and irritability before menstruation ever since she was young. For the previous three months due to work being busier, one or two days before menstruation she had epistaxis. It stopped after menstruation. Even though she used medicine to stop the bleeding, she always had epistaxis before menstruation. The patient's general health was good. She was smallish in stature. Her pulse was wiry, thready and fast. Her tongue was red with a thin white coating. This time, she followed the doctor's instruction to visit four days before menstruation. The syndrome belonged to inverted menstruation due to penetrating vessel stagnation and obstruction. The penetrating vessel's stagnated fire flared upward to cause it. The patient was thin and her pulse was thready and fast. She had a yang excess body. If yang is excessive, it is hot and easily transforms into fire. The penetrating vessel stagnation and obstruction then caused rebellious qi to travel upward accompanied by heart or liver fire moving upward to induce epistaxis. The treatment plan was to regulate the penetrating vessel, and descend rebellion, aided by soothing the liver and invigorating the blood to eliminate the fire and heat pathogen from the lower *jiao*.

Prescription

麦门冬	*mài mén dōng*	30g	Radix Ophiopogonis
党参	*dǎng shēn*	15g	Radix Codonopsis
半夏	*bàn xià*	9g	Rhizoma Pinelliae
赤芍	*chì sháo*	12g	Radix Paeoniae Rubra
川牛膝	*chuān niú xī*	15g	Radix Cyathulae
当归	*dāng guī*	12g	Radix Angelicae Sinensis
香附	*xiāng fù*	12g	Rhizoma Cyperi
柴胡	*chái hú*	12g	Radix Bupleuri
代赭石	*dài zhě shí*	15g	Haematitum
薄荷	*bò he*	9g	Herba Menthae
甘草	*gān cǎo*	6g	Radix Glycyrrhizae

Water was used to decoct, then was divided into two servings, one pack per day.

Second Visit

May 23rd: After taking three packs of the formula, during menstruation, she did not have epistaxis but she still had the breast distention and irritability symptoms, however, they were not as severe as before. She was instructed to continue taking the original formula for three more packs, one week before menstruation.

Third Visit

August 30th: The patient had continuously taken the formula for three menstrual periods. After this menstrual period, all the symptoms had gone and there was no epistaxis.

The above examples have very obvious differences in their clinical expressions, but to analyze them from their aetiologies, there is only one i.e. rebellious qi of the penetrating vessel. Hence, the treatments were mainly to regulate the penetrating vessel and descend rebellious qi and both had effective results.

Women's premenstrual syndrome (PMS) is when, before menstruation, emotional changes appear, such as headache, irritability, easy to anger, insomnia, somnolence, oedema, nausea and vomiting, epistaxis, breast distention, abdominal pain, abnormal libido, etc. It usually starts around one week before menstruation and becomes severe gradually, as well as gradually disappearing after menstruation. Traditional Chinese medicine thought is that this syndrome is related to disharmony of the penetrating vessel's function. The *Divine Pivot - On the Seas* said: "*The penetrating vessel is the sea of blood.*" This means that the penetrating vessel has the function to govern the "*sea of blood*". The sea of blood is directly regulated by the penetrating vessel. If the penetrating vessel is full then the sea of blood will be flourishing and menstruation comes on time. If the penetrating vessel is deficient then the sea of blood is also deficient. There will be irregular menstruation or amenorrhoea, etc. Hence, Zhang Jing-yue said: "*The penetrating vessel is the root of menstruation.*" The *Basic Questions - Treatise of Heavenly Truth from Remote Antiquity* (素问·上古天真论 , *Sù Wèn - Shàng Gǔ Tiān Zhēng Lùn*) said: "*For girls at the age of 14, the dew of heaven (menstruation) arrives...the penetrating vessel is flourishing, menstruation comes regularly.*" The penetrating vessel is our body's twelve meridians gathering place. The penetrating vessel's excess, deficiency, flourishing or shortage rely on the five *zang* organs and six *fu* organs flourishing to regulate it. The *Divine Pivot - Treating the Fat and Thin According to Circumstance* (灵枢·逆顺肥瘦 , *Líng Shū - Nì Shùn Féi Shòu*) said: "*The penetrating vessel is the sea for the five zang organs and six fu organs. All of the five zangs and six fus generate from it.*" In the *Divine Pivot - Movement and Transport* (灵枢·动输 , *Líng Shū - Dòng Shū*) it said: "*The penetrating vessel is the sea for the twelve meridians*", hence, in clinical practice, if there is sickness in the five *zang*s or six *fu*s, it will cause penetrating vessel disharmony. Conversely, any disharmony of the penetrating vessel will also cause abnormality in the five *zang*s and six *fu*s. However, due to the penetrating vessel's special physiology and its meridian's route, the penetrating vessel sickness is mostly related to menstruation. Most symptoms are expressed as qi rebelling upward, abdominal pain or abdominal urgency. The *Classic of Difficulties - Twenty-Ninth Difficulty* pointed out that: "*The penetrating vessel sickness is rebellious qi and internal urgency.*" The *Classic of Difficulties - Twenty-Eighth Difficulty* (难经·二十八难 , *Nàn Jīng - Èr Shí Bā Nàn*) said: "*The penetrating vessel starts from ST30 (qì jiē) and travels with the stomach channel of the foot yangming, it passes around the umbilicus and travels upward to the chest and disperses.*" Hence, if the penetrating vessel is disharmonious it will cause qi to rebel upward and qi rebellion causes contraction of the internal organs and abdominal pain. We observed PMS clinical manifestations, which indicated that there are more symptoms in the upper *jiao* that have the qi rebelling upward syndrome such as headache, dizziness, cough and asthma, nausea, vomiting, epistaxis, irritability, breast distention, insomnia, etc. as well as rebellious qi induced internal urgency. These symptoms are similar to the *Classic of Difficulties'* (难经 , *Nàn Jīng*) description of penetrating vessel sickness. Hence, the treatment principle mainly is to regulate the penetrating vessel and descend the rebellious. When the penetrating vessel is regulated and qi is smooth, the syndrome will be eliminated by itself. The author uses his own personal formula *Jiā Wèi Mài Mén Dōng Tāng* (Supplemented Ophiopogon Decoction, 加味麦门冬汤) to treat this sickness with good results. Using *Mài Mén Dōng Tāng* (Ophiopogon Decoction, 麦门冬汤) to treat this disease was inspired by Chen Xiu-yuan and Zhang Xi-chun two medical experts. *Mài Mén Dōng Tāng* is Zhang Zhong-jing's *Essentials From the Golden Cabinet* formula. It mainly treats lung atrophy. The *Essentials From the Golden Cabinet* said: "*Severe rebellious qi going upward, and throat discomfort, can be treated by Mài Mén Dōng Tāng* (Ophiopogon Decoction) *to stop rebellion and descend the qi.*" Chen Xiu-yuan was the first one to use this formula that has the "*stopping rebellion and descending qi*" function to treat inverted menstruation. Later on, Zhang Xi-chun used *Jiā Wèi Mài Mén Dōng Tāng* (Supplemented Ophiopogon Decoction) to treat inverted menstruation. Zhang added *dān shēn* (Radix et Rhizoma Salviae Miltiorrhizae), *táo rén* (Semen Persicae), *chì sháo* (Radix Paeoniae Rubra), *shān yào* (Rhizoma Dioscoreae), etc. into the *Mài Mén Dōng Tāng* (Ophiopogon Decoction), to invigorate blood, regulate menstruation and tonify the kidney, which has a stronger effect than *Mài Mén Dōng Tāng* (Ophiopogon Decoction). The author in his practice thought that inverted menstruation belongs to qi rebellion before menstruation, though its symptoms are different with the other premenstrual syndromes, their aetiologies are the same. He used this formula to treat it, and it was really effective. The original formula has less strength to regulate the penetrating vessel and descend rebellious qi, hence, *chuān niú xī* (Radix Cyathulae), and *dài zhě shí* (Haematitum), etc. were added in. The author's own personal formula *Jiā Wèi Mài Mén Dōng Tāng* (Supplemented Ophiopogon Decoction) ingredients, applications and modifications are as follows:

Ingredients

麦门冬	*mài mén dōng*	15~30g	Radix Ophiopogonis
半夏	*bàn xià*	10~15g	Rhizoma Pinelliae
党参	*dǎng shēn*	10~15g	Radix Codonopsis
川牛膝	*chuān niú xī*	15~20g	Radix Cyathulae
代赭石	*dài zhě shí*	15~20g	Haematitum
当归	*dāng guī*	12g	Radix Angelicae Sinensis
杭白芍	*háng bái sháo*	15g	Radix Paeoniae Lactiflorae
甘草	*gān cǎo*	6g	Radix Glycyrrhizae

Application

After immersing herbs in water for half an hour, use a strong heat to decoct until boiling. Then simmer to decoct for half an hour. Filter off the decoction. To the residual herbs add more water to decoct for 20 minutes. Filter off this decoction. Mix the two decoctions together, divide into two servings to be taken in the morning and evening. Take four to six packs continuously one week before menstruation.

Modifications

1. For headache, and dizziness, add *chuān xiōng* (Rhizoma Chuanxiong), *jú huā* (Flos Chrysanthemi), *hé yè* (Folium Nelumbinis), *gōu téng* (Ramulus Uncariae Cum Uncis), etc.

2. For irritability, insomnia, a lot of dreams, add *zǎo rén* (Semen Ziziphi Spinosae), *yè jiāo téng* (Caulis Polygoni Multiflori), *chuān lián* (川连 , Rhizoma Coptidis), etc.

3. For palpitations and insomnia, add *shēng lóng gǔ* (raw Os Draconis) and *shēng mǔ lì* (raw Concha Ostreae).

4. For breast distended pain or distention of both hypochondria with fullness and pain, add *yán hú suǒ* (Rhizoma Corydalis), *chái hú* (Radix Bupleuri), *jú yè* (橘叶 , *Folium Citri*) and *chuān liàn zǐ* (*Fructus Toosendan*), etc.

5. For vomiting and epistaxis, add *xiān hè cǎo* (Herba Agrimoniae), *zhī zǐ* (栀子 , Fructus Gardeniae) and *máo gēn* (茅根 , Rhizoma Imperatae), etc.

6. For mouth or tongue ulcers, add *yīn chén* (Herba Artemisiae Scopariae), *shān zhī zǐ* (山栀子 , Fructus Gardeniae) and *qīng dài* (青黛 , Indigo Naturalis), etc.

7. For cough or asthma, add *xìng rén* (Semen Armeniacae Dulcis), *hòu pò* (厚朴 , Cortex Magnoliae Officinalis) and *chuān bèi* (Bulbus Fritillariae Cirrhosae), etc.

8. For facial eczema or wind rash, add *fáng fēng* (Radix Saposhnikoviae) and *huò xiāng* (Herba Agastachis).

9. For vomiting, and or nausea, add large doses of *bàn xià* (Rhizoma Pinelliae) and *dài zhě shí* (Haematitum).

10. For abdominal pain, add *yán hú suǒ* (Rhizoma Corydalis), *pú huáng* (Pollen Typhae) and *líng zhī* (Faeces Trogopterori).

11. For diarrhoea or oedema, eliminate *zhě shí* (Haematitum), reduce *mài dōng*'s (Radix Ophiopogonis) dosage, add *fú líng* (Poria), *bái zhú* (Rhizoma Atractylodis Macrocephalae), *zé xiè* (Rhizoma Alismatis) and *shān yào* (Rhizoma Dioscoreae), etc.

Formula Explanation

In the formula, *chuān niú xī* (Radix Cyathulae), *dāng guī* (Radix Angelicae Sinensis) and *háng bái sháo* (Radix Paeoniae Lactiflorae) enter the blood level to regulate the penetrating vessel and descend rebellious qi. *Bàn xià* (Rhizoma Pinelliae), *mài mén dōng* (Radix Ophiopogonis), *dǎng shēn* (Radix Codonopsis), and *dài zhě shí* (Haematitum) harmonise the stomach and tonify qi, and descend rebellious qi. Due to the fact that the penetrating vessel belongs to the *yangming*, to treat the *yangming* is to treat the penetrating vessel. If the *yangming* stomach qi is sufficient then the stomach is harmonious and qi descends, and the penetrating vessel qi also descends. *Háng bái sháo* (Radix Paeoniae Lactiflorae) and *gān cǎo* (Radix Glycyrrhizae) relax urgency and alleviate the pain. This formula, all together, achieves functions of harmonising the stomach, descending rebellion, regulating the penetrating vessel and alleviating the contraction of the internal organs.

The author thought that if it was effective for PMS, it may also had certain functions to regulate female progesterone and relax the nerves. Modern pharmacological research has discovered the calming and analgesic functions of herbs such as *háng bái sháo* (Radix Paeoniae Lactiflorae), *gān cǎo* (Radix Glycyrrhizae) and *dāng guī* (Radix Angelicae Sinensis), etc. and that there is also a certain treatment effect in alleviating mental strain. *Dǎng shēn* (Radix Codonopsis), *mài dōng* (Radix Ophiopogonis), *háng bái sháo* (Radix Paeoniae Lactiflorae) and *gān cǎo* (Radix Glycyrrhizae) contain ingredients like carbohydrate, starch, and vitamins. They have a certain function to alleviate menstrual toxins. *Chuān niú xī* (Radix Cyathulae), *dāng guī* (Radix Angelicae Sinensis) and *háng bái sháo* (Radix Paeoniae Lactiflorae) are the main herbs to regulate menstruation. Further study is required to discover whether they have a regulating function on female progesterone.

3. *Chuān Niú Xī* (Radix Cyathulae) ——— *Tǔ Biē Chóng* (土鳖虫, Eupolyphaga seu Steleophaga)

Function

To tonify the liver and kidney, to strengthen the lower back and knees, to transform stasis and alleviate pain.

Application

This is used to treat lower back, knees and leg pain that belong to liver and kidney deficiency; and stasis obstructed meridians and collaterals. The author always uses this compatibility to treat lumbar muscle strain or lumbar vertebral hyperosteogeny.

Compatibility Analysis

Niú xī (Radix Cyathulae) and *tǔ biē chóng* (土鳖虫 , Eupolyphaga seu Steleophaga) all enter the liver channel of the foot *jueyin* meridian's blood level. The *Guide to Clinical Practice* (临症指南 , *Lín Zhèng Zhǐ Nán*) said of *niú xī* (Radix Cyathulae): "*It can tonify the liver and kidney, strengthen sinews and bones, and treat lumbar and knee pain.*" *Tǔ biē chóng* (Eupolyphaga seu Steleophaga) is a blood invigorating insect herb. It is good at invigorating blood and transforming stasis, and connecting sinews and bone. The two herbs have the effect to tonify the liver and kidney, to transform stasis and alleviate pain, as well as to tonify but not leave stasis, to transform but not injure the upright qi. Clinically, if lumbar and knee pain is more due to liver and kidney deficiency, *huái niú xī* (Radix Achyranthis Bidentatae) can be used, if it is more due to blood stasis, then use *chuān niú xī* (Radix Cyathulae) in a large dosage.

Case Study

Initial Visit

A male patient, 48 years old, March 5th, 1988.

The patient had low back pain for two years. It was more obvious when he was fatigued or at night. The x-ray showed that there was hyperosteogeny at the anterior and lateral edges of 1 ~ 5 lumbar vertebrae. Diagnosis: degenerated lumbar vertebral arthritis. He had treatment in a hospital but did not see a remarkable result and so came to visit. On inspection his lumbar muscle was stiff, and he had pain when flexing his lumbar spine. After pressing or massage it felt comfortable. The patient had good physical development. His urine and bowel movements were normal. His pulse was wiry, thready and when pressed deeply, it was forceless. His tongue was red and his coating was thin and white. The lumbar region is the kidneys' house, if there is lumbar pain in the night and it ameliorates on pressing it is due to kidney deficiency. The liver governs the sinews, if the liver is deficient it cannot nourish and so the lumbar muscle will be stiff. Prolonged sickness will enter the collaterals and cause the blood vessels to stagnate and obstruct. The treatment principle was to tonify the liver and kidney, to invigorate blood and unblock the collaterals.

Prescription

川牛膝	*chuān niú xī*	20g	Radix Cyathulae
土鳖虫	*tŭ biē chóng*	10g	Eupolyphaga seu Steleophaga
杜仲	*dù zhòng*	20g	Cortex Eucommiae
川断	*chuān duàn*	20g	Radix Dipsaci
鹿角霜	*lù jiăo shuāng*	10g	Cornu Cervi Degelatinatum
桑寄生	*sāng jì shēng*	20g	Herba Taxilli
延胡索	*yán hú suŏ*	15g	Rhizoma Corydalis
木瓜	*mù guā*	15g	Fructus Chaenomelis
杭白芍	*háng bái sháo*	15g	Radix Paeoniae Lactiflorae
透骨草	*tòu gŭ căo*	10g	Herba Glechomae
甘草	*gān căo*	6g	Radix Glycyrrhizae

Water was used to decoct, one pack per day.

Second Visit

March 10th: After taking five packs of the above formula, his low back pain was obviously reduced. However, when flexing his lower back he still felt pain and his lumbar area was still stiff and numb. As the formula was effective it was not changed and he continued on with it.

Third Visit

March 17th: The low back pain basically disappeared. The lumbar stiffness was also improved. He took more than 20 packs of the original formula and recovered.

4. *Chuān Niú Xī* (Radix Cyathulae) —
Jié Gĕng (桔梗, Radix Platycodonis), *Chái Hú* (柴胡, Radix Bupleuri)

Function

To raise the clear and descend the turbid, to soothe and regulate qi.

Application

It may be used for qi stagnation and blood stasis inducing chest impediment syndrome or qi disharmony inducing diarrhoea syndrome. It also can be used for stubborn hiccoughs.

Compatibility Analysis

Chuān niú xī (Radix Cyathulae) enters the blood level. Its property is going downward and descending the turbid. *Jié gěng* (Radix Platycodonis) enters the qi level. Its property is good at moving upward to raise the clear qi. *Chái hú* (Radix Bupleuri) enters the liver channel of the foot *jueyin* meridian and can soothe the liver and disperse qi to unblock and regulate the raising and descending route, the three herbs combined together have functions to raise the clear, descend the turbid and regulate qi. Clinically, if combined with invigorating blood and eliminating stasis herbs, they can enhance the power to promote qi movement and invigorate blood, make qi move and unblock stasis, and harmonise qi and blood. It is commonly used for chest impediment as with Wang Qing-ren's *Xuè Fǔ Zhú Yū Tāng* (House of Blood Stasis-Expelling Decoction, 血府逐瘀汤), which is the application of this compatibility. If combined with the herbs that strengthen the spleen, harmonise the stomach and soothe the liver, it can enhance the effect for the spleen and stomach to raise clear qi and descend the turbid. If combined with *Sháo Yào Gān Cǎo Tāng* (Peony and Liquorice Decoction, 芍药甘草汤), it can regulate qi and enhance its alleviating urgency and relieving spasm functions, as well as treat stubborn hiccough (diaphragm spasm).

Case 1

Chronic colitis.

Initial Visit

A female patient, 35 years old, March 20th, 1987.

The patient had a history of chronic colitis for six years. Coldness or tiredness or anger would aggravate the condition. Two months previously due to eating cold and raw food, she had abdominal pain and diarrhoea which became severe and her stools were thin with bubbles. Every day she had diarrhoea 3 ~ 4 times. As soon as she had abdominal pain, she had diarrhoea. After the diarrhoea, the pain did not reduce. Her appetite was alright, but other accompanying symptoms included abdominal distention, lassitude, yellowish face, a pale tongue, a thin white tongue coating, and a deep and slightly wiry pulse. A routine stool laboratory test did not show any abnormality. The syndrome belonged to liver stagnation and spleen deficiency, with dampness arthralgia and qi sinking. The treatment plan was to soothe the liver and strengthen the spleen, raise the clear and descend the turbid.

Prescription

茯苓	fú líng	15g	Poria
杭白芍	háng bái sháo	24g	Radix Paeoniae Lactiflorae
陈皮	chén pí	10g	Pericarpium Citri Reticulatae
白术	bái zhú	12g	Rhizoma Atractylodis Macrocephalae

防风	*fáng fēng*	10g	Radix Saposhnikoviae
桔梗	*jié gěng*	4g	Radix Platycodonis
柴胡	*chái hú*	12g	Radix Bupleuri
枳壳	*zhǐ qiào*	9g	Fructus Aurantii
川牛膝	*chuān niú xī*	12g	Radix Cyathulae
丹参	*dān shēn*	15g	Radix et Rhizoma Salviae Miltiorrhizae
乌药	*wū yào*	9g	Radix Linderae
甘草	*gān cǎo*	6g	Radix Glycyrrhizae

Second Visit

March 25th: After taking five packs of the above formula, the abdominal pain reduced and the stool was formed, with movements 1 ~ 2 times per day. She continued to take the above formula.

Third Visit

March 31st: After taking another five packs, the abdominal pain and diarrhoea disappeared. Her bowel movements were once a day, but she still had lassitude. The formula was changed to *Guī Pí Tāng* (Spleen-Returning Decoction, 归脾汤) to treat this.

Case 2

Diaphragm spasm.

Initial Visit

A male patient, 78 years old, May 7th, 1992.

The patient was old but healthy and strong. Recently he visited his son and stayed there. Due to the life, diet and environment change, he suddenly developed frequent hiccoughs that he could not control by himself. He had a hospital check up and it was diagnosed as diaphragm spasm and they gave him chlorpromazine to treat it. However, it did not work and the hiccough became even worse, his appetite was also affected. It had gone on for four days. He had a wiry pulse, a thin white tongue coating and a loud hiccough sound. The treatment plan was to use a formula to soothe and regulate qi, and to alleviate spasm.

Prescription

川牛膝	*chuān niú xī*	15g	Radix Cyathulae
桔梗	*jié gěng*	8g	Radix Platycodonis
柴胡	*chái hú*	12g	Radix Bupleuri
枳壳	*zhǐ qiào*	12g	Fructus Aurantii
红花	*hóng huā*	9g	Flos Carthami
桃仁	*táo rén*	9g	Semen Persicae
杭白芍	*háng bái sháo*	50g	Radix Paeoniae Lactiflorae
甘草	*gān cǎo*	20g	Radix Glycyrrhizae
丹参	*dān shēn*	20g	Radix et Rhizoma Salviae Miltiorrhizae
当归	*dāng guī*	12g	Radix Angelicae Sinensis

He was instructed to use water to decoct and to drink it very frequently, one pack per day.

Second Visit

May 10th: His son said that after he took one third of a pack of the above formula, there was no nausea and vomiting, and the hiccough calmed a little bit. He continued with the rest of the decoction, and the hiccough stopped. After he took another pack, he was back to normal and there was no more hiccoughing.

5. *Chuān Niú Xī* (Radix Cyathulae) ——
Shān Jiǎ (山甲, Squama Manis)

Function

To free stranguria and expel stones.

Application

This is used for urolithic stranguria syndrome (urinary system stone).

Compatibility Analysis

Niú xī (Radix Cyathulae) is good at moving downward it can promote urination and free painful urinary syndrome. *Shān jiǎ* (山甲 , Squama Manis) invigorates blood, transforms stasis and is good at opening the orifices. The two herbs combined together can open the lower orifices, invigorate blood, clear stranguria and expel stones. In promoting urination and easing stranguria formulae, adding these two herbs can increase the effectiveness of expelling the stones.

Case Study

Initial Visit

A female patient, 37 years old, June 22nd, 1986.

The patient had left side low back pain for a couple of months. She had a B-ultrasound test at a hospital. On the superior and inferior ends of the kidney system, a 0.5cm × 0.7cm stone could be seen. The patient had good physical development. With the exception of low back pain, she had no other discomfort. Her pulse was deep and thready. Her tongue coating was thin and white and her tongue was red. The treatment plan was to use a formula to promote urination and clear stranguria, to transform stasis and eliminate stones.

Prescription

金钱草	*jīn qián cǎo*	20g	Herba Lysimachiae
海金沙	*hǎi jīn shā*	15g	Spora Lygodii (wrapped up)
川牛膝	*chuān niú xī*	20g	Radix Cyathulae
山甲珠	*shān jiǎ zhū*	10g	Squama Manis
王不留行	*wáng bù liú xíng*	15g	Semen Vaccariae

车前子	*chē qián zǐ*	15g	Semen Plantaginis (wrapped up)
仙灵脾	*xiān líng pí*	15g	Radix Clematadis
白花蛇舌草	*bái huā shé shé cǎo*	20g	Herba Hedyotis Diffusae
甘草梢	*gān cǎo shāo*	12g	Radix Glcyrrhizae

Water was used to decoct, then it was divided into two servings. One pack per day. She was also instructed to drink more water and do more exercise.

Third Visit

July 9th: After she continuously took fourteen packs of the above formula, she suddenly had an episode of severe pain in the left side kidney area. A routine urine test showed: RBC +++. This was the sign of the stone moving downward to the urethra. The patient was told not to be startled. Using the original formula, *chuān niú xī* (Radix Cyathulae) was changed to 30g, and *máo gēn* (Rhizoma Imperatae) 30g, and *yì mǔ cǎo* (Herba Leonuri) 20g, were also added. She continued taking it.

Fourth Visit

July 20th: After taking ten packs of the above formula, she eliminated a couple of pieces of sand like stone, and the low back pain reduced. She continued on with the original formula.

Fifth Visit

July 27th: After taking the formula, the low back pain disappeared. The B-ultrasound test showed: There were no stones in the kidney, urethra or urinary bladder.

6. *Chuān Niú Xī* (Radix Cyathulae) —— *Jī Nèi Jīn* (鸡内金, Endothelium Corneum Gigeriae Galli), *Hǔ Pò* (琥珀, Succinum)

Function

To free stranguria and transform stasis, to dissolve and expel stones.

Application

This is used for urinary system stones, large kidney stones that are difficult to expel.

Compatibility Analysis

When a kidney stone is big or at a location from which it is not easily expelled, or there are multiple stones, or the medicinal treatment is ineffective, the melting stone and reducing stone method can be applied.

Chuān niú xī (Radix Cyathulae) can free stranguria, invigorate blood, transform stasis and guide herbs downward. *Jī nèi jīn* (Endothelium Corneum Gigeriae Galli) has functions to transform stasis and expel stones. *Hǔ pò* (琥珀 , Succinum) is sweet and neutral. It has functions to promote urination and transform stasis. The *Essentials of the Materia Medica* said: "……*unblock the urinary bladder, treat five lin syndromes.*" It is an important herb to invigorate blood, disperse stasis, promote urination, and clear stranguria. These three herbs all have the function to invigorate blood and transform stasis. *Niú xī* (Radix Cyathulae) and *hǔ pò* (Succinum) each can enter the kidney channel of the foot *shaoyin* and the urinary bladder channel of the foot *taiyang* meridians and can promote kidney qi and blood circulation to assist expelling stones. If combined with *jī nèi jīn* (Endothelium Corneum Gigeriae Galli), then the stone expelling and dissolving power is even stronger. The author in his clinical practice, when seeing stubborn multiple kidney calculi or a large stone, most of the time he uses this compatibility to treat it, sometimes, with good results.

Case Study

Initial Visit

A female patient, 45 years old, July 21st, 1992.

The patient had a physical exam at a hospital because of low back discomfort, and discovered there were a couple of stones, in both kidneys. She used Chinese medicine to eliminate stones for more than a month, but it did not work and so she came to visit. The patient had a good constitution but was slightly obese. The B-ultrasound report: Both kidneys had a couple of stones. The biggest one was 0.9cm. She was diagnosed with multiple kidney calculi. Her pulse was deep and wiry. Her tongue coating was thin and white. The treatment plan was to clear stranguria, transform stasis and dissolve the stones.

Prescription

川牛膝	*chuān niú xī*	150g	Radix Cyathulae
鸡内金	*jī nèi jīn*	150g	Endothelium Corneum Gigeriae Galli
琥珀	*hǔ pò*	50g	Succinum
石韦	*shí wéi*	80g	Folium Pyrrosiae
海金沙	*hǎi jīn shā*	100g	Spora Lygodii (wrapped up)
滑石	*huá shí*	100g	Talcum
车前子	*chē qián zǐ*	80g	Semen Plantaginis
穿山甲	*chuān shān jiǎ*	30g	Roasted Manitis Squama
茯苓	*fú líng*	80g	Poria
泽泻	*zé xiè*	80g	Rhizoma Alismatis
甘草	*gān cǎo*	60g	Radix Glycyrrhiza

The above herbs were all ground into a fine powder, to be taken 5g each time, with warm water, three times a day. Meanwhile, she was also instructed to take Vitamin B 680mg, three times a day.

Second Visit

September 16th: She took the above formula for more than 50 days, her low back pain and uncomfortable feeling had disappeared. Her B-ultrasound test showed that the left kidney stones had disappeared and the right kidney had two stones which had reduced in size. The biggest one was 0.5cm. As it was effective, the formula did not change. She continued with the same formula for treatment.

Note

The patient in this case took Vitamin B6 to treat the kidney stone, the theory was based on the article *"Vitamin B6 to Prevent Kidney Stones"* in *Health News*, March 15th, 1999, p. 2.

Chuān xiōng's (川芎 , Rhizoma Chuanxiong) taste and property are acrid and warm. It enters the liver channel of the foot *jueyin*, the gallbladder channel of the foot *shaoyang* and the pericardium channel of the hand *jueyin* meridians. It has functions to invigorate the blood, promote movement of qi, expel wind and relieve pain. It is most often used for irregular menstruation, abdominal pain, wind inside the head, headache, cold pain and spasm, etc. during menstruation. In Zhang Yuan-su's theses, there was the comment: *"Upward, it travels through head and eye. Downward, it travels through the sea of blood."* These words sufficiently express *chuān xiōng*'s (Rhizoma Chuanxiong) function. *Chuān xiōng* (Rhizoma Chuanxiong) is acrid and frees the movement of qi. It is the qi herb in the blood. It has the strongest pain relieving power and goes upward to stop headache, hence, Li Dong-yuan stated: *"Headache ought to use chuān xiōng* (Rhizoma Chuanxiong)." However, it is properly used for exterior syndrome's headache and blood deficiency type headache, for liver yang rising, patients need to use it carefully. In travelling downward to relieve abdominal pain it is better used for patients with blood stasis and menstruation disharmony.

1. *Chuān Xiōng* (Rhizoma Chuanxiong) —— *Tiān Má* (天麻, Rhizoma Gastrodiae)

Function

To transform stasis and extinguish wind, to relieve spasm and alleviate pain.

Application

This is used for all kinds of headache.

Compatibility Analysis

Chuān xiōng (Rhizoma Chuanxiong) is acrid and warm. It is an important herb to invigorate blood, expel wind and stop headache. *Tiān má*'s (天麻 , Rhizoma Gastrodiae) property and taste are neutral and acrid. It enters the liver channel of the foot *jueyin* meridian and can extinguish wind and relieve spasm. It is the main herb to treat migraine. Zhang Yuan-su said that *tiān má* (Rhizoma Gastrodiae) can *"treat dizziness and headache"*. Li Dong-yuan said: *"For liver yin deficiency patients it is proper to use tiān má* (Rhizoma Gastrodiae) *and xiōng qióng* (芎藭 , Rhizoma Chuanxiong) *to tonfiy. It even treats wind-heat headache, or wind spasm, or palpitations."* *Tiān Má Wán* (Gastrodia Pill, 天麻丸) in *Prescriptions of Universal Benefit* (普济方 , *Pǔ Jì Fāng*) used the compatibility of *tiān má* (Rhizoma Gastrodiae) and *chuān xiōng* (Rhizoma Chuanxiong) to treat migraine, headache, blurry vision, or coma. We can see that these two herbs have been an effective compatibility to treat headache since ancient times. The *Divine Husbandman's Classic of the Materia Medica* said of *tiān má* (Rhizoma Gastrodiae): *"Taking it for a long time can tonify qi and strength, generate yin and health."* This means that *tiān má* (Rhizoma Gastrodiae) has the function to tonify yin. It combined with *chuān xiōng* (Rhizoma Chuanxiong) can invigorate blood, transform stasis, expel wind, but without injury to yin and blood. Hence, Li Dong-yuan said: *"liver deficiency patients are suitable for má* (Rhizoma Gastrodiae) *and xiōng* (Rhizoma Chuanxiong)." The two herbs combined can treat all kinds of headache.

According to syndrome differentiation, adding these two herbs always produces a good result.

Case Study

Initial Visit

A female patient, 20 years old, December 23rd, 1985.

The patient had a history of headache for two years. When she was anxious or she had insomnia, the headache became severe. This time, she had already had the headache for a week, due to having a final exam and anxiety over studying. She took pain killers but they did not work and so she came to visit. On observation, her sickness and pain were evident from her facial appearance, she had no spirit, and she had a wiry, thready, and fast pulse, a red tongue with a thin white coating. The patient complained of severe pain on the left side of her head which made her lose interest in studying and appetite. Her menstruation came on time, but with less amount and dark colour. The diagnosis was blood stasis induced headache syndrome. The treatment plan was to invigorate blood, transform stasis, calm spasm and stop the pain.

Prescription

川芎	*chuān xiōng*	15g	Rhizoma Chuanxiong
天麻	*tiān má*	12g	Rhizoma Gastrodiae
葛根	*gé gēn*	20g	Radix Puerariae Lobatae
丹参	*dān shēn*	24g	Radix et Rhizoma Salviae Miltiorrhizae
双钩	*shuāng gōu*	30g	Ramulus Uncariae Cum Uncis
全蝎	*quán xiē*	6g	Buthus Martensi
柴胡	*chái hú*	10g	Radix Bupleuri
当归	*dāng guī*	12g	Radix Angelicae Sinensis
甘草	*gān cǎo*	6g	Radix Glycyrrhizae

Water was used to decoct, then it was divided into two servings, to be taken in the morning and at night, one pack per day.

Second Visit

December 25th: She took three packs of the above formula, the headache reduced but she still had a poor appetite and insomnia. *Yè jiāo téng* (Caulis Polygoni Multiflori) 30g and *zǎo rén* (Semen Ziziphi Spinosae) 15g were added. She continued to take it.

Third Visit

December 29th: After taking another three packs of the formula, the headache basically disappeared, but her sleep was still not good. She used the above formula and added medicinals to nourish the heart and calm the spirit to be taken again.

川芎	*chuān xiōng*	15g	Rhizoma Chuanxiong
天麻	*tiān má*	12g	Rhizoma Gastrodiae
丹参	*dān shēn*	24g	Radix et Rhizoma Salviae Miltiorrhizae

石菖蒲	*shí chāng pú*	10g	Rhizoma Acori Tatarinowii
生龙骨	*shēng lóng gǔ*	15g	Os Draconis (raw)
远志	*yuǎn zhì*	10g	Radix Polygalae
龟甲	*guī jiǎ*	10g	Carapax et Plastrum Testudinis
川连	*chuān lián*	6g	Rhizoma Coptidis
肉桂	*ròu guì*	3g	Cortex Cinnamomi
甘草	*gān cǎo*	6g	Radix Glycyrrhizae

Water was used to decoct, then it was divided into two servings.

Fourth Visit

January 3rd, 1986: Her sleep was good, and her headache was slight. She followed the formula for three more packs for prevention.

Note

Headache is a common disease in clinical practice. Chinese Medicine differentiates it into two big categories of external invasion headache and internal injury headache. The external invasion headache is more due to careless living, either sitting or sleeping under a wind attack, invaded by wind, cold, dampness, heat, etc. external pathogens cause it, but especially the wind pathogen. The internal injury headache is more related to *zang* or *fu* disharmony and can be further divided into liver yang rising, kidney deficiency, blood deficiency, qi deficiency, turbid phlegm, and blood stasis patterns. The aetiologies can be covered using four words: deficiency; stagnation; phlegm; and stasis. Qi deficiency cannot transform or blood deficiency cannot moisten, both can prevent the blood flow from flowing smoothly then stasis or stagnation appears. Once the stasis or stagnation blocks the flow, it will cause pain, hence, the headache syndrome can use stasis as a word to cover it. Internal headache corresponds nowadays to the biomedical migraine (angioneurotic headache), cluster headache, nervous headache, neurasthenia, neurosis, etc.

In practice, the author uses his own personal formula *Jiā Wèi Tiān Má Wán* (Supplemented Gastrodia Pill, 加味天麻丸) to treat headache and has had satisfactory results. It is detailed below for reference.

Ingredients

天麻	*tiān má*	15g	Rhizoma Gastrodiae
川芎	*chuān xiōng*	15~30g	Rhizoma Chuanxiong
丹参	*dān shēn*	30g	Radix et Rhizoma Salviae Miltiorrhizae
葛根	*gé gēn*	30g	Radix Puerariae Lobatae
当归	*dāng guī*	15g	Radix Angelicae Sinensis
双钩	*shuāng gōu*	30g	Ramulus Uncariae Cum Uncis
全蝎	*quán xiē*	6g	Buthus Martensi
荷叶	*hé yè*	10g	Folium Nelumbinis
甘草	*gān cǎo*	6g	Radix Glycyrrhizae

Water is used to decoct, then it is divided into two servings, one pack per day. Or, use the above herbs in proportions, ground into very fine powder and use honey to pack into pills. Each pill should weigh 9g. Take one pill three times a day.

Modifications

1. For migraine, add *chái hú* (Radix Bupleuri).

2. For frontal headache, add *bái zhǐ* (Radix Angelicae Dahuricae).

3. For vertex headache, add *gǎo běn* (藁本 , Rhizoma et Radix Ligustici).

4. For occipital headache, add *qiāng huó* (羌活 , Rhizoma et Radix Notopterygii).

5. For whole headache, add *wú gōng* (蜈蚣 , Scolopendra).

6. For qi deficiency headache, add *huáng qí* (Radix Astragali) and *dǎng shēn* (Radix Codonopsis), etc.

7. For blood stasis headache, add *Táo Hóng Sì Wù Tāng* (Peach Kernel and Carthamus Four Agents Decoction, 桃红四物汤).

8. For blood deficiency headache, add *Sì Wù Tāng* (Four Agents Decoction) and *huáng qí* (Radix Astragali).

9. For phlegm or damp heavy headache, add *bàn xià* (Rhizoma Pinelliae), *nán xīng* (南星 , Arisaema cum Bile), *chén pí* (Pericarpium Citri Reticulatae) and *fú líng* (Poria), etc.

10. If accompanied by neurasthenia, insomnia, add *yè jiāo téng* (Caulis Polygoni Multiflori) and *zǎo rén* (Semen Ziziphi Spinosae), etc.

11. If with external wind-cold, add *jīng jiè* (Herba Schizonepetae), *fáng fēng* (Radix Saposhnikoviae) and *sū yè* (苏叶 , Folium Perillae), etc.

12. If with external wind-heat, add *sāng yè* (Folium Mori) and *jīn yín huā* (金银花 , Flos Lonicerae Japonicae), etc.

13. If with external wind-damp, add *qiāng huó* (Rhizoma et Radix Notopterygii) and *shēng yì rén* (raw Semen Coicis), etc.

Compatibility Analysis

In the formula, *tiān má* (Rhizoma Gastrodiae) and *chuān xiōng* (Rhizoma Chuanxiong) are used to relieve spasm and expel wind, to transform stasis and stop pain as a king herb. *Dān shēn* (Radix et Rhizoma Salviae Miltiorrhizae), *dāng guī* (Radix Angelicae Sinensis) and *gé gēn* (Radix Puerariae Lobatae) have the function to invigorate blood and transform stasis, to nourish blood and unblock the collaterals. It unblocks blood flow and eliminates stasis, then the pain stops spontaneously, these are the assistant herbs. *Shuāng gōu* (Ramulus Uncariae Cum Uncis) and *quán xiē* (Buthus Martensi) have functions to calm the spirit, stop pain and alleviate spasm. They can assist *tiān má* (Rhizoma Gastrodiae) and *chuān xiōng* (Rhizoma Chuanxiong), to more effectively alleviate the pain acting as the deputy herbs. *Hé yè* (Folium Nelumbinis) is a product to arouse clear yang, raise clear qi, descend turbid qi and guide all herbs to the head, acting as the envoy herb. This formula is gentle but is more effective for stopping the pain. With modification this can be used for all kinds of headache.

2. *Chuān Xiōng* (Rhizoma Chuanxiong) ——
Gōu Téng (钩藤, Ramulus Uncariae Cum Uncis)

Function

To invigorate blood, expel wind and alleviate the pain.

Application

This is used for many kinds of headache such as vasospastic headache, nervous headache, hypertension headache, and the external invasion headache, etc. According to syndrome differentiation, adding this compatibility can increase the effectiveness of treatment.

Compatibility Analysis

Chuān xiōng (Rhizoma Chuanxiong) is acrid and warm. It can invigorate blood, promote qi movement, alleviate pain, and clear wind. *Gōu téng* (Ramulus Uncariae Cum Uncis) is sweet, bitter and cold. It can clear heat and extinguish wind, calm the liver and relieve spasm. The two herbs are both the *jueyin* channel's herbs. *Gōu téng* (Ramulus Uncariae Cum Uncis) based on the modern pharmacological research contains rhynchophylline, which has functions to expand blood vessels and relieve blood vessel's spasm. It also enhances the power of *chuān xiōng* (Rhizoma Chuanxiong) to promote qi movement and invigorate blood. Meanwhile, it can use its coldness to balance the warmth and dryness of *chuān xiōng* (Rhizoma Chuanxiong). The ligustrazine in *chuān xiōng* (Rhizoma Chuanxiong) also has a good ability to expand peripheral blood vessels. Its extract functions better to alleviate pain and to be an analgesic. The two herbs combined together can mutually enhance each other and they then have a stronger power to expel wind and alleviate pain.

Case Study

Initial Visit

A female patient, 7 years old, January 10th, 1987.

The patient had a frontal headache for more than half a year. Every morning until 9:00a.m. and at night before sleep, it was aggravated. It was accompanied by nausea, depression, fatigue and lassitude. Her appetite was normal. She had no convulsion history. She was investigated by the Neurology and Otolaryngology Departments in a hospital, but no abnormalities were found. Biomedicine diagnosed it as an angioneurotic headache (migraine). On observation, the patient had good physical development. Her tongue coating was thin and white and her tongue was slightly red. Her mother said that after midday, the patient experienced feelings of cold and heat with head distention. She was diagnosed with liver and stomach disharmony syndrome. When the liver is stagnated, there is cold and heat. The stomach lost its harmony and its descending functions so the clear yang could not rise up inducing headache and head distention, this coupled with the turbid yin not being able to descend downward, induced the nausea and fatigue. The treatment plan was to use a formula to soothe the liver and harmonise the stomach, to raise the clear and descend the turbid.

Prescription

川芎	*chuān xiōng*	15g	Rhizoma Chuanxiong
钩藤	*gōu téng*	24g	Ramulus Uncariae Cum Uncis (decoct later)
白芷	*bái zhǐ*	9g	Radix Angelicae Dahuricae
柴胡	*chái hú*	12g	Radix Bupleuri
荷叶	*hé yè*	9g	Folium Nelumbinis
升麻	*shēng má*	3g	Rhizoma Cimicifugae
半夏	*bàn xià*	9g	Rhizoma Pinelliae
黄芩	*huáng qín*	9g	Radix Scutellariae
丹参	*dān shēn*	15g	Radix et Rhizoma Salviae Miltiorrhizae
苍术	*cāng zhú*	9g	Rhizoma Atractylodis
甘草	*gān cǎo*	5g	Radix Glycyrrhizae

Water was used to decoct, then it was divided several times, to drink often, one pack per day.

Second Visit

January 16th: After taking five packs of the above formula, the headache reduced a lot, the head distention, and cold and hot feeling also became lighter. She took five more packs of the original formula.

Third Visit

January 21st: Her headache and head distention basically disappeared. Five more packs of the original formula were used to strengthen the treatment effect.

Later at a follow up visit, there had been no reoccurrence.

3. *Chuān Xiōng* (Rhizoma Chuanxiong) ——
Cāng Ěr Zǐ (苍耳子, Fructus Xanthii)

Function

To expel wind and alleviate pain.

Application

This can be used for external wind-cold invasion headaches, with a stuffy and running nose. It also can be used for chronic rhinitis and nasosinusitis, and for wind-dampness arthralgia syndrome, etc.

Compatibility Analysis

Chuān xiōng (Rhizoma Chuanxiong) is acrid and warm. It invigorates blood and promotes movement of qi; it expels wind and alleviates pain. *Cāng ěr zǐ* (Fructus Xanthii) is acrid, bitter and warm. It especially enters the lung through the nostrils. Its dispersing power is the strongest. Upwards it can reach the vertex, downwards it can reach the knees and feet, inwards it can reach the bone marrow and outwards it penetrates the skin. It has the function to expel wind and transform dampness. *Chuān xiōng* (Rhizoma Chuanxiong) with the help of *cāng ěr zǐ* (Fructus Xanthii) enters the nostrils and can invigorate blood to promote the local circulation. *Cāng ěr zǐ* (Fructus Xanthii) with the help of the promoting qi movement and invigorating blood power of *chuān xiōng* (Rhizoma Chuanxiong), can increase its expelling wind and alleviating pain effect.

Case Study

Initial Visit

A female patient, 36 years old, March 30th, 1987.

The patient had flu for more than a month without recovery. It was sometimes better and sometimes worse. Her right supraobital bone felt painful. If pressed on, it became severely painful. She had a stuffy nose with a yellow turbid discharge, anorexia, nausea, a bitter mouth and no appetite, slight body heat and aversion to cold. Her pulse was wiry, slippery, thready and fast. Her tongue was red with a white tongue coating. Prolonged external wind-cold had transformed to heat, and heat stagnated in the upper *jiao* lung orifices. Hence, she had the stuffy nose with the turbid discharge and headache. The external pathogen was not relieved and entered the *shaoyang* half external and half internal phase. This produced the bitter mouth, nausea, hot and cold body sensations. The treatment plan was to harmonise the *shaoyang* by clearing heat, expelling wind and opening the orifices.

Prescription

苍耳子	*cāng ěr zǐ*	12g	Fried Fructus Xanthii
川芎	*chuān xiōng*	15g	Rhizoma Chuanxiong
辛夷	*xīn yí*	9g	Flos Magnoliae
薄荷	*bò he*	9g	Herba Menthae
柴胡	*chái hú*	15g	Radix Bupleuri
半夏	*bàn xià*	9g	Rhizoma Pinelliae
黄芩	*huáng qín*	9g	Radix Scutellariae
金银花	*jīn yín huā*	20g	Flos Lonicerae Japonicae
蒲公英	*pú gōng yīng*	15g	Herba Taraxaci
僵蚕	*jiāng cán*	10g	Bombyx Batryticatus
羌活	*qiāng huó*	9g	Rhizoma et Radix Notopterygii
甘草	*gān cǎo*	6g	Radix Glycyrrhizae

Water was used to decoct, then it was divided into two servings, served warm, one pack per day.

Second Visit

April 3rd: After taking four packs of the above formula, the hot and cold symptoms were eliminated; her appetite became normal, her headache reduced; and her stuffy nose with discharge improved by more than half. The original formula was used eliminating

Xiǎo Chái Hú Tāng (Minor Bupleurum Decoction) and it was taken again.

Prescription

苍耳子	*cāng ěr zǐ*	15g	Fried Fructus Xanthii
川芎	*chuān xiōng*	15g	Rhizoma Chuanxiong
辛夷	*xīn yí*	9g	Flos Magnoliae
薄荷	*bò he*	9g	Herba Menthae
金银花	*jīn yín huā*	20g	Flos Lonicerae Japonicae
黄芩	*huáng qín*	9g	Radix Scutellariae
蒲公英	*pú gōng yīng*	15g	Herba Taraxaci
羌活	*qiāng huó*	9g	Rhizoma et Radix Notopterygii
僵蚕	*jiāng cán*	10g	Bombyx Batryticatus
甘草	*gān cǎo*	6g	Radix Glycyrrhizae

Water was used to decoct, one pack per day.

Third Visit

April 8th: After taking another five packs, the headache, and the stuffy nose with discharge were a lot better. The right supraobital bone still had a pressing pain, but was better than before. The formula was effective so remained unchanged. The patient was instructed to follow the formula, to take it every other day continuously. Later, at a follow up visit, the patient had taken more than twenty packs, and recovered.

4. *Chuān Xiōng* (Rhizoma Chuanxiong) —— *Bǎn Lán Gēn* (板蓝根, Radix Isatidis)

Function

To release the exterior and clear heat.

Application

It is used for viral influenza, aches and pains all over the body, headache and fever. the cause is not important, it can be due to exterior wind-cold or wind-heat, because it can be differentiated first and then applied. Meanwhile, it is also effective for reducing fever.

Compatibility Analysis

Chuān xiōng (Rhizoma Chuanxiong) is acrid and warm. It can expel wind and alleviate pain. It is a commonly used herb to treat exterior wind-cold induced headache. *Bǎn lán gēn* (Radix Isatidis) is bitter and cold. It can clear heat, cool blood and detoxify toxins. It has better treatment effectiveness for exterior wind-heat induced swollen and sore throat, and headache with fever. These two herbs combined together, the acridity and warmth of *chuān xiōng* (Rhizoma Chuanxiong) can prevent the cold arthralgia from *bǎn lán gēn* (Radix Isatidis). *Bǎn lán gēn* (Radix Isatidis) can reduce the warmness and dryness from *chuān xiōng* (Rhizoma Chuanxiong). The acrid and dispersing nature of *chuān xiōng* (Rhizoma Chuanxiong) can help *bǎn lán gēn* (Radix Isatidis) to clear heat in order to disperse it thoroughly. The bitterness and coldness of *bǎn lán gēn* (Radix Isatidis) can help *chuān xiōng* (Rhizoma Chuanxiong) to release the exterior and drop the fever. These two herbs used together can neutralise and regulate cold and heat as well as releasing the exterior and dropping the fever effectively.

Case Study

Initial Visit

A male patient, 23 years old, labour worker, December 7th, 1988.

The patient had an exterior invasion with fever for three days continuously without relief. He took tablets for the fever but the temperature still did not drop. His highest temperature reached 39℃. He had aversion to cold, fever, body aches, a dry mouth, a sore throat, brownish urine, a floating and fast pulse, a red tongue, with a thin white tongue coating, and a routine blood test showed no abnormalities. The diagnosis was exterior wind-cold invasion with pathogens stagnated in the muscle and skin levels, with the interior heat being unable to disperse. The treatment plan was to release the exterior and clear heat.

Prescription

板蓝根	*bǎn lán gēn*	30g	Radix Isatidis
川芎	*chuān xiōng*	15g	Rhizoma Chuanxiong
金银花	*jīn yín huā*	15g	Flos Lonicerae Japonicae
连翘	*lián qiáo*	15g	Fructus Forsythiae
薄荷	*bò he*	9g	Herba Menthae
荆芥	*jīng jiè*	9g	Herba Schizonepetae
柴胡	*chái hú*	15g	Radix Bupleuri
黄芩	*huáng qín*	9g	Radix Scutellariae
桔梗	*jié gěng*	9g	Radix Platycodonis
甘草	*gān cǎo*	6g	Radix Glycyrrhizae

This was decocted with water and divided in two, to be taken twice, one pack per day.

Second Visit

December 10th: After taking one pack of the above formula, he had sweating all over his body and his fever dropped a little bit. After taking two packs, his fever was gone and the pain in all his joints was also reduced a lot. After taking three packs, all of his symptoms basically disappeared except the dry mouth and sore throat. Based on the above formula, *lú gēn* (芦根 , Rhizoma Phragmitis) 20g, and *shè gān* (射干 , Rhizoma Belamcandae) 9g, were added. He continued taking the formula.

Third Visit

December 13th: His dry mouth and sore throat had reduced a lot. He continued to take the same formula for another three packs and recovered.

Mù Tōng (木通 , Caulis Akebiae)

Mù Tōng's (木通 , Caulis Akebiae) taste and property are bitter and slightly cold. It enters the heart channel of the hand *shaoyin*, the lung channel of the hand *taiyin*, the small intestine channel of the hand *shaoyang* and the urinary bladder channel of the foot *taiyang* meridians. It has functions to clear heat, promote urination and free stranguria. It is a commonly used herb for mouth ulcers, hot painful urination, sore throat and red eyes. This product is beneficial for promoting urination. It can unblock thè qi stagnation, invigorate the blood vessels, open the nine orifices and the mammary glands. It also can be used for blocked lactation, amenorrhoea, etc. The earliest record for *mù tōng* (木通 , Caulis Akebiae) is in the *Divine Husbandman's Classic of the Materia Medica*, but it only mentioned that *mù tōng* (Caulis Akebiae) can *"freely open the nine orifices, blood vessels and joints."* It did not mention that it can promote urination and free stranguria. Until Zheng Quan started to develop it, in the *Treasury of Words on the Materia Medica* (本草汇言 , *Běn Cǎo Huì Yán*), he states that *"mù tōng (Caulis Akebiae) opens the nine orifices, eliminates stagnated heat, guides the small intestine, treats turbid painful urination, calms palpitations, convulsions and mania. It is an important herb to treat the heart and the small intestine........"* The author uses this herb in three ways. The first one is to clear the heart fire in order to treat mouth ulcers and red eyes. The second is to free stranguria to treat painful urination. The third is to unblock the blood vessels and the qi stagnation in order to treat blocked lactation and amenorrhea. Due to the bitter and cold nature of this herb, it can unblock and drain downward, therefore, it can treat the syndromes belonging to the yang deficiency and qi weakness category. It should be used with caution in pregnant women.

1. *Mù Tōng* (Caulis Akebiae) ──
Shēng Dì (生地, Radix Rehmanniae)

Function

To nourish yin and clear the heart, to sedate heat and free stranguria.

Application

In clinical practice, if due to yin deficiency with excess fire, we can see excess heat in the heart channel of the hand *shaoyin* meridian, irritability and heat in the heart and chest, thirst and a red face, a preference for cold drinks, or mouth and/or tongue ulcers, or the heart heat transferred to the small intestine manifesting in red or painful urination, or red eyes and brownish urine. All of these can use this compatibility for treatment.

Compatibility Analysis

If the kidney yin is deficient then heart fire will be out off control and flare upward. Therefore, the heart and chest will feel irritable and hot. *Mù tōng* (Caulis Akebiae) is bitter and cold. It is beneficial for promoting urination and can guide the heart fire downward to expel it from the body via the small intestine. These two herbs combined together can nourish the kidney yin to clear the heart and sedate heart fire through promoting urination. Qian Yi's *Dǎo Chì Sǎn* (Red-Abducting Powder, 导赤散) uses this compatibility, which uses large doses of *shēng dì* (Radix Rehmanniae). Wu Qian said that *"it is pertinent to use it for those patients with wood deficiency but without excess fire."* He also said that *"it promotes urination but does not injure yin and sedates fire but does not injure the stomach."* This compatibility is the best one for heat in the heart channel of the hand *shaoyin* meridian.

Initial Visit

A male patient, 60 years old, July 6th, 1985.

The patient had long term family problems that could not be resolved easily which induced emotional stress and irritability. His prolonged constraints transferred into fire. His heart was irritable and he easily got angry, he had thirst with a little saliva. He had mouth and tongue ulcers that were hard to bear. He took *Niú Huáng Jiě Dú Piàn* (Bovine Bezoar Toxin-resolving Tablet, 牛黄解毒片) orally and *Zhū Huáng Sǎn* (Pearl and Bezoar Powder, 珠黄散) externally to treat it but without effect and so he came to visit. On examination, the patient's tongue was red with a scanty tongue coating. There were two soy bean sized ulcers on the left side of his tongue. He had slippery and forceful pulse, and brownish urine. The diagnosis was qi stagnation transformed to fire with heart fire flaring upward. The treatment plan was to clear the heart, sedate fire and nourish yin.

Prescription

生地	*shēng dì*	20g	Radix Rehmanniae
木通	*mù tōng*	9g	Caulis Akebiae
竹叶	*zhú yè*	6g	Herba Lophatheri
栀子	*zhī zǐ*	10g	Fructus Gardeniae
茵陈	*yīn chén*	12g	Herba Artemisiae Scopariae
藿香	*huò xiāng*	10g	Herba Agastachis
甘草	*gān cǎo*	6g	Radix Glycyrrhizae

This was decocted with water and divided into three servings, two servings taken orally in the morning and night, and the other one was divided many times to be used as a mouth wash. One pack per day.

Second Visit

July 12th: After continuously taking six packs of the above formula, his irritability and thirst were eliminated. The ulcers on his tongue also disappeared and there was no pain. Three more packs of the same formula were used to treat him and he recovered.

Initial Visit

A female patient, 45 years old, February 7th, 1991.

The patient had a history of urinary tract infections. Recently, due to overexertion, it manifested as frequent urination; the urination was short, painful and not smooth, with red urine and irritability, a routine urine test showed: WBC++，RBC++. On observation she had a red tongue, with a thin yellow coating, and a thready and fast pulse. The diagnosis was yin deficiency with exuberant heart fire. The urination being short, painful, and not flowing smoothy with red urine was due to heart fire moving to the small intestine. The treatment plan was to treat by using a nourishing yin, clearing fire, sedating fire and freeing painful urination formula. A modified *Dǎo Chì Sǎn* (Red-Abducting Powder) was used for treatment.

Prescription

生地	*shēng dì*	30g	Radix Rehmanniae
木通	*mù tōng*	10g	Caulis Akebiae
竹叶	*zhú yè*	9g	Herba Lophatheri
金银花	*jīn yín huā*	20g	Flos Lonicerae Japonicae
车前子	*chē qián zǐ*	12g	Semen Plantaginis (wrapped up)
石韦	*shí wéi*	15g	Folium Pyrrosiae
甘草	*gān cǎo*	6g	Radix Glycyrrhizae

Water was used to decoct, then it was divided into two servings, one pack per day.

Second Visit

February 10 th: After taking three packs of the above formula, her irritability, dry mouth, and painful urination were all reduced, and the urine quantity increased. She continued taking the original formula again.

Third Visit

February 15 th: After taking five more packs, all her symptoms disappeared, and her routine urine test was normal.

Mù Hú Dié's (木蝴蝶 , Semen Oroxyli) taste and property are bitter and neutral. It enters the lung channel of the hand *taiyin* and the liver channel of the foot *jueyin* meridians. It functions to moisten the lung and stop cough, to clear heat and open up the voice. It is a common herb to treat cough and dysphonia.

1. *Mù Hú Dié* (Semen Oroxyli) ——
Chán Tuì (蝉蜕, Periostracum Cicadae),
Fèng Huáng Yī (凤凰衣, Membrana Follicularis Ovi)

Function

To moisten the lung and open up the voice.

Application

This is used for dry heat yin injury causing dysphonia and lost voice.

Compatibility Analysis

A hoarse sound, or voice loss is a disease of the throat or larynx. In the *Inner Classic* (内经 , *Nèi Jīng*) it was called *yin* (喑). The *Compendium of Medicine* (医学纲目 , *Yī Xué Gāng Mù*) called it "*she yin* (舌喑)", to differentiate it from that of a stiff tongue after a stroke and speaking without fluency. Dysphonia or lost voice is clinically classified into two types, deficiency and excess. The excess is more due to external wind-cold or wind-heat invasion. The external pathogen attacks the lung, the lung qi loses its dispersing function, and the voice becomes hoarse, and the pharynx does not open and close smoothly, causing sudden dysphonia or loss of voice. In clinical practice it is also called metal excess causing loss of voice. The deficiency syndrome is more due to lung dryness and kidney deficiency, fluid is burned and injured, the larynx loses moisture inducing dysphonia. For the excess syndrome it is better to release the exterior and disperse the lung. For the deficiency syndrome, it is better to nourish the yin and moisten the dryness.

Mù hú dié (Semen Oroxyli) is bitter and neutral. It can clear heat, moisten the lung, stop cough and open up the voice. *Chán tuì* (蝉蜕 , Periostracum Cicadae) is salty, sweet and cold. It enters the lung channel of the hand *taiyin* and the liver channel of the foot *jueyin* meridians. It can clear heat and disperse lung qi, treat cough and dysphonia. Chen Zhong-qi said: "*Grind one qian into powder, take with water, this can treat mute disease.*" The *Grand Materia Medica* said it treats "*adult voice loss*". *Fèng huáng yī* (凤凰衣 , Membrana Follicularis Ovi) is sweet and neutral. Its colour is white and it enters the lung channel of the hand *taiyin* meridian. It can nourish yin and moisten the lung, it can treat chronic cough, induce shortness of breath and voice loss. *Fèng huáng yī* (Membrana Follicularis Ovi) is the egg shell membrane after a chick hatches. In clinical practice the raw egg's internal membrane can be used as a substitute and its power to moisten the lung and open up the voice is even better. The above three herbs used together have the power to clear heat and disperse lung qi, nourish yin and moisten dryness, open up the voice and stop cough. Using syndrome differentiation, it can be added in for wind-heat attacking the lung or dry heat injuring yin, inducing dysphonia and voice loss, and this can increase the effectiveness for opening the voice up.

Case 1

External invasion dysphonia.

Initial Visit

A male patient, 6 years old, March 4 th, 1989.

The child came to visit due to an external invasion causing cough and dysphonia. His mother stated that the child had an external invasion and fever. After she administered a children's flu powder, his fever was slightly reduced but he had relentless coughing with dysphonia, and speech was difficult and unclear. On observation the child had a red throat, an enlarged tonsil, a hoarse breath sound in both lungs, a red tongue with a thin white coating, and a thready and fast pulse. The diagnosis: An external pathogen had invaded the lung, lung heat had then stagnated inside, so the lung qi could not disperse nor descend, with rough inhalation and exhalation inducing dysphonia and voice loss. The treatment plan was to clear heat and disperse the lung, to open up the voice and stop cough. A modified *Zhǐ Sòu Sǎn* (Cough-Stopping Powder, 止嗽散) was used to treat him.

Prescription

百部	*bǎi bù*	9g	Radix Stemonae
桔梗	*jié gěng*	9g	Radix Platycodonis
紫菀	*zǐ wǎn*	9g	Radix et Rhizoma Asteris
蝉蜕	*chán tuì*	9g	Periostracum Cicadae
木蝴蝶	*mù hú dié*	10g	Semen Oroxyli
凤凰衣	*fèng huáng yī*	2pcs	Membrana Follicularis Ovi
荆芥	*jīng jiè*	6g	Herba Schizonepetae
川贝	*chuān bèi*	8g	Bulbus Fritillariae Cirrhosae
杏仁	*xìng rén*	8g	Semen Armeniacae Dulcis
板蓝根	*bǎn lán gēn*	10g	Radix Isatidis
甘草	*gān cǎo*	6g	Radix Glycyrrhizae

Water was used to decoct, then it was divided into two servings, one pack per day.

Second Visit

March 9 th: After taking five packs of the above formula, the cough stopped, and the dysphonia went. He took another two packs of the original formula and recovered.

Case 2

Lung dryness dysphonia.

Initial Visit

A male patient, 49 years old, July 6 th, 1994.

The patient had dysphonia and voice loss for half a month. He had treatments but did not recover so he came to visit. The patient's business was slow which induced worry and depression, and stagnated fire generated internally. He also talked too much during meetings injuring his lung qi, gradually his voice became hoarse, until he finally lost his voice. He took lubricating throat tablets and *Cǎo Shān Hú* (Sarcandrae Ramulus et Folium, 草珊瑚), but they did not work. On observation, the patient had a slightly red throat (he did not state that he had a sore throat), a red tongue with a thin white coating, a thready, fast and slightly wiry pulse, this was accompanied by a dry mouth, irritability, tinnitus, etc. The treatment plan was to clear heat, moisten the lung, open up the voice and eliminate depression.

Prescription

木蝴蝶	*mù hú dié*	20g	Semen Oroxyli
蝉蜕	*chán tuì*	10g	Periostracum Cicadae
凤凰衣	*fèng huáng yī*	4pcs	Membrana Follicularis Ovi
桔梗	*jié gěng*	10g	Radix Platycodonis
薄荷	*bò hé*	9g	Herba Menthae
杭白芍	*háng bái sháo*	15g	Radix Paeoniae Lactiflorae
柴胡	*chái hú*	12g	Radix Bupleuri
茯苓	*fú líng*	15g	Poria
百合	*bǎi hé*	24g	Bulbus Lilii
甘草	*gān cǎo*	6g	Radix Glycyrrhizae

Water was used to decoct, then it was divided into two servings, one pack per day.

In addition, *mù hú dié* (Semen Oroxyli) 10g, and *pàng dà hǎi* (胖大海 , Semen Sterculiae Lychnophora) two pieces, were decocted with water, to be taken as an herbal tea.

Second Visit

July 13 th: The patient's throat felt easier, it was not as difficult to speak as before, his dry mouth and irritability also reduced. He took the same formula again.

Third Visit

July 20th: After taking seven more packs he could speak but was still hoarse. He continued to take the same formula again.

Fourth Visit

July 28th: All his symptoms were basically eliminated. The patient was instructed to take the same formula every other day for five more packs to prevent it reoccurring.

16 **Dān Pí (丹皮 , Cortex Moutan)**

Dān pí's (丹皮 , Cortex Moutan) taste and property are acrid, bitter and slightly cold. It enters the heart channel of the hand *shaoyin*, the liver channel of the foot *jueyin*, the kidney channel of the foot *shaoyin* and the pericardium channel of the hand *jueyin* meridians. It functions to cool blood, expel stasis and clear heat. The *Divine Husbandman's Classic of the Materia Medica* said: *"It mainly treats cold and heat, stroke inducing convulsion, anxiety and pathogenic qi, it eliminates abdominal abscesses, accumulations, stasis in the intestines and stomach, it calms the five organs, and treats abscesses and ulcers."* Currently *dān pí* (Cortex Moutan) roughly has two clinical applications. One is clearing heat and cooling blood, used for heat in the blood level inducing rashes with fever, or heat pushing blood to move restlessly inducing haemoptysis, epistaxis, haematochezia, etc. The second one is invigorating blood and dispelling stasis. It can be used for blood stasis, abdominal abscesses, amenorrhoea, intestinal abscesses, etc.

1. *Dān Pí* (Cortex Moutan) ——
Guì Zhī (桂枝, Ramulus Cinnamomi)

Function

To invigorate blood, promote menstruation, and to eliminate stasis and alleviate pain.

Application

This is used for blood stasis inducing female irregular menstruation, menstrual pain, abdominal abscesses, accumulations, etc. Modern clinics use it for uterine fibroids, chronic pelvic inflammation, ovarian cysts, etc. with good effect.

Compatibility Analysis

Guì zhī (Ramulus Cinnamomi) is acrid, warm, red in colour and enters the blood level. It can warm up yang qi and the blood vessels; and it can expel cold stagnation in the blood vessels. *Dān pí* (Cortex Moutan) is acrid, bitter and cold. It can invigorate blood; expel stasis and heat clumping in the blood vessels. The two herbs together can invigorate blood, promote menstruation, and eliminate stagnation and impediment in the blood vessels. Zhou Shu said: *"Dān pí (Cortex Moutan) enters the heart channel of the hand shaoyin meridian to unblock stagnation and impediment in the blood vessels and is very similar to guì zhī (Ramulus Cinnamomi). The qi of guì zhī (Ramulus Cinnamomi) is warm, hence it unblocks cold obstruction, in the blood vessels. The qi of mǔ dān (Cortex Moutan) is cold, hence, it unblocks heat clumping in the blood vessels."* These two herbs used together, neutrally regulate heat and cold. Aside from enhancing functions to invigorate blood and transform blood stasis, they also can relax the bias of each other's property, hence, in clinical practice; their application scope can be expanded. It is of no consequence whether the condition is due to deficiency, excess, cold or heat; if there is blood stasis, it can be used for all. *Guì Zhī Fú Líng Wán* (Cinnamon Twig and Poria Pill, 桂枝茯苓丸) and *Wēn Jīng Tāng* (Channel-Warming Decoction, 温经汤) are both examples of the use of this compatibility of *mǔ dān* (Cortex Moutan) and *guì zhī* (Ramulus Cinnamomi).

Case Study

Initial Visit

A female patient, 38 years old, January 10ᵗʰ, 1989.

The patient had low back pain and lower right abdominal pain for three months. She went to the Gynaecology Department of an hospital to get checked and was diagnosed with chronic adnexitis. She had acupuncture and herbal treatment for a week, but it did not have a remarkable effect, so she came to visit our hospital clinic. Her chief complaint was a mild pain in her lower right abdomen, a sinking and distending sensation, lower back soreness and pain, almost like it was broken, symptoms were worse after exertion, her menstruation was early, each cycle being almost twenty days, the menstrual flow was dark in colour, average in quantity, with a lot of leukorrhoea, etc. On palpation, there was a mass in the right lower abdomen, about the size of a date, which refused pressure. She had a wiry, thready and deep pulse, and a thin white tongue coating with a dark tongue (qi and blood deficiency inducing blood stasis type). The treatment plan was to tonify qi, invigorate blood and transform stasis.

Prescription

桂枝	guì zhī	10g	Ramulus Cinnamomi
丹皮	dān pí	12g	Cortex Moutan
桃仁	táo rén	9g	Semen Persicae
当归	dāng guī	12g	Radix Angelicae Sinensis
赤芍	chì sháo	15g	Radix Paeoniae Rubra
生黄芪	shēng huáng qí	20g	Radix Scutellariae (raw)
海螵蛸	hǎi piāo xiāo	15g	Endoconcha Sepiae
茜草	qiàn cǎo	9g	Radix et Rhizoma Rubiae
茯苓	fú líng	15g	Poria
甘草	gān cǎo	6g	Radix Glycyrrhizae

Water was used to decoct, then it was divided into two servings, served warm, one pack per day.

Second Visit

January 13ᵗʰ: After taking three packs of the above formula, the abdominal pain had slightly reduced, and the leukorrhoea was also reduced. She continued taking the original formula with yán hú suǒ (Rhizoma Corydalis) 15g, added to it.

Third Visit

January 20ᵗʰ: She took another six packs, her low back and abdominal pain had all reduced significantly. On palpation, the mass had also reduced and there was no pain on pressure. She continued with the same formula.

Fourth visit

February 15ᵗʰ: After another twenty more packs of the above formula modification, the mass disappeared along with all the symptoms, and she recovered.

Dān Shēn (丹参 , Radix et Rhizoma Salviae Miltiorrhizae)

Dān shēn's (丹参 , Radix et Rhizoma Salviae Miltiorrhizae) taste and property are bitter and slightly cold. It enters the heart channel of the hand *shaoyin* and the liver channel of the foot *jueyin* meridians. It is red in colour and enters the blood level. It functions to invigorate blood, promote menstruation, nourish blood and tranquilise the spirit. It can be used for irregular menstruation in women, blood obstructed amenorrhoea, postpartum blood stasis and abdominal pain, abdominal abscesses, and accumulations, also for falling or fighting injuries, sores, abscesses and pain due to swellings, etc. It has functions to expand blood vessels and reduce blood pressure because it contains tanshinone therefore, it is also used for cardiac heart disease and liver or spleen enlargement. Overall, *dān shēn* (Radix et Rhizoma Salviae Miltiorrhizae) is an effective treatment for treating many kinds of disease, it is irrelevant whether its due to blood stasis, or to blood not flowing smoothly. Aside from that, *dān shēn* (Radix et Rhizoma Salviae Miltiorrhizae) also functions to nourish blood, and tranquilise and calm the spirit. The *Clarification of the Theory for Women Disease* (妇人明理论 , *Fù Rén Míng Lǐ Lùn*) said: "*Dān shēn (Radix et Rhizoma Salviae Miltiorrhizae), a single herb, has the same functions as Sì Wù Tāng (Four Agents Decoction, 四物汤)*", this indicates that *dān shēn* (Radix et Rhizoma Salviae Miltiorrhizae) has functions to tonify, nourish and invigorate blood. It is an herb to invigorate blood but does not injure blood.

1. *Dān Shēn* (Radix et Rhizoma Salviae Miltiorrhizae) —— *Dāng Guī* (当归, Radix Angelicae Sinensis)

Function

To invigorate blood, promote menstruation, transform stasis and eliminate tumours.

Application

This is used for qi and blood stagnation and obstruction, heart and abdominal pain, leg pain, arm pain, internal or external sores and abscesses as well as all kinds of *zang* or *fu* organ's accumulations.

Compatibility Analysis

Dān shēn (Radix et Rhizoma Salviae Miltiorrhizae) is bitter and slightly cold. It enters the heart channel of the hand *shaoyin* and the liver channel of the foot *jueyin* meridians, at the blood level, and is an important herb for blood invigoration and promoting menstruation. *Dāng guī* (Radix Angelicae Sinensis) is acrid, sweet, bitter and warm. It also enters the heart channel of the hand *shaoyin*, the liver channel of the foot *jueyin* and the spleen channel of the foot *taiyin* meridians at the blood level. It is sweet, warm, moistening, acrid, aromatic and is good at unblocking and moving. It is an extraordinary herb to tonify and invigorate blood. Both herbs have effects on tonifying and invigorating blood. *Dān shēn* (Radix et Rhizoma Salviae Miltiorrhizae) has a cold property, and can cool and invigorate blood; whereas, *dāng guī* (Radix Angelicae Sinensis) has a warm property and can warm blood vessels. The two herbs combined together mutually enhance their functions to move and invigorate blood, and promote menstruation. Meanwhile, the coldness of *dān shēn* (Radix et Rhizoma Salviae Miltiorrhizae) can balance the warmth of *dāng guī* (Radix Angelicae Sinensis), and the warmth of *dāng guī* (Radix Angelicae Sinensis) can lower the coldness of *dān shēn* (Radix et Rhizoma Salviae

Miltiorrhizae), to neutrally regulate cold and heat. It is most suitable for all kinds of qi and blood stagnation and obstruction diseases. This compatibility originated from Zhang Xichun's *Huó Luò Xiào Líng Dān* (Network-Quickening Miraculous Effect Elixir, 活络效灵丹) with *rŭ xiāng* (乳香 , Olibanum) and *mò yào* (没药 , Myrrha) eliminated. Clinically, with modifications, its treatment applications are very broad. To treat lower limb pain, add *niú xī* (Radix Cyathulae). For upper limb pain, add *lián qiáo* (连翘 , Fructus Forsythiae). For female blood stasis inducing abdominal pain, add *táo rén* (桃仁 , Semen Persicae) and *wŭ líng zhī* (Faeces Trogopterori). For ulcers, abscesses and painful swellings, add *jīn yín huā* (Flos Lonicerae Japonicae) and *lián qiào* (Fructus Forsythiae). They all have effective results.

Case 1

Mammary gland hypertrophy.

Initial Visit

A female patient, 37 years old, March 8th, 1986.

The patient had hard tumours on both breasts with pain, for more than half a month. On palpation, the tumours were the size of a chicken's egg and refused pressure, the superficial skin was neither red, nor swollen. Her diet, urination and bowel movement were normal. The Oncology Department diagnosed it as mammary gland hypertrophy. Her pulse was wiry and slippery. Her tongue coating was thin and white, and her tongue was slightly red. There were no cold or heat symptoms. The diagnosis was the syndrome of qi and blood stagnation and obstruction. The treatment plan was to invigorate blood, dredge the collaterals and expel tumours.

Prescription

丹参	*dān shēn*	30g	Radix et Rhizoma Salviae Miltiorrhizae
当归	*dāng guī*	15g	Radix Angelicae Sinensis
玄参	*xuán shēn*	15g	Radix Scrophulariae
柴胡	*chái hú*	10g	Radix Bupleuri
橘叶	*jú yè*	10g	Folium Citri
王不留行	*wáng bù liú xíng*	10g	Semen Vaccariae
陈皮	*chén pí*	9g	Pericarpium Citri Reticulatae
连翘	*lián qiáo*	10g	Fructus Forsythiae
薄荷	*bò he*	6g	Herba Menthae
生牡蛎	*shēng mŭ lì*	30g	Concha Ostreae (raw)
甘草	*gān cǎo*	6g	Radix Glycyrrhizae

Water was used to decoct, then it was divided into two servings, to be taken in the morning and at night, one pack per day.

Second Visit

March 14th: After taking six packs of the above formula, the pain was eliminated, the tumours had reduced by more than half, and there was no pain when pressed on. She continued taking the original formula.

Third Visit

March 21th: After taking another five packs, the tumour basically disappeared. She was instructed to take three more packs to complete the treatment.

Case 2

Ovarian cyst.

Initial Visit

A female patient, 28 years old, December 14th, 1990.

Half a year previously, the patient had an operation due to an ectopic pregnancy. Two months after her operation, her right abdomen became painful, with the pain extending to her waist and back. It was aggravated during activities. On inspection of her lower abdomen, close to her right side, there was a tumour. A B-ultrasound showed a 5.6cm × 4.5cm × 3.9cm dark cyst area on the right side of her uterus and diagnosed it as a right side uterine cyst. She was afraid to have another operation and so, came to visit. Her pulse was wiry and deep. Her tongue was red and her tongue coating was thin. The ectopic pregnancy operation injured her qi and blood as well as inducing bad blood to stagnate and obstruct, clumping in the lower abdomen and accumulating into an abdominal abscess. The treatment plan was to use a formula to invigorate blood, transform stasis, reduce abscesses and disperse tumours.

Prescription

丹参	*dān shēn*	24g	Radix et Rhizoma Salviae Miltiorrhizae
当归	*dāng guī*	12g	Radix Angelicae Sinensis
制乳香	*zhì rǔ xiāng*	6g	Olibanum (processed)
制没药	*zhì mò yào*	6g	Myrrha (processed)
桂枝	*guì zhī*	10g	Ramulus Cinnamomi
丹皮	*dān pí*	12g	Cortex Moutan
桃仁	*táo rén*	9g	Semen Persicae
杭白芍	*háng bái sháo*	15g	Radix Paeoniae Lactiflorae
生牡蛎	*shēng mǔ lì*	30g	Concha Ostreae (raw)
三棱	*sān léng*	8g	Rhizoma Sparganii
莪术	*é zhú*	8g	Rhizoma Curcumae
皂角刺	*zào jiǎo cì*	10g	Spina Gleditsiae
甘草	*gān cǎo*	6g	Radix Glycyrrhizae

Water was used to decoct, one pack per day.

After following the formula with modifications for more than ninety packs, the abdominal pain disappeared, a gynaecological examination showed up normal, and a B-ultrasound showed no cyst.

Case 3

Mastitis.

Initial Visit

A female patient, 29 years old, August 16th, 1986.

The patient had breast-fed normally for the first two months after delivery. Four days previously, she started feeling distention and pain in the left breast, her lactation reduced, and was followed by aversion to cold and a fever, pain in the limbs and irritability, nausea and vomiting. On inspection, the patient's upper left breast skin was red, swollen, and hot. On palpation, there was a mass that was painful on pressure and hot to the touch, an enlarged lymph node could be felt inferior to her left axilla, her body temperature was 38°C. Her pulse was wiry, fast and forceful. Her tongue coating was white and greasy, and the tongue was red. This was the initial stage of a breast abscess. The treatment plan was to clear heat, cool blood, and invigorate blood and the collaterals.

Prescription

丹参	dān shēn	24g	Radix et Rhizoma Salviae Miltiorrhizae
当归	dāng guī	15g	Radix Angelicae Sinensis
玄参	xuán shēn	24g	Radix Scrophulariae
穿山甲	chuān shān jiǎ	10g	Roasted Manitis Squama
金银花	jīn yín huā	24g	Flos Lonicerae Japonicae
连翘	lián qiáo	15g	Fructus Forsythiae
赤芍	chì sháo	12g	Radix Paeoniae Rubra
橘络	jú luò	10g	Vascular Aurantii
柴胡	chái hú	12g	Radix Bupleuri
瓜蒌	guā lóu	10g	Fructus Trichosanthis
制乳香	zhì rǔ xiāng	6g	Olibanum (processed)
制没药	zhì mò yào	6g	Myrrha (processed)
甘草	gān cǎo	9g	Radix Glycyrrhizae

Water was used to decoct, then divided into servings, one pack per day.

Second Visit

August 19th: After three packs of the above formula, the redness and swelling reduced, and the pain stopped, there was no reoccurrence of the aversion to cold or fever, and the body temperature dropped back to normal. As it was so effective, the formula was not changed. She continued taking the original formula and after three packs she recovered completely.

2. *Dān Shēn* (Radix et Rhizoma Salviae Miltiorrhizae) —— *Pú Gōng Yīng* (蒲公英, Herba Taraxaci)

Function

To clear heat, transform stasis, reduce swellings and expel nodules.

Application

This can be used for chronic gastritis, with symptoms of stomach distention, fullness or pain, reduced appetite, nausea, vomiting, belching, etc.

Chronic gastritis belongs to Chinese medicine's "stomachache" or "stuffiness" classification. It is a disease where there is chronic gastric mucosal inflammation and pathological change, with many inducing factors, and the aetiology is still not clear. The related factors include: The sequelae of acute gastritis, injury from spicy food or medicine, lack of stomach acid, malnutrition, endocrine disorders, auto-immune reactions, bile regurgitation, pylorus spiral bacillus infection, etc. Under endoscope observation, the initial stage of gastritis can show gastric mucosal hyperaemia, erosion, and red and white colours are visible, but red is more prevalent. Then at the mid and late stage the mucous membrane gland atrophies and becomes white in colour; and the mucosal layer becomes thinner or flat. The blood vessels under the mucosa are prominent or hypertrophied, or there is intestinal epidermis metaplasia.

The aetiology of gastritis, analysed from a Chinese medicine point of view, is due to blood stasis and collaterals obstruction as well as damp-heat stagnating internally. It can be due to irregular diet or many other factors. The stomach collaterals qi and blood function becomes disturbed inducing qi stagnation and blood stasis. Prolonged stagnation transforms to heat, heat burns the stomach body and induces this disease. Hence, this disease treatment, aside from strengthening the stomach and harmonising the middle *jiao*, the primary way for treatment is to treat the root, transform stasis, clear the collaterals, clear heat and soothe the liver.

Dān shēn (Radix et Rhizoma Salviae Miltiorrhizae) invigorates blood, transforms stasis, eliminates stagnation and frees the collaterals. It can improve local blood circulation. If stasis is eliminated and collaterals are unblocked, then pain can be stopped. *Pú gōng yīng* (蒲公英, Herba Taraxaci) is sweet, bitter and cold. It enters the spleen channel of the foot *taiyin* and the stomach channel of the foot *yangming* meridians. It can clear heat and detoxify toxins. Its qi is clear and it functions to soothe the liver and clear stagnated heat. Modern medical research has proved that: *Pú gōng yīng* (Herba Taraxaci) can increase the potential voltage difference of the stomach mucosal membrane; therefore, it increases the gastric mucosal barrier function. Meanwhile, *pú gōng yīng* (Herba Taraxaci) has a stronger anti-bacterial function. *Dān shēn* (Radix et Rhizoma Salviae Miltiorrhizae) and *pú gōng yīng* (Herba Taraxaci) combined together, function to invigorate blood, transform stasis, free the collaterals, clear stagnated heat and detoxify toxins. This correlates with the aetiology of gastritis, so they are an effective treatment modality for it.

Initial Visit

A male patient, 63 years old, November 20th, 1995.

The patient had a history of stomach distention and stomachache for three years. Recently due to increased distention and pain, he came to visit. He complained that he had stomach distention and stomachache in the upper abdomen, which was worse after eating; he had frequent hiccough, dry mouth, poor appetite, anorexia, and a pressing pain. The endoscope report showed the stomach pylorus membrane had severe erosion, hyperaemia, red and white colours were visible, but mostly red, and scattered bleeding points could be seen. He was diagnosed with gastritis. His pulse was deep and wiry. His tongue was red and slightly dark. His tongue coating was thin white and slightly yellow. His Chinese medicine diagnosis was stomachache. Through syndrome differentiation it was classified as stomach deficiency, qi stagnation, and blood stasis transforming to heat, with stomach deficiency preventing qi from descending, causing the hiccough. The qi stagnation induced the stomach and abdomen distention and fullness, and caused the blood stasis; with the blood stasis and the collaterals obstructed it caused the stomachache. Prolonged qi and blood stagnation and obstruction transformed into heat and injured the yin, hence the dry mouth, and thin, white, slightly yellowish tongue coating. The treatment plan was to harmonise the stomach and nourish yin, to clear heat and transform stasis.

Prescription

百合	*bǎi hé*	30g	Bulbus Lilii
丹参	*dān shēn*	24g	Radix et Rhizoma Salviae Miltiorrhizae
乌药	*wū yào*	12g	Radix Linderae
砂仁	*shā rén*	8g	Fructus Amomi
蒲公英	*pú gōng yīng*	15g	Herba Taraxaci
香附	*xiāng fù*	12g	Rhizoma Cyperi
杭白芍	*háng bái sháo*	15g	Radix Paeoniae Lactiflorae
白术	*bái zhú*	12g	Rhizoma Atractylodis Macrocephalae
枳壳	*zhǐ qiào*	12g	Fructus Aurantii
延胡索	*yán hú suǒ*	12g	Rhizoma Corydalis
甘草	*gān cǎo*	6g	Radix Glycyrrhizae

Water was used to decoct, then it was divided into two servings, to be taken in the morning and at night, one pack per day.

Second Visit

November 27th: After taking seven packs of the above formula, the stomachache stopped, and the appetite increased. The original formula was used for treatment again.

Third Visit

December 5th: The patient felt better, all his symptoms were basically eliminated. He again used five packs of the above formula and was instructed to take it every other day to complete the treatment.

3. *Dān Shēn* (Radix et Rhizoma Salviae Miltiorrhizae) —— *Gé Gēn* (葛根, Radix Puerariae Lobatae)

Function

To invigorate blood and promote menstruation, to expand coronary arteries and reduce blood pressure.

Application

This can be used for qi stagnation and blood stasis inducing coronary heart disease, and insufficient blood supply to the cardiac muscle. It also can be used for high blood pressure inducing dizziness or angioneurotic headache and cervical vertebrae type dizziness. This also can be used for allergic purpura.

Compatibility Analysis

Dān shēn (Radix et Rhizoma Salviae Miltiorrhizae) invigorates blood, promotes menstruation and functions to expand blood vessels and reduce blood pressure. *Gé gēn* (Radix Puerariae Lobatae) is acrid, sweet and neutral and also has the function to expand blood vessels of the heart and brain. As well as the invigorating blood function of *dān shēn* (Radix et Rhizoma Salviae Miltiorrhizae), it also can tonify blood, so it invigorates blood but without injuring it. With *gé gēn* (Radix Puerariae Lobatae), our ancestors used it more for relaxing muscles, releasing the exterior, raising clear yang, and stopping irritability and thirst. Modern research has discovered that *gé gēn* (Radix Puerariae Lobatae) contains flavonoids, and it can expand blood vessels. It has better treatment effectiveness for improving headache, dizziness, cervical ankylosis, tinnitus, body and limbs numbness, etc. Though *gé gēn* (Radix Puerariae Lobatae) can expand blood vessels, its power to invigorate blood and promote menstruation is not enough; if *dān shēn* (Radix et Rhizoma Salviae Miltiorrhizae) is used to help, it can increase its blood supply power. *Dān shēn* (Radix et Rhizoma Salviae Miltiorrhizae) combined with *gé gēn* (Radix Puerariae Lobatae) can even enhance its invigorating blood and promoting menstruation powers. The two herbs also have functions to nourish blood and generate fluid. As well as invigorating blood and promoting menstruation, they can tonify and nourish blood without injury to the upright qi. Regardless of deficiency or excess, it can be used for all. It is a neutral and safe herbal compatibility to expand coronary arteries and reduce blood pressure.

Case 1

Chest impediment.

Initial Visit

A male patient, 50 years old, February 28 th, 1986.

The patient had coronary heart disease and high blood pressure for the previous seven to eight years. Recently it became more severe, and he had to rest at home all the time. His current symptoms included: Left chest pain, stuffiness in the chest, a suffocating sensation, shortness of breath and lassitude, asthma on exertion, the chest pain expanded to the left shoulder and arm, as well as the back, causing discomfort; after exertion or climbing stairs it became severe and was accompanied by a headache and dizziness. The EKG showed: The V5 T wave was inverted, the Ⅰ, Ⅱ, aVF T waves were all low and flat. On inspection his lower limbs were slightly swollen, his facial colour was dark without lustre, his tongue body was dark with stasis spots, his tongue coating was thick and peeled, his pulse was deep and slow, and his blood pressure was 140/100mm Hg. His diagnosis was qi deficiency and blood stasis inducing chest impediment syndrome. The treatment plan was to tonify qi, invigorate blood, and unblock the collaterals.

Prescription

丹参	dān shēn	20g	Radix et Rhizoma Salviae Miltiorrhizae
葛根	gé gēn	20g	Radix Puerariae Lobatae
生黄芪	shēng huáng qí	40g	Radix Astragali (raw)
党参	dǎng shēn	15g	Radix Codonopsis
桂枝	guì zhī	9g	Ramulus Cinnamomi
赤芍	chì sháo	18g	Radix Paeoniae Rubra
川芎	chuān xiōng	10g	Rhizoma Chuanxiong
三七粉	sān qī fěn	4g	Radix et Rhizoma Notoginseng Powder (dissolved)
川牛膝	chuān niú xī	18g	Radix Cyathulae
红花	hóng huā	9g	Flos Carthami
降香	jiàng xiāng	15g	Lignum Dalbergiae Odoriferae
甘草	gān cǎo	6g	Radix Glycyrrhizae

Water was used to decoct, then it was divided into two servings, one pack per day.

Second Visit

April 10 th: After taking more than thirty packs of modifications of this formula, his stuffiness in his chest, his chest pain and feelings of suffocation all basically disappeared. Going up stairs was also easier than before. He only had slight shortness of breath and could work for half a day. The EKG reported that the cardiac muscle still had insufficient blood supply, however, the T-wave was better than before. The patient was instructed to take the same formula again.

Case 2

Headache.

Initial Visit

A male patient, 38 years old, June 3rd, 1988.

The patient said that half a year ago, due to an emergency, he had to ride a motorcycle to work. It was in cold weather and he was not wearing a helmet, and so he suffered a wind-cold invasion. After he returned, he had a left sided migraine which gradually got worse daily. He took pain killers and had acupuncture treatments, but they still did not resolve the root. The pain was sometimes better and sometimes quite severe. When severe, it went from the back of the skull, GB20 (fēng chí) acupoint, along the left side to the vertex where he felt a needle sensation pain. His pulse was deep and wiry. His diet, urination and bowel movement were normal. He was diagnosed with prolonged wind-cold pathogen invading the taiyang channel and collaterals inducing meridians and blood vessels stagnation and obstruction syndrome. The treatment should invigorate blood, collaterals and expel cold.

Prescription

丹参	*dān shēn*	30g	Radix et Rhizoma Salviae Miltiorrhizae
葛根	*gé gēn*	20g	Radix Puerariae Lobatae
当归	*dāng guī*	15g	Radix Angelicae Sinensis
川芎	*chuān xiōng*	20g	Rhizoma Chuanxiong
羌活	*qiāng huó*	12g	Rhizoma et Radix Notopterygii
桃仁	*táo rén*	9g	Semen Persicae
红花	*hóng huā*	9g	Flos Carthami
柴胡	*chái hú*	10g	Radix Bupleuri
川牛膝	*chuān niú xī*	12g	Radix Cyathulae
红花	*hóng huā*	9g	Flos Carthami
甘草	*gān cǎo*	6g	Radix Glycyrrhizae
葱白	*cōng bái*	1piece	Bulbus Allii Fistulosi (as a guide)

Water was used to decoct, then it was divided into two servings, served warm, one pack per day.

Second Visit

June 6th: After taking three packs, the headache reduced, the pulse changed to float up. He sweated more after taking the herbs. Using the original formula *cōng bái* (葱白 , Bulbus Allii fistulosi), and *qiāng huó* (Rhizoma et Radix Notopterygii) were eliminated, and *dì lóng* (地龙 , Pheretima) 15g, was added. He continued to to take it again.

Third Visit

June 10th: He took four more packs, the sweating stopped, the headache reduced, and the needle sensation pain was also eliminated. He continued with the same formula.

Note

This patient totally took more than twenty packs of the formula, and the headache recovered. After three months follow up there was no reoccurrence.

Case 3

Allergic purpura.

Initial Visit

A male patient, 14 years old, January 22nd, 1991.

Two weeks previously, the patient had flu, cold, anorexia, and lassitude, but after treatments he became better. Five days previously, he discovered his skin had blood stagnation spots on his four limbs, more pronounced on his lower limbs and hips. It started off small, but later gradually developed and blended into patches. A hospital diagnosed it as allergic purpura. A routine blood test showed all, including platelet count, was normal. His tongue was red and the tongue coating was thin and white. His pulse was thin and fast. The diagnosis was external pathogens attacking the superfices, heat toxin stagnating internally, burning and injuring yin and the collaterals, with blood overflowing into the muscle and skin. The treatment plan was to cool blood, stop bleeding, transform stasis, and eliminate spots.

Prescription

丹参	dān shēn	20g	Radix et Rhizoma Salviae Miltiorrhizae
当归	dāng guī	15g	Radix Angelicae Sinensis
葛根	gé gēn	20g	Radix Puerariae Lobatae
紫草	zǐ cǎo	15g	Radix Arnebiae
丹皮	dān pí	10g	Cortex Moutan
生地	shēng dì	15g	Radix Rehmanniae
杭白芍	háng bái sháo	15g	Radix Paeoniae Lactiflorae
玄参	xuán shēn	15g	Radix Scrophulariae
防风	fáng fēng	9g	Radix Saposhnikoviae
金银花	jīn yín huā	15g	Flos Lonicerae Japonicae
甘草	gān cǎo	9g	Radix Glycyrrhizae

Water was used to decoct, one pack per day.

Second Visit

January 17 th: After taking five packs of the above formula, the purpura ceased to expand. The original spots were disappearing or were eliminated, except for on the lower limbs, which had a few residual spots. The rest of the skin cleared up to be as normal. He was instructed to take five packs of the same formula to complete the treatment.

Case 4

Dizziness.

Initial Visit

A female patient, 36 years old, June 24th, 1986.

A month previously, on May 21st, the patient had sudden dizziness and fell down, after which she had secondary vertigo, head distention, nausea and vomiting. When symptoms were severe, she did not want to open her eyes and perceived surrounding objects as moving or spinning. Her frontal cervical spine x-ray report stated: The fifth cervical vertebral process was fractured; the sixth cervical vertebra was slightly hypertrophied. She was treated with biomedicine but it did not improve and so she transferred to the Chinese medicine clinic for treatment. The patient had vertigo, wanted to vomit, and had accompanying neck pain, if exerted the pain got worse, she had a poor appetite, a thin tongue coating, and a wiry and deep pulse. The treatment plan was to invigorate blood, transform stasis, harmonise the stomach and descend the rebellious.

Prescription

丹参	dān shēn	30g	Radix et Rhizoma Salviae Miltiorrhizae
葛根	gé gēn	20g	Radix Puerariae Lobatae
茯苓	fú líng	15g	Poria

当归	*dāng guī*	12g	Radix Angelicae Sinensis
陈皮	*chén pí*	9g	Pericarpium Citri Reticulatae
半夏	*bàn xià*	9g	Rhizoma Pinelliae
钩藤	*gōu téng*	30g	Ramulus Uncariae Cum Uncis
竹茹	*zhú rú*	9g	Caulis Bambusae in Taenia
枳实	*zhǐ shí*	9g	Fructus Aurantii Immaturus
川断	*chuān duàn*	15g	Radix Dipsaci
苏梗	*sū gěng*	6g	Caulis Perillae
甘草	*gān cǎo*	6g	Radix Glycyrrhizae

The patient was instructed to decoct with water and divide into two servings to be taken often, until the vomiting stopped.

Second Visit

June 28th: After taking four packs of the above formula, she had no vertigo, and her appetite increased, but there was no improvement in her neck pain. The prescription was changed to:

Prescription

葛根	*gé gēn*	40g	Radix Puerariae Lobatae
丹参	*dān shēn*	20g	Radix et Rhizoma Salviae Miltiorrhizae
川断	*chuān duàn*	20g	Radix Dipsaci
当归	*dāng guī*	15g	Radix Angelicae Sinensis
杭白芍	*háng bái sháo*	30g	Radix Paeoniae Lactiflorae
透骨草	*tòu gǔ cǎo*	15g	Caulis Impatientis
寄生	*jì shēng*	20g	Herba Taxilli
生地	*shēng dì*	15g	Radix Rehmanniae
土鳖虫	*tǔ biē chóng*	9g	Eupolyphaga seu Steleophaga
甘草	*gān cǎo*	6g	Radix Glycyrrhizae

Water was used to decoct, then it was divided into two servings. Meanwhile, she also took one bottle of *Qī Lí Sǎn* (Seven Pinches Powder, 七厘散).

Third Visit

July 15th: After she took more than twelve packs, her neck pain basically disappeared, her appetite was normal, and she had no vertigo. She was instructed to take one bottle of *Qī Lí Sǎn* (Seven Pinches Powder) per day and to stop after one week.

4. *Dān Shēn* (Radix et Rhizoma Salviae Miltiorrhizae) —— *Biē Jiǎ* (鳖甲, Carapax Trionycis)

Function

To invigorate blood, transform stasis, soften hardness and dispel nodules.

Application

This is used for liver or spleen enlargement and liver cirrhosis.

Compatibility Analysis

Biē jiǎ (鳖甲 , Carapax Trionycis) is salty and cold. It enters the liver channel of the foot *jueyin* meridian. It can soften hardness and expel nodules, nourish yin and reduce deficiency heat. It is a commonly used herb for abdominal abscesses and accumulations. *Dān Shēn* (Radix et Rhizoma Salviae Miltiorrhizae) eliminates stasis, invigorates blood, and promotes blood circulation to make stagnated and obstructed blood flow freely. It also has the function to reduce an enlarged liver or spleen. The two herbs both have the function to treat liver or spleen enlargement. However, *biē jiǎ* (Carapax Trionycis) is more often used for softening hardness and dispelling nodules; and *dān shēn* (Radix et Rhizoma Salviae Miltiorrhizae) is more often used for invigorating blood and eliminating stasis. The two herbs applied together, can mutually enhance each other. With the liver or spleen stagnated and obstructed with blood, *biē jiǎ* (Carapax Trionycis) can be used to soften hardness and expel nodules, and then *dān shēn* (Radix et Rhizoma Salviae Miltiorrhizae) can invigorate blood and eliminate stasis better. Once blood is invigorated and stasis eliminated, then they can further soften the hardness and expel nodules. Besides treating liver or spleen enlargement and improving liver function, *dān shēn* (Radix et Rhizoma Salviae Miltiorrhizae) and *biē jiǎ* (Carapax Trionycis) can also function well to prevent liver fibroses. For liver cirrhosis patients, they can soften the liver and shrink the spleen, and improve clinical symptoms. They also have a certain effectiveness for treating steatorrhoeic hepatosis.

Case 1

Early stage liver cirrhosis.

Initial Visit

A male patient, 53 years old, October 25th, 1996.

The patient had a history of hepatitis B. Over the previous months, the liver area felt painful and uncomfortable, he had anorexia, lassitude, stomach and abdominal distention and fullness, which was worse after eating, his face was lustreless, and he sometimes had gum bleeding. A liver function test showed: ALT 120u/L, total protein 70g/L, albumin 36.3g/L, globulin 33.7g/L. Of the five indicators for hepatitis B, the first, second and fifth were positive. A B-ultrasound showed: The liver physical echo had increased, light spots were rough and big, the hepatic portal vein was 1.45cm, the spleen had enlarged, the spleen thickness was 5.9cm, and the splenic vein was 0.92cm, ascites (-). The tongue coating was white and turbid with teeth marks. The tongue body was dark. The pulse was wiry. The diagnosis was liver blood stagnation and obstruction, with disorder of the dispersing function. Liver blood stasis caused liver area pain. If the dispersing function is in disorder then there will be spleen and stomach disharmony, hence, the anorexia and abdominal distention. Liver blood stasis meant that the spleen did not transform and transport, hence the lustreless face. The treatment plan was to invigorate blood, transform stasis, soothe the liver and strengthen the spleen.

Prescription

丹参	*dān shēn*	20g	Radix et Rhizoma Salviae Miltiorrhizae
鳖甲	*biē jiǎ*	10g	Carapax Trionycis
当归	*dāng guī*	15g	Radix Angelicae Sinensis
桃仁	*táo rén*	9g	Semen Persicae
生黄芪	*shēng huáng qí*	24g	Radix Astragali (raw)
鸡血藤	*jī xuè téng*	24g	Caulis Spatholobi
泽兰	*zé lán*	15g	Herba Lycopi
虎杖	*hǔ zhàng*	15g	Rhizoma et Radix Polygoni Cuspidati
白花蛇舌草	*bái huā shé shé cǎo*	30g	Herba Hedyotis Diffusae
蒲公英	*pú gōng yīng*	15g	Herba Taraxaci
枳壳	*zhǐ qiào*	12g	Fructus Aurantii
白术	*bái zhú*	12g	Rhizoma Atractylodis Macrocephalae
猪苓	*zhū líng*	15g	Polyporus
厚朴	*hòu pò*	9g	Cortex Magnoliae Officinalis
仙鹤草	*xiān hè cǎo*	20g	Herba Agrimoniae
甘草	*gān cǎo*	6g	Radix Glycyrrhizae

Water was used to decoct to 300ml to be divided into two servings, to be taken in the morning and night, one pack per day.

Second Visit

November 14th: After taking twenty packs of the above formula in total, the lassitude and abdominal distention reduced. He still had anorexia and liver discomfort. *Shā rén* (Fructus Amomi) 8g, and *yán hú suǒ* (Rhizoma Corydalis) 15g, were added and he continued to take it.

Third Visit

December 4th: After taking another twenty packs of the above formula, all symptoms remarkably reduced. A liver function test showed: All normal. Total protein 68.0g/L, albumin 49.4g/L, globulin 28.6g/L. He continued to take the above formula.

Fourth Visit

December 27th: After taking yet another twenty more packs, he felt all symptoms had disappeared. A B-ultrasound showed: The spleen was smaller than before. As it was not convenient to decoct on the farm, his prescription was changed to *Gān Níng Sān Hào Jiāo Náng* (Liver-Quietening #III Capsules, 肝宁三号胶囊) for treatment.

Note

Gān Níng Sān Hào Jiāo Náng (Liver-Quietening #III Capsules) is one of a series of Chinese herbal formulae which our hospital's Liver Diseases Research Centre developed for treatment of liver diseases. It mainly treats early stage liver cirrhosis and chronic persistent hepatitis. It is composed of more than ten herbs including *dān shēn* (Radix et Rhizoma Salviae Miltiorrhizae), *biē jiǎ* (Carapax Trionycis), *dōng chóng xià cǎo* (Cordyceps), *sān qī* (Radix et Rhizoma Notoginseng) and *zé lán* (泽兰 , Herba Lycopi), etc.

Case 2

Steatorrhoeic hepatosis (fatty liver).

Initial Visit

A female patient, 46 years old, April 10th, 1997.

The patient was obese. In recent times, she had lower limbs lassitude, liver area discomfort, distention and slight pain. Her appetite, urination and bowel movement were normal. She suspected she had liver disease and came for a visit. Her liver function test and hepatitis B five indicators were all normal. A B-ultrasound report showed: The liver physical echo was fine and enhanced, later the sound was weaker. She was diagnosed with steatorrhoeic hepatosis. Her blood lipid profile showed: Triglyceride 2.3mmol/L, cholesterol 6.7mmol/L. Her pulse was deep and wiry. Her tongue coating was thin and white. Patients with body obesity have more dampness; dampness stagnation becomes phlegm; phlegm and dampness stagnation obstruct qi and blood preventing them from flowing freely. The liver collaterals obstruction caused her to feel the stuffiness, distention, pain and discomfort. The treatment plan was to invigorate blood, transform stasis, eliminate dampness and transform phlegm.

Prescription

丹参	*dān shēn*	20g	Radix et Rhizoma Salviae Miltiorrhizae
鳖甲	*biē jiǎ*	10g	Carapax Trionycis
生山楂	*shēng shān zhā*	20g	Fructus Crataegi (raw)
茯苓	*fú líng*	30g	Poria
泽泻	*zé xiè*	20g	Rhizoma Alismatis
郁金	*yù jīn*	12g	Radix Curcumae
泽兰	*zé lán*	15g	Herba Lycopi
柴胡	*chái hú*	15g	Radix Bupleuri
枳壳	*zhǐ qiào*	12g	Fructus Aurantii
白术	*bái zhú*	12g	Rhizoma Atractylodis Macrocephalae
延胡索	*yán hú suǒ*	15g	Rhizoma Corydalis
砂仁	*shā rén*	8g	Fructus Amomi
三棱	*sān léng*	6g	Rhizoma Sparganii
莪术	*é zhú*	6g	Rhizoma Curcumae
甘草	*gān cǎo*	6g	Radix Glycyrrhizae

Water was used to decoct.

Second Visit

May 30th: The above formula was modified and a total of more than forty packs was taken. The clinical symptoms disappeared and her body weight reduced by 5kg. A B-ultrasound test showed: The liver echo did not seem abnormal, triglyceride 1.7mmol/L, cholesterol 4.3mmol/L.

5. *Dān Shēn* (Radix et Rhizoma Salviae Miltiorrhizae) —— *Yè Jiāo Téng* (夜交藤, Caulis Polygoni Multiflori)

Function

To nourish blood, calm and tranquilise the spirit.

Application

This is used for heart or liver blood deficiency inducing palpitations and insomnia. It is particularly good for liver disease patients with insomnia.

Compatibility Analysis

Dān shēn (Radix et Rhizoma Salviae Miltiorrhizae) functions to nourish blood, invigorate blood and calm the spirit. The *Materia Medica of South Yunnan* (滇南本草 , *Diān Nán Běn Cǎo*) said that *dān shēn* (Radix et Rhizoma Salviae Miltiorrhizae) can "*tonify the heart, firm the will, treat forgetfulness, fearful throbbing, fright palpitations and insomnia*". Modern pharmacological research and clinical observation discovered that *dān shēn* (Radix et Rhizoma Salviae Miltiorrhizae) has functions to tranquilise, act as an analgesic and calm the spirit. *Yè jiāo téng* (Caulis Polygoni Multiflori) is the vine stem of *hé shǒu wū* (Radix Polygoni Multiflori). It is sweet and neutral. It enters the heart channel of hand *shaoyin* and the liver channel of the foot *jueyin* meridians. It is a commonly used herb for nourishing blood and calming the spirit. It also can nourish blood and the collaterals. *Dān shēn* (Radix et Rhizoma Salviae Miltiorrhizae) with *yè jiāo téng* (Caulis Polygoni Multiflori), together have the characteristic to nourish blood and calm the spirit, they also can invigorate blood and the collaterals, eliminate stasis and generate new blood. Although it tonifies, it does not obstruct. The author used this compatibility to treat hepatitis patients with insomnia and had good results. The main aetiology of hepatitis is the liver losing its dispersing function, qi stagnation and blood stasis. If heavy settling and astringing the spirit herbs are used, it may obstruct the liver channel of foot *jueyin* meridian's dispersing function. Using the biomedicine Diazepam, etc. can damage the liver, hence, if choosing *dān shēn* (Radix et Rhizoma Salviae Miltiorrhizae) with *yè jiāo téng* (Caulis Polygoni Multiflori) it is more gentle and effective, and it also benefits liver diseases.

Case Study

Initial Visit

A male patient, 40 years old, May 14th, 1997.

The patient was a carrier of the hepatitis B virus for a couple of years. Recently due to being busier, he did not have good rest and presented with symptoms of whole body lassitude, stomach stuffiness and anorexia, abdominal distention, and insomnia with a lot of dreams, etc. A liver function test showed: ALT 437U/L; and the seven indicators for hepatitis B showed: HBs Ag (+), Anti Hbe (+), Anti Hbc (+). His pulse was wiry and thready, and his tongue coating was thin and white. The treatment plan was to nourish the liver, strengthen the spleen, harmonise the stomach and calm the spirit.

Prescription

丹参	dān shēn	20g	Radix et Rhizoma Salviae Miltiorrhizae
夜交藤	yè jiāo téng	30g	Caulis Polygoni Multiflori
柴胡	chái hú	15g	Radix Bupleuri
泽兰	zé lán	15g	Herba Lycopi
虎杖	hǔ zhàng	15g	Rhizoma et Radix Polygoni Cuspidati
枳壳	zhǐ qiào	12g	Fructus Aurantii
白术	bái zhú	12g	Rhizoma Atractylodis Macrocephalae
蒲公英	pú gōng yīng	15g	Herba Taraxaci
五味子	wǔ wèi zǐ	9g	Fructus Schisandrae Chinensis
砂仁	shā rén	8g	Fructus Amomi
猪苓	zhū líng	15g	Polyporus
乌药	wū yào	15g	Radix Linderae
土茯苓	tǔ fú líng	20g	Rhizoma Smilacis Glabrae
生黄芪	shēng huáng qí	20g	Radix Astragali (raw)
白花蛇舌草	bái huā shé shé cǎo	30g	Herba Hedyotis Diffusae
甘草	gān cǎo	6g	Radix Glycyrrhizae

Water was used to decoct, then it was divided into two servings, one pack per day.

Second Visit

May 21st: After taking seven packs of the above formula, his sleeping had improved, and all of his symptoms were getting better. As it was effective, the prescription was not altered, and he continued to take it again.

Third Visit

May 28th: His sleep was good, and all his symptoms had reduced. *Yè jiāo téng* (Caulis Polygoni Multiflori) was eliminated and he continued to take it.

Fourth visit

August 18th: The patient totally took more than ninety packs. His liver function returned to normal.

Bèi mǔ (贝母 , Bulbus Fritillariae), clinically divides into two kinds: *Chuān bèi mǔ* (川贝母 , Bulbus Fritillariae Cirrhosae) and *zhè bèi mǔ* (浙贝母 , Bulbus Fritillariae Thunbergii). *Chuān bèi mǔ* (Bulbus Fritillariae Cirrhosae) is bitter, sweet and slightly cold. *Zhè bèi mǔ* (Bulbus Fritillariae Thunbergii) is bitter and cold. They both enter the lung channel of the hand *taiyin* and the heart channel of the hand *shaoyin* meridians. They can clear heat, expel nodules, transform phlegm and stop cough. Clinically, *chuān bèi mǔ* (Bulbus Fritillariae Cirrhosae) is sweet, cold and moist. Its power to stop cough and moisten the lung is stronger than that of *zhè bèi mǔ* (Bulbus Fritillariae Thunbergii). *Zhè bèi mǔ* (Bulbus Fritillariae Thunbergii) is bitter and cold. Its power to clear heat and expel nodules is better than that of *chuān bèi mǔ* (Bulbus Fritillariae Cirrhosae), hence, *chuān bèi mǔ* (Bulbus Fritillariae Cirrhosae) moistens the lung, stops cough, descends qi and transforms phlegm. It is properly used for chronic bronchitis inducing difficult phlegm expectoration or lung atrophy, lung abscess inducing a heated cough, a dry cough and a chronic cough. *Zhè bèi mǔ* (Bulbus Fritillariae Thunbergii) transforms phlegm, and expels nodules. It is appropriately used for phlegm nodules, scrofula, goitres, abscesses, etc.

1. *Chuān Bèi Mǔ* (Bulbus Fritillariae Cirrhosae) —— *Xìng Rén* (杏仁, Semen Armeniacae Dulcis)

Function

To stop cough, moisten the lung, and calm asthma.

Application

It may be used for many kinds of cough and/or asthma syndromes. It is unimportant whether it is an external invasion of wind-cold or wind-heat or internal injured cough or asthma. Through syndrome differentiation, they can all be used with additions to enhance the stopping cough and calming asthma results.

Compatibility Analysis

Chuān bèi mǔ (Bulbus Fritillariae Cirrhosae) is sweet, bitter and slightly cold. *Xìng rén* (Semen Armeniacae Dulcis) tastes bitter and is slightly warm. The two herbs both taste bitter and enter the lung channel of the hand *taiyin* meridian. Bitterness can sedate the lung, descend qi and calm asthma. *Chuān bèi mǔ* (Bulbus Fritillariae Cirrhosae) is sweet, cold and moist and can moisten the lung to stop cough. *Xìng rén* (Semen Armeniacae Dulcis) is bitter and warm. It is a lipid kind of herb and so also can moisten the lung to stop cough. Of the two herbs, one is cold and the other is warm. Used together they can neutrally regulate cold and heat, mutually controlling to gently achieve balance, and can increase the power of the moistening the lung and stopping cough effect, as well as being suitable for all kinds of cough and/or asthma.

Case 1

Yin deficiency cough.

Initial Visit

A female patient, 52 years old, July 23rd, 1987.

The patient had cough and asthma without phlegm, and could not lie down for the previous half month. If she lay down, then she would have chest tightness, accompanied by a salty taste inside the mouth, an itchy dry throat, and constipation. Her pulse was wiry and thready. Her tongue was red with a scanty tongue coating. Her EKG showed up normal. A chest x-ray showed: Both lungs' interstitial markings were rough. Through Chinese medicine it was differentiated as yin deficiency with the lung not dispersing syndrome. If there is yin deficiency then there is no phlegm, a dry throat, and a thready pulse. With yin deficiency, the lung lacks moisture, and the regulation of the dispersing and descending function is lost, hence the cough, asthma and constipation. The treatment plan was to nourish yin, moisten the lung and calm asthma.

Prescription

百部	bǎi bù	15g	Radix Stemonae
紫菀	zǐ wǎn	10g	Radix et Rhizoma Asteris
冬花	dōng huā	10g	Flos Farfarae
川贝	chuān bèi	10g	Bulbus Fritillariae Cirrhosae
杏仁	xìng rén	10g	Semen Armeniacae Dulcis
麦冬	mài dōng	12g	Radix Ophiopogonis
桔梗	jié gěng	10g	Radix Platycodonis
当归	dāng guī	12g	Radix Angelicae Sinensis
桑叶	sāng yè	12g	Folium Mori
杷叶	pá yè	12g	Folium Eriobotryae
侧柏叶	cè bǎi yè	10g	Cacumen Platycladi
甘草	gān cǎo	6g	Radix Glycyrrhizae

Second Visit

July 27th: The above prescription used water to decoct and after taking four packs, the cough and asthma reduced a lot. Only the salty taste in the mouth and the itchy throat had not been eliminated. The salty mouth sensation was due to kidney deficiency and water overflowing syndrome. The prescription was changed to modified *Jīn Shuǐ Liù Jūn Jiān* (Six Gentlemen Metal and Water Brew, 金水六君煎).

Prescription

熟地	shú dì	15g	Radix Rehmanniae Praeparata
当归	dāng guī	15g	Radix Angelicae Sinensis
茯苓	fú líng	15g	Poria
陈皮	chén pí	9g	Pericarpium Citri Reticulatae
川贝	chuān bèi	10g	Bulbus Fritillariae Cirrhosae
杏仁	xìng rén	10g	Semen Armeniacae Dulcis
半夏	bàn xià	9g	Rhizoma Pinelliae
桔梗	jié gěng	10g	Radix Platycodonis
百部	bǎi bù	15g	Radix Stemonae
侧柏叶	cè bǎi yè	15g	Cacumen Platycladi
甘草	gān cǎo	6g	Radix Glycyrrhizae

Water was used to decoct and then it was taken.

Third Visit

July 30th: After three packs of the above formula the salty taste in the mouth and the itchy throat reduced a lot, the bowel movements were also free and smooth, and there was no reoccurrence of the cough and asthma. Two more packs of the same formula were taken to clear it up completely.

Case 2

Initial Visit

A male patient, 4 years old, July 4th, 1984.

The sick child had pneumonia with fever, two months previously. After treatment, the fever reduced but the cough and asthma did not stop, and were worse at night. After ineffective treatment from Chinese medicine and biomedicine treatment with injections and herbs he came to visit. On observation the sick child had developed well. On auscultation of both lungs: Wet rales could be heard, and respiration sounded rough. A chest x-ray revealed: Both lungs' interstitial markings had increased. He was diagnosed with bronchitis. A routine blood test was normal. His tongue coating was thin and white; and his pulse was thready and fast. The diagnosis: After pneumonia, the residual pathogen did not clear, and the lung was unable to descend syndrome. The treatment plan was to use a formula to disperse the lung, stop cough, clear heat and calm asthma.

Prescription

川贝	*chuān bèi*	7g	Bulbus Fritillariae Cirrhosae
杏仁	*xìng rén*	5g	Semen Armeniacae Dulcis
百部	*bǎi bù*	7g	Radix Stemonae
桑皮	*sāng pí*	7g	Cortex Mori
桔梗	*jié gěng*	5g	Radix Platycodonis
侧柏叶	*cè bǎi yè*	9g	Cacumen Platycladi
陈皮	*chén pí*	5g	Pericarpium Citri Reticulatae
炙麻黄	*zhì má huáng*	3g	Herba Ephedrae (processed)
鱼腥草	*yú xīng cǎo*	10g	Herba Houttuyniae
甘草	*gān cǎo*	3g	Radix Glycyrrhizae

Water was used to decoct, then it was divided into three servings, to be served warm.

Second Visit

July 7th: After three packs, the cough and asthma reduced, and the appetite increased. However, the respiratory sounds were still rough and wet rales could be heard in both lungs. The original formula was used with *dì lóng* (Pheretima) 6g, added and he continued to take it again.

Third Visit

July 12th: After another five packs, all symptoms basically disappeared, except that the respiratory sounds were still slightly rough. Three further packs of the same formula were administered again and he recovered.

2. *Zhè Bèi Mǔ* (Bulbus Fritillariae Thunbergii) —— *Wū Zéi Gǔ* (乌贼骨, Os Sepiellae Se Sepiae)

Function

To hold, restrain and heal ulceration, to control acidity and alleviate pain.

Application

This can be used for gastric and duodenal ulcers, and colon and rectum ulcers. It is also used for hyperacidity, acid regurgitation, stomachache and gastric haemorrhage, etc.

Compatibility Analysis

Wū zéi gǔ (乌贼骨 , Os Sepiellae Se Sepiae) has a remarkable function to absorb pepsin and harmonise stomach acidity. Hence, it is effective for treating all kinds of ulcer diseases, gastric haemorrhages, acid regurgitation and pain. However, the properties of *wū zéi gǔ* (Os Sepiellae Se Sepiae) are warm and astringent. If taken for a long time it can cause constipation. *Zhè bèi mǔ* (Bulbus Fritillariae Thunbergii) is bitter and cold. It can clear heat, disperse the lung, descend qi and also has the function to moisten the intestines. It can reduce the side effects of *wū zéi gǔ* (Os Sepiellae Se Sepiae) drying the intestines. Moreover, *zhè bèi mǔ* (Bulbus Fritillariae Thunbergii) is a good product to expel nodules and treat abscesses. It also can enhance *zhè bèi mǔ* (Bulbus Fritillariae Thunbergii) to effectively treat ulcers. This compatibility, is called *Wū Bèi Sǎn* (Cuttlefish Bone and Fritillaria Powder, 乌贝 散) in clinic. If, based on syndrome differentiation, this compatibility is added, it will increase the treatment effectiveness.

Case Study

Initial Visit

A female patient, 27 years old, November 26th, 1987.

The patient had a history of stomachache for many years which was sometimes better and sometimes worse. Recently, due to cold it became worse. She had stomachache which refused pressure accompanied by heartburn and acid reflux. An upper GI tract barium radiography diagnosed it as: Stomach pylorus region and duodenal ulcers. Her pulse was wiry and thready. Her tongue coating was thin and white and her tongue was red. The treatment plan was to regulate qi, harmonise the stomach, control acidity and alleviate pain. Professor Jiao Shu-de's *Sān Hé Tāng* (Triple Combination Decoction, 三合 汤) was modified for treatment.

Prescription

百合	*bǎi hé*	24g	Bulbus Lilii
乌药	*wū yào*	10g	Radix Linderae
檀香	*tán xiāng*	6g	Lignum Santali Albi

砂仁	*shā rén*	3g	Fructus Amomi
乌贼骨	*wū zéi gǔ*	15g	Os Sepiellae Se Sepiae
浙贝母	*zhè bèi mǔ*	10g	Bulbus Fritillariae Thunbergii
丹参	*dān shēn*	20g	Radix et Rhizoma Salviae Miltiorrhizae
枳壳	*zhǐ qiào*	10g	Fructus Aurantii
白术	*bái zhú*	12g	Rhizoma Atractylodis Macrocephalae
杭白芍	*háng bái sháo*	15g	Radix Paeoniae Lactiflorae
甘草	*gān cǎo*	6g	Radix Glycyrrhizae

Water was used to decoct, then it was divided into two servings, to be taken in the morning and at night, one pack per day.

Second Visit

December 3rd: After taking seven packs of the above formula, the stomachache had gone, and there was no acid regurgitation or heartburn sensation. As it was effective, the prescription was not changed and after he continuously took it for seven packs more, all symptoms were eliminated.

Wū yào's (乌药 , Radix Linderae) taste and property are acrid and warm. It enters the spleen channel of the foot *taiyin,* the stomach channel of the foot *yangming*, the lung channel of the hand *taiyin* and the kidney channel of the foot *shaoyin* meridians. It functions to promote qi movement, expel cold and alleviate pain. Its tastes are acrid and aromatic to free movement in the meridians and collaterals. It is an important herb to regulate qi and alleviate pain. Huang Gong-xiu said: "*It is appropriate for use for all diseases that belong to qi rebellion, as well as where there is chest and abdomen discomfort.*" For disease in the upper with stagnation, clumping, qi rebellion, and in severe asthma patients, a formula such as the *Sì Mò Tāng* (Four Milled Ingredients Decoction, 四磨汤) can regulate the upper *jiao* stagnation and clumping. For middle *jiao* cold stagnation and clumping inducing cold pain, a formula such as the *Wū Chén Tāng* (Lindera and Aquilaria Decoction, 乌沉汤) can warm the middle *jiao* and alleviate the pain. For lower *jiao* cold, urinary bladder qi deficiency inducing urinary incontinence, a formula such as the *Suō Quán Wán* (Stream-Reducing Pill, 缩泉丸) can warm the kidney to stop enuresis. *Wū yào* (乌药 , Radix Linderae) stops enuresis because it can warm the kidney. If kidney qi is sufficient, then the kidney gate is secured and enuresis stops. Meanwhile, for kidney yang deficiency cold inducing dysuria, *wū yào* (Radix Linderae) has a urine promotion function. In clinical practice, liver cirrhosis with ascitics and dysuria belongs to the lower *jiao* deficiency cold, if large amounts of *wū yào* (Radix Linderae) are used it can increase the quantity of urination.

1. *Wū Yào* (Radix Linderae) ——
Xiāng Fù (香附, Rhizoma Cyperi)

Function

To soothe the liver, regulate qi, expel cold and alleviate pain.

Application

It is mainly used for gastro-intestinal (GI) tract functional disorders of the stomach and intestines. It is better for clinical stomach distention, belching, hiccough, anorexia or borborygmus, abdominal distention, and umbilical and abdominal pain that are due to cold.

Compatibility Analysis

Wū yào (Radix Linderae) is acrid and warm. It functions to promote qi movement, expel cold and alleviate pain. Based on modern research, *wū yào* (Radix Linderae) has dual regulation to excite or restrain the smooth muscles of the stomach and intestines to help the GI tract function recover. Meanwhile, it can promote GI tract secretions and has the function to help digestion. *Xiāng fù* (香附 , Rhizoma Cyperi) is acrid, bitter and neutral. It enters the liver channel of the foot *jueyin* meridian. It is an important herb to soothe the liver, regulate qi and alleviate pain. Zhang Shan-lei said: "*Xiāng fù's (Rhizoma Cyperi) acrid taste is very strong and its aromatic property is very thick. They both deal with qi. Hence, it particularly treats qi stagnation.*" Li Shi-zhen said "*xiāng fù* (Rhizoma Cyperi) is *the overall governor of qi diseases.*" Based on modern research, it has a direct inhibitory action on the smooth muscle of the intestinal tract and it has analgesic functions. *Wū yào* (Radix Linderae) and *xiāng fù* (Rhizoma Cyperi) both are classified as qi regulating herbs, both have functions to regulate the qi mechanism disorder in the stomach and intestinal tract as well as regulate qi and stop pain. However, *wū yào* (Radix Linderae) along with its regulating qi function, also functions to warm the kidney and expel cold, and warm the lower *jiao* to help the middle *jiao* transform and transport, as in adding firewood under the cauldron. *Xiāng fù* (Rhizoma Cyperi) along with its regulating qi function, it also can soothe the liver and relieve constraint, and soothe the liver constraint to regulate the middle *jiao*, and harmonise the liver and spleen. The two herbs combined together function to mutually enhance each other and can supplement each other's disadvantages as well as expand their application scope. In clinical practice, it can be used for stomach and intestines neurosis syndrome, gastritis, stomach spasm inducing stomachache, stomach and intestines functional disorder and intestinal adhesions after surgery, etc. This compatibility is acrid and warm. It is suitable for cold, inducing pain. If matched with clearing heat, regulating the intestines and harmonising the stomach herbs, then patients that belong to the heat type also can use it. It can use their function to regulate the stomach and intestines.

Case 1

Stomach spasm.

Initial Visit

A female patient, 35 years old, April 13th, 1988.

The patient had a history of stomachache. The day before she visited, she had an acute episode of stomach distention, fullness and pain. It was worse after eating and was accompanied by acid regurgitation and growling. An upper GI tract barium radiograph showed: The stomach peristalsis was weak, with little tension, the pylorus opened slowly and it was diagnosed as stomach spasm. Her pulse was deep and thready. Her tongue coating was thin and white. The syndrome belonged to the middle *jiao* deficiency cold. The treatment plan was to regulate qi, warm the middle *jiao*, expel cold and alleviate pain.

Prescription

香附	*xiāng fù*	12g	Rhizoma Cyperi
乌药	*wū yào*	12g	Radix Linderae
砂仁	*shā rén*	3g	Fructus Amomi
檀香	*tán xiāng*	6g	Lignum Santali Albi
白术	*bái zhú*	10g	Rhizoma Atractylodis Macrocephalae
枳壳	*zhǐ qiào*	10g	Fructus Aurantii
丹参	*dān shēn*	20g	Radix et Rhizoma Salviae Miltiorrhizae
党参	*dǎng shēn*	12g	Radix Codonopsis
杭白芍	*háng bái sháo*	15g	Radix Paeoniae Lactiflorae
海螵蛸	*hǎi piāo xiāo*	15g	Endoconcha Sepiae
甘草	*gān cǎo*	6g	Radix Glycyrrhizae

Water was used to decoct and then it was taken.

Second Visit

May 28th: He came to visit due to another illness and informed us that after taking four packs of the above formula the stomachache and distention were eliminated and there was no reoccurrence.

Case 2

Dysentery and stomachache.

Initial Visit

A male patient, 49 years old, June 11th, 1986.

The patient complained that ever since he had dysentery during his adolescence, for a couple of decades, around early June every year, he had dysentery, tenesmus, abdominal pain and diarrhoea, for more than ten times every day, each episode went on for more than a month without recovery. Although each time he had treatment, the next year around early June, it still reoccurred. For the previous few days, he had abdominal pain and dysentery again. His stool contained pus and he had a tenesmus sensation. His pulse was wiry. His tongue coating was slippery, greasy and turbid. His tongue was dark. This was the condition of qi and blood stagnation and obstruction, with turbid dampness staying internally. Li Shi-zhen said: "*On regulating qi, then abdominal pain will self eliminate. On invigorating blood then pus in the stool will be eliminated.*" The treatment plan was to regulate qi, invigorate blood, and transform dampness.

Prescription

白芍	*bái sháo*	24g	Radix Paeoniae Alba
赤芍	*chì sháo*	15g	Radix Paeoniae Rubra
桃仁	*táo rén*	6g	Semen Persicae
红花	*hóng huā*	6g	Flos Carthami
乌药	*wū yào*	10g	Radix Linderae
香附	*xiāng fù*	10g	Rhizoma Cyperi
川连	*chuān lián*	9g	Rhizoma Coptidis
木香	*mù xiāng*	9g	Radix Aucklandiae
防风	*fáng fēng*	9g	Radix Saposhnikoviae
甘草	*gān cǎo*	6g	Radix Glycyrrhizae

Water was used to decoct, then it was divided into two servings, and taken in the morning and at night.

Second Visit

June 14th: After taking three packs of the above formula, the diarrhoea had stopped and the bowel movements became normal, but he still had pain around the umbilicus. The same formula was used for treatment again.

Third Visit

June 21st: After taking another three packs of the above formula, there was no abdominal pain or dysentery. He stopped the formula for three days after which he still felt a dull abdominal pain. His pulse was still deep and hesitant, and although his tongue coating had already turned a thin white, the tongue was still dark. The eliminating dampness herbs were removed from the above formula, and the herbs to invigorate blood, regulate qi and alleviate pain were reinforced and he continued taking it again.

Prescription

白芍	*bái sháo*	15g	Radix Paeoniae Alba
赤芍	*chì sháo*	15g	Radix Paeoniae Rubra
乌药	*wū yào*	12g	Radix Linderae
香附	*xiāng fù*	10g	Rhizoma Cyperi
延胡索	*yán hú suǒ*	15g	Rhizoma Corydalis
丹皮	*dān pí*	10g	Cortex Moutan
桂枝	*guì zhī*	10g	Ramulus Cinnamomi
桃仁	*táo rén*	9g	Semen Persicae
红花	*hóng huā*	9g	Flos Carthami
枳壳	*zhǐ qiào*	10g	Fructus Aurantii
黄芪	*huáng qí*	30g	Radix Astragali
蒲黄	*pú huáng*	9g	Pollen Typhae
五灵脂	*wǔ líng zhī*	9g	Faeces Trogopterori
甘草	*gān cǎo*	9g	Radix Glycyrrhizae

Water was used to decoct, then it was divided into two servings, one pack per day.

Fourth Visit

June 24th: After taking another three packs, the abdominal pain disappeared and all symptoms were eliminated. The patient requested treatment of the root problem to prevent reoccurrence the following year. Prolonged sickness is usually in the blood and in the collaterals. For a couple of decades of chronic disease, it should be treated by invigorating blood and transforming stasis. The above formula was modified and changed to a powder and treatment was attempted.

Prescription

白芍	*bái sháo*	60g	Radix Paeoniae Alba
赤芍	*chì sháo*	60g	Radix Paeoniae Rubra
乌药	*wū yào*	50g	Radix Linderae
香附	*xiāng fù*	50g	Rhizoma Cyperi
丹皮	*dān pí*	30g	Cortex Moutan
桂枝	*guì zhī*	30g	Ramulus Cinnamomi
桃仁	*táo rén*	30g	Semen Persicae
红花	*hóng huā*	30g	Flos Carthami
桔梗	*jié gěng*	20g	Radix Platycodonis
枳壳	*zhǐ qiào*	30g	Fructus Aurantii
柴胡	*chái hú*	30g	Radix Bupleuri
川牛膝	*chuān niú xī*	30g	Radix Cyathulae
当归	*dāng guī*	30g	Radix Angelicae Sinensis
甘草	*gān cǎo*	30g	Radix Glycyrrhizae

The above herbs were all ground together to form a fine powder. Each time, he was told to take 5g, twice a day, to be taken with warm water.

Note

The above formula was modified from *Xuè Fǔ Zhú Yū Tāng* (House of Blood Stasis-Expelling Decoction). It aims to invigorate blood and transform stasis to regulate the stomach and intestinal function. At a follow up visit in the second year, the patient reported that there was no reoccurrence.

Wǔ líng zhī's (五灵脂 , Faeces Trogopterori) taste and property are sweet, warm and fishy. Its qi is turbid. It enters the liver channel of the foot *jueyin* meridian. Its functions can promote qi movement, unblock stasis and alleviate pain. The liver governs blood. All kinds of pain belong to the liver. *Wǔ líng zhī* (Faeces Trogopterori) enters the liver channel of the foot *jueyin* meridian. It expels stagnation and obstruction in the blood. Hence, it is good at treating blood diseases and alleviating all kinds of pain. In the *Grand Materia Medica*, it said it can: *"Stop menstrual overflow, persistent leukorrhoea with a reddish discharge, before or after delivery, qi and blood inducing all kinds of pain in the male or female heart, abdomen, hypochondria, lower abdomen…all kinds of pain.... and injuries from snake, scorpion or centipede bites."* It is evident that *wǔ líng zhī* (Faeces Trogopterori) is an analgesic medicine.

1. *Wǔ Líng Zhī* (Faeces Trogopterori) ——
Pú Huáng (蒲黄, Pollen Typhae)

Function

To invigorate blood and promote movement of stasis, to expel nodules and alleviate pain.

Application

It can be used for any blood stasis accumulations and obstructions inducing all kinds of diseases such as blood stasis in females inducing irregular menstruation, painful menstruation, amenorrhoea, etc. It also can be used for heart blood vessel diseases inducing angina pectoral pain. In clinic it has been effective for treating neurosis type pain and stomach or intestinal spasm inducing stomachache or abdominal pain.

Compatibility Analysis

If all is freely moving then there is no pain; if there is pain, then there is a blockage. All kinds of blood stasis stay internally, the meridian route is obstructed or stagnated, and blood flow does not run smoothly, all of this can induce pain. *Wǔ líng zhī* (Faeces Trogopterori) is sweet and warm. It enters the liver channel of the foot *jueyin* meridian. It is good at invigorating blood and expelling stasis. It is a qi herb within blood herbs, hence, its power to stop pain is strong. Modern pharmacological research for this product found that it functions to relax smooth muscle spasm and can increase the permeability of blood vessels to improve blood circulation. *Pú huáng* (Pollen Typhae) is acrid and neutral. It enters the heart channel of the hand *shaoyin* and the liver channel of the foot *jueyin* meridians. It has the effect to invigorate blood and stop bleeding. If combined with *wǔ líng zhī* (Faeces Trogopterori), they can mutually supplement and achieve a mutual enhancement function. The analgesic power is even stronger. The two herbs in combination is called *Shī Xiào Sǎn* (Sudden Smile Powder, 失 笑 散). It is a formula from the *Formulary of the Bureau of Medicines of the Taiping Era* Volume 9, also called the *Duàn Gōng Xuán Sǎn* (Bow String-broken Powder, 断弓弦散). Laboratory research has demonstrated: This formula can increase an organism's tolerance for reduced pressure and oxygen shortage. It has an antagonistic action on pituitrin inducing acute heart muscle blood shortage in the rat, and has a remarkable sedative function on spontaneous activities of the mouse. It also has a reducing blood pressure function. The *Golden Mirror of the Medical Tradition* (医 宗 金 鉴 , *Yī Zōng Jīn Jiàn*) said: "*Formulae using líng zhī (Faeces Trogopterori) that is acrid, warm and enters the liver channel of the foot jueyin meridian, if used raw then it promotes blood circulation. Pú huáng (Pollen Typhae) is acrid, neutral and enters the liver channel of the foot jueyin meridian. Used raw it can break blood. It also has the function to eliminate stasis and generate new blood. Sweetness does not injure the spleen, acridity can expel stasis. Without notice, all symptoms can be eliminated. One can laugh and put it aside.*" This explains that the two herbs have very good treatment effects for stopping pain, invigorating blood and transforming stasis.

Case 1

Painful menstruation.

Initial Visit

A female patient, 19 years old, June 27th, 1987.

The patient had painful menstruation for two years. The menstrual cycle was normal. Each menstrual period was five to seven days. The first menstruation was at age fourteen. For the previous two years, one day before, or the first day of menstruation, she had abdominal pain with a dragging downwards sensation. When the pain was severe, it was accompanied by cold limbs, cold abdomen, and scanty menstruation which was black in colour. Her pulse was deep and wiry. Her tongue coating was thin and white and her tongue was dark. The diagnosis was qi stagnation and blood stasis inducing painful menstruation. She visited during her menstrual period. The treatment plan used modified *Shào Fù Zhú Yū Tāng* (Lesser Abdomen Stasis-Expelling Decoction, 少腹逐瘀汤) to invigorate blood and transform stasis.

Prescription

蒲黄	*pú huáng*	9g	Pollen Typhae
五灵脂	*wǔ líng zhī*	9g	Faeces Trogopterori
泽兰	*zé lán*	20g	Herba Lycopi
小茴香	*xiǎo huí xiāng*	6g	Fructus Foeniculi
炮姜	*páo jiāng*	6g	Rhizoma Zingiberis Praeparatum
赤芍	*chì sháo*	10g	Radix Paeoniae Rubra
肉桂	*ròu guì*	6g	Cortex Cinnamomi
当归	*dāng guī*	12g	Radix Angelicae Sinensis
川芎	*chuān xiōng*	9g	Rhizoma Chuanxiong
川牛膝	*chuān niú xī*	15g	Radix Cyathulae
延胡索	*yán hú suǒ*	15g	Rhizoma Corydalis
甘草	*gān cǎo*	6g	Radix Glycyrrhizae

Water was used to decoct, then it was divided into two servings, served warm, one pack per day.

Second Visit

July 4th: After taking four packs of the above formula continuously, her abdominal pain was remarkably reduced. The menstruation amount was still scanty and black in colour. She was instructed to take the same formula three days before the next menstruation continuously for four packs.

Note

The patient continuously took the decoction for three menstrual periods for a total of twelve packs. After stopping the formula she did not have another painful menstruation episode.

Case 2

Chest impediment.

Initial Visit

A female patient, 54 years old, October 6th, 1988.

The patient presented with left chest pain expanding to the shoulder and back causing discomfort which was accompanied by chest tightness, throat blockage and oedema of the four limbs for the previous few months. She had a check up in a Beijing hospital and was diagnosed with an acute inferior and posterior myocardial infarction. Her pulse was slow with a heart rate of 40 beats/minute. Her tongue coating was turbid and greasy. She had symptoms of shortness of breath, anorexia, and asthma after exertion. The diagnosis was phlegm and dampness obstructing internally, with blood stasis and unsmooth flow of blood syndrome. When phlegm and dampness obstruct internally then there is chest stuffiness and oedema in the limbs, a greasy tongue coating and anorexia. When there is blood stasis with unsmooth flow, then chest pain, shortness of breath and a slow pulse manifest. The treatment plan was to use a formula to invigorate blood, transform stasis, and to warm in order to transform phlegm and dampness.

Prescription

丹参	*dān shēn*	30g	Radix et Rhizoma Salviae Miltiorrhizae
葛根	*gé gēn*	24g	Radix Puerariae Lobatae
蒲黄	*pú huáng*	10g	Pollen Typhae
五灵脂	*wǔ líng zhī*	10g	Faeces Trogopterori
桂枝	*guì zhī*	10g	Ramulus Cinnamomi
茯苓	*fú líng*	20g	Poria
白术	*bái zhú*	12g	Rhizoma Atractylodis Macrocephalae
瓜蒌	*guā lóu*	15g	Fructus Trichosanthis
半夏	*bàn xià*	9g	Rhizoma Pinelliae
太子参	*tài zǐ shēn*	20g	Radix Pseudostellariae
麦冬	*mài dōng*	12g	Radix Ophiopogonis
五味子	*wǔ wèi zǐ*	8g	Fructus Schisandrae Chinensis
桃仁	*táo rén*	9g	Semen Persicae

| 红花 | *hóng huā* | 9g | Flos Carthami |
| 炙甘草 | *zhì gān cǎo* | 6g | Radix et Rhizoma Glycyrrhizae Praeparata cum Melle |

Water was used to decoct, then it was divided into two servings, to be taken in the morning and at night, one pack per day.

Second Visit

October 13th: After taking seven packs of the above formula, the oedema was eliminated, the chest pain, stuffiness and suffocating sensation also improved. However, she still had the slow pulse and chest stuffiness. The formula was changed to the following:

Prescription

生黄芪	*shēng huáng qí*	24g	Radix Astragali (raw)
麦冬	*mài dōng*	15g	Radix Ophiopogonis
五味子	*wǔ wèi zǐ*	8g	Fructus Schisandrae Chinensis
太子参	*tài zǐ shēn*	20g	Radix Pseudostellariae
丹参	*dān shēn*	24g	Radix et Rhizoma Salviae Miltiorrhizae
葛根	*gé gēn*	24g	Radix Puerariae Lobatae
桂枝	*guì zhī*	10g	Ramulus Cinnamomi
茯苓	*fú líng*	20g	Poria
半夏	*bàn xià*	9g	Rhizoma Pinelliae
厚朴花	*hòu pò huā*	9g	Flos Magnoliae Officinalis
玉竹	*yù zhú*	30g	Rhizoma Polygonati Odorati
蒲黄	*pú huáng*	9g	Pollen Typhae
五灵脂	*wǔ líng zhī*	9g	Faeces Trogopterori
炙甘草	*zhì gān cǎo*	6g	Radix et Rhizoma Glycyrrhizae Praeparata cum Melle

Water was used to decoct and taken as before.

Third Visit

October 21st: The chest pain was eliminated, the heart rate reached 50 beats/minute and she could do light house chores. To the above formula, *pú huáng* (Pollen Typhae) and *wǔ líng zhī* (Faeces Trogopterori) were eliminated, and it was taken again.

Fourth Visit

October 29th: The patient felt well and only had slight shortness of breath and chest stuffiness. An EKG reported that there was still an inferior and posterior myocardial infarction. She was told to continuously take the same formula.

Bái sháo (白芍 , Radix Paeoniae Alba) is bitter, sour and slightly cold. It enters the liver channel of the foot *jueyin* meridian. It can nourish the liver and retain yin by astringing, calm the liver, harmonise blood and alleviate pain. *Sháo yào* (芍药 , Radix Paeoniae Lactiflorae) in the classic formulae did not differentiate between red or white. It was only since the *Illustrated Classic of the Materia Medica* (本草图经 , *Běn Cǎo Tú Jīng*) that they started to differentiate it. Thereafter, every doctor thought of the white one as tonifying and the red one as sedating, the white one as astringing and red one as dispersing. *Bái sháo* (Radix Paeoniae Alba) has functions to nourish blood, retain yin by astringing, soften the liver and alleviate pain; *chì sháo* (Radix Paeoniae Rubra) is effective for invigorating the blood and moving stagnation. The two herbs have different clinical applications. *Bái sháo* (Radix Paeoniae Alba) has a broader application scope. This is roughly summarised as: 1. To calm the liver, used for liver yin deficiency, liver yang rising inducing dizziness, tinnitus, headache and head distention, etc.; 2. To nourish the blood and harmonise blood, used for blood deficiency syndrome or female irregular menstruation; 3. To soften the liver and alleviate pain, used for all kinds of spasmodic pain. In addition, *bái sháo* (Radix Paeoniae Alba) has many compatibilities, if used correctly, it is very effective.

1. *Bái Sháo* (Radix Paeoniae Alba) ——
Guì Zhī (桂枝, Ramulus Cinnamomi)

Function

To harmonise nutritive and defensive qi, release muscles and the exterior, warm the meridians and the blood vessels.

Application

This compatibility is the main one in *Guì Zhī Tāng* (Cinnamon Twig Decoction, 桂枝汤) from the *Discussion on Cold Damage*. It can be used for exterior deficiency inducing external invasion of wind-cold and cold causing abdominal pain, postpartum spontaneous sweating and wind-damp arthralgia, etc.

Compatibility Analysis

Guì zhī (Ramulus Cinnamomi) is acrid and warm. It can warm the meridians and move blood, free yang qi to release the muscles. It not only can regulate nutritive qi but also harmonise defensive qi and release muscle and skin stagnated with wind-cold pathogens. *Bái sháo* (Radix Paeoniae Alba) nourishes blood and retains yin by astringing. It can prevent *guì zhī* (Ramulus Cinnamomi) from being over acrid and dispersing thus injuring yin. The *Golden Mirror of the Medical Tradition* said that: "*Guì zhī (Ramulus Cinnamomi) governing sháo yào (Radix Paeoniae Lactiflorae) achieves the goal of retaining too much sweat. Sháo yào (Radix Paeoniae Lactiflorae) helping guì zhī (Ramulus Cinnamomi) harmonises nutritive qi and has a regulating defensive qi function.*" *Guì zhī* (Ramulus Cinnamomi) and *sháo yào* (Radix Paeoniae Lactiflorae) compatibility allows sweating but does not induce injury to the nutritive qi and blood, it stops sweating but does not trap the pathogen. One to open and the other to close can release the exterior and harmonise the interior. Hence, an external invasion syndrome of the exterior deficiency type with spontaneous sweating, could not be released without this compatibility. In addition, *bái sháo* (Radix Paeoniae Alba) nourishes blood, softens the liver and alleviates pain. If there is blood deficiency with cold it induces all kinds of pain, *guì zhī* (Ramulus Cinnamomi) is needed to warmly unblock in order to be effective.

Case 1

Postpartum body ache.

Initial Visit

A female patient, 34 years old, April 9th, 1986.

Three months after the patient had a baby, she came to visit due to whole body joint pain. Her chief complaint was due to lack of quality rest after delivery inducing whole body joint pain, lassitude, and sweating, which was getting worse by the day. She could not exert herself, once she exerted, she had persistent sweating. Her pulse was deep, slow and forceless. Her tongue was pale and the tongue coating was thin and white. The diagnosis was postpartum qi and blood deficiency with invasion of wind-cold. During the postpartum stage there is more deficiency. The wind pathogens invade organisms during deficiency, obstructing and stagnating meridians and collaterals, and so it induces the body joint aches. The wind pathogen attacked the exterior, and then nutritive qi and defensive qi were in disharmony, hence, there was persistent sweating. The treatment plan was to use a formula to warm the meridians and unblock the blood vessels, tonify qi and nourish blood, harmonise nutritive qi and eliminate wind.

Prescription

白芍	*bái sháo*	15g	Radix Paeoniae Alba
桂枝	*guì zhī*	10g	Ramulus Cinnamomi
当归	*dāng guī*	20g	Radix Angelicae Sinensis
丹参	*dān shēn*	30g	Radix et Rhizoma Salviae Miltiorrhizae
生黄芪	*shēng huáng qí*	24g	Radix Astragali (raw)
鸡血藤	*jī xuè téng*	30g	Caulis Spatholobi
白术	*bái zhú*	12g	Rhizoma Atractylodis Macrocephalae
防风	*fáng fēng*	9g	Radix Saposhnikoviae
牛膝	*niú xī*	15g	Radix Cyathulae
甘草	*gān cǎo*	6g	Radix Glycyrrhizae

Water was used to decoct, one pack per day.

Second Visit

April 29th: After taking the above formula modification for more than twenty packs, the sweating stopped, the lassitude was eliminated, and the joint pain was relieved, except for the upper limbs, which still had lassitude, soreness and heaviness. Again, the original formula was used, adding *jiāng huáng* (Rhizoma Curcumae Longae) 9g, and *shān zhū yú* (Fructus Corni) 10g, she continued to take it and recovered.

Case 2

Sweating.

Initial Visit

A male patient, 47 years old, March 1st, 1986.

The patient had always had spontaneous sweating. A slight exertion or during a meal, caused sweat to drip and it was accompanied by shortness of breath, lassitude, insomnia, lack of concentration, and dizziness. He had the symptoms for more than half a year. He had biomedicine treatments but they were ineffective and so he came to visit our hospital's clinic. His pulse was wiry, thready and fast. His tongue was red and tongue coating was thin and white. His case history of this condition was taken in detail, the patient recalled: Half a year ago he had flu and fever. After treatments, he had sweats and the fever reduced, he had lassitude, and had deficiency sweats easily. The diagnosis was the nutritive qi and defensive qi disharmony inducing sweating. Sweat is the heart's fluid. Prolonged sweating will surely injure heart yin. With the yin injured, the qi also becomes deficient and induces both qi and blood deficiency of the heart channel of the hand *shaoyin* meridian, causing insomnia and palpitations. The treatment plan was to regulate nutritive and defensive qi, tonify heart qi, nourish the heart yin, and calm the heart and spirit.

Prescription

杭白芍	*háng bái sháo*	15g	Radix Paeoniae Lactiflorae
桂枝	*guì zhī*	10g	Ramulus Cinnamomi
百合	*bǎi hé*	30g	Bulbus Lilii
茯苓	*fú líng*	15g	Poria
党参	*dǎng shēn*	15g	Radix Codonopsis
生龙骨	*shēng lóng gǔ*	15g	Os Draconis (raw)
生牡蛎	*shēng mǔ lì*	15g	Concha Ostreae (raw)
甘草	*gān cǎo*	6g	Radix Glycyrrhizae

A pinch of *xiǎo mài* (Fructus Tritici) was added as a guiding herb. Water was used to decoct, then it was divided into two servings, to be taken warm in the morning and at night.

Second Visit

March 7th: After taking six packs of the above formula, the sweating stopped, and all symptoms reduced, but he still had insomnia, and lack of concentration. To the original formula, *zǎo rén* (Semen Ziziphi Spinosae) 15g, and *yè jiāo téng* (Caulis Polygoni Multiflori) 30g, were added. He continued to take six more packs and recovered.

Case 3

Gushing (Running Piglet) Syndrome.

Initial Visit

A female patient, 40 years old, April 13th, 1985.

Four years previously, the patient's mother passed away, she was overcome with grief and suddenly felt suffocated. After waking up, she felt distention, fullness and discomfort in the lower abdomen, and qi from the lower abdomen was rebelling upward. Later symptoms gradually deteriorated. It was very obvious after she got angry. She noticed that qi from the lower abdomen counterflowed to the chest and head. She had cervical ankylosis and could not turn her head with both eyes closed tight, she disliked hearing others' voices, her throat got blocked like there was something caught in it. The treatment plan was aimed at phlegm and qi knotted together inducing plum-stone qi (globus hystericus) syndrome. Though the throat blockage improved a little bit, the qi was still as rebellious as before. Her pulse was deep, thready and forceless. Her tongue was pale and the coating was thin. The diagnosis was lower *jiao* deficiency cold, gushing qi rebelling upwards inducing Gushing (Running Piglet) Syndrome. The formula was changed to Zhong-jing's *Guì Zhī Jiā Guì Tāng* (Cinnamon Twig Decoction With Extra Cinnamon, 桂枝加桂汤).

Prescription

白芍	*bái sháo*	30g	Radix Paeoniae Alba
桂枝	*guì zhī*	15g	Ramulus Cinnamomi
半夏	*bàn xià*	4g	Rhizoma Pinelliae
厚朴	*hòu pò*	10g	Cortex Magnoliae Officinalis
茯苓	*fú líng*	15g	Poria
紫苏	*zǐ sū*	10g	Folium Perillae
荷叶	*hé yè*	10g	Folium Nelumbinis
沉香	*chén xiāng*	5g	Lignum Aquilariae Resinatum
甘草	*gān cǎo*	6g	Radix Glycyrrhizae
生姜	*shēng jiāng*	3pcs	Rhizoma Zingiberis Recens (as a guide)
大枣	*dà zǎo*	3pcs	Fructus Jujubae (as a guide)

Shēng jiāng (Rhizoma Zingiberis Recens) 3 pieces，and *dà zǎo* (Fructus Jujubae) 3 pieces were added as guiding herbs.

Water was used to decoct, then it was divided into two servings, one pack per day.

Second Visit

April 16th: After taking three packs of the above formula, she felt qi was not rebelling, immediately after taking the formula, but after one hour, the qi was rebellious again and gushed upward. It was an indication that the main herb dosage was too little. The same formula was used again with *guì zhī* (Ramulus Cinnamomi) increased to 60g, and *bái zhú* (Rhizoma Atractylodis Macrocephalae) increased to 30g.

Third Visit

April 23rd: After taking four packs, the symptoms reduced a lot. The qi stopped rebelling upwards. She only felt something in her lower abdomen, and had insomnia. As it was effective, the prescription was not changed. To the original formula *shēng lóng gǔ* (raw Os Draconis) and *shēng mǔ lì* (raw Concha Ostreae) 15g each were added, and she continued to take it again.

Fourth Visit

April 28th: All symptoms were eliminated; the pulse floated up and became slow, however, there still was occassional insomnia. The formula was changed to *Guī Pí Tāng* (Spleen-Returning Decoction, 归脾汤) to complete clearing the symptoms.

2. *Bái Sháo* (Radix Paeoniae Alba) ——
Gōu Téng (钩藤, Ramulus Uncariae Cum Uncis)

Function

To nourish the blood, soften the liver, extinguish wind and stop spasm.

Application

Used for symptoms of liver yin deficiency or liver blood deficiency or liver deficiency and liver yang rising up inducing headache, dizziness, etc.

Compatibility Analysis

Bái sháo (Radix Paeoniae Alba) enters the liver channel of the foot *jueyin* meridian to nourish blood and soften the liver. *Gōu téng* (Ramulus Uncariae Cum Uncis) also enters the liver channel of the foot *jueyin* meridian. It can calm the liver, clear heat and extinguish wind. In clinic, the root problem of liver yang rising and liver wind moving internally, is due to the liver channel of the foot *jueyin* meridian's blood being inadequate and being unable to restrain liver yang, therefore, it rose upward. If there is more liver yang rising, it would consume the liver yin and induce more liver yin deficiency. *Bái sháo* (Radix Paeoniae Alba) nourishes blood and softens the liver to treat the root problem. *Gōu téng* (Ramulus Uncariae Cum Uncis) clears heat and calms the liver to treat its manifestations. The two herbs combined treat both manifestations and the root problem. They are mutually enhanced.

Case 1

Initial Visit

A female patient, November 26th, 1986.

The patient had dizziness for more than a year and it was accompanied by uncontrollable shaking of her head, left and right. When tired or worried, it was the most severe. Although, she got treatments from many places, they were not effective and she came to visit the Chinese medicine doctor for treatment. On observation, the patient had good physical development, her blood pressure was 140/90mm Hg, and her appetite, urination and bowel movement were all normal. Her pulse was wiry and thready. Her tongue was red and tongue coating was thin and white. Syndrome differentiation: All kinds of wind inducing trembling and dizziness belong to the liver. The patient's head shaking and dizziness was a sign that her liver was sick. Her wiry and thready pulse with red tongue was due to liver yin deficiency. If yin is deficient then yang will rise and generate wind, therefore she presented with head shaking and dizziness syndrome. The treatment plan was to tonify the liver and kidney, nourish the blood and extinguish wind.

Prescription

杭白芍	*háng bái sháo*	20g	Radix Paeoniae Lactiflorae
枸杞	*gǒu qǐ*	12g	Fructus Lycii
熟地	*shú dì*	15g	Radix Rehmanniae Praeparata
山萸肉	*shān yú ròu*	15g	Fructus Corni
钩藤	*gōu téng*	30g	Ramulus Uncariae Cum Uncis (decoct later)
全蝎	*quán xiē*	6g	Scorpio
僵蚕	*jiāng cán*	10g	Bombyx Batryticatus
川牛膝	*chuān niú xī*	15g	Radix Cyathulae
丹参	*dān shēn*	20g	Radix et Rhizoma Salviae Miltiorrhizae
甘草	*gān cǎo*	6g	Radix Glycyrrhizae

Water was used to decoct, and divided into two servings.

Second Visit

January 7th, 1987: Her son came to visit and said that after taking six packs of the above formula, the head shaking and dizziness stopped and there was no reoccurrence for more than a month. Recently due to arranging her son's wedding, she did not get enough rest, and the head shaking and dizziness appeared again, and so he came to get a formula to treat her. Four packs of the same formula was prescribed again and the patient was informed that if she recovered, she should take another four packs to strengthen the effectiveness.

Case 2

Initial Visit

A male patient, 59 years old, April 30th, 1986.

The patient had flu and fever two months previously. After recovery, dizziness and nausea appeared which were sometimes better and sometimes worse. He had been diagnosed by an hospital with meniere's disease. He had treatments but with no remarkable effect so he visited this clinic for treatment. On inspection his blood pressure was normal and his cerebral blood flow chart was also normal. His pulse was wiry, thready and fast. His tongue was red with a scanty coating. His symptoms were also accompanied by tinnitus and an ear blockage. This occurred after external invasion with injured yin. When yin is injured then the pulse is thready and fast, and there is tinnitus. With yin injured the deficiency fire flared upward and induced dizziness, and nausea. The treatment plan was to nourish yin, clear heat, extinguish wind and descend the rebellion.

Prescription

钩藤	*gōu téng*	30g	Ramulus Uncariae Cum Uncis (decoct later)
杭白芍	*háng bái sháo*	30g	Radix Paeoniae Lactiflorae

麦冬	*mài dōng*	30g	Radix Ophiopogonis
半夏	*bàn xià*	15g	Rhizoma Pinelliae
代赭石	*dài zhě shí*	15g	Haematitum
丹参	*dān shēn*	24g	Radix et Rhizoma Salviae Miltiorrhizae
生地	*shēng dì*	30g	Radix Rehmanniae
川牛膝	*chuān niú xī*	15g	Radix Cyathulae
葛根	*gé gēn*	30g	Radix Puerariae Lobatae
茯苓	*fú líng*	20g	Poria
甘草	*gān cǎo*	6g	Radix Glycyrrhizae

Water was used to decoct, and divided many times. The patient was informed that if he vomited after taking the formula, then he should wait for another half an hour before taking it again, and it should be fine. He should not take it all at once. One pack per day.

Second Visit

May 10th: After taking three packs of the above formula, the dizziness reduced a lot, the nausea and rebellious symptoms disappeared except for when raising or shaking the head, there was still a dizziness sensation. Three more packs of the original formula were taken. Later on he revealed that he had recovered.

3. *Bái Sháo* (Radix Paeoniae Alba) ——
Chái Hú (柴胡, Radix Bupleuri)

Function

To harmonise blood, soften the liver, eliminate stagnation and alleviate pain.

Application

This can be used for a constrained liver with qi stagnation, and liver and spleen disharmony inducing the liver channel of the foot *jueyin* and the gallbladder channel of the foot *shaoyang* meridians' disease syndromes, such as hepatitis, cholecystitis, pancreatitis, pleurodynia, etc. Also used for female abnormal menstruation, mastitis that belongs to the constrained liver with qi stagnation.

Compatibility Analysis

The liver is a strong *zang* organ, physically, it is yin but its functions are yang. Its qi easily rebels horizontally. Once it rebels horizontally, it insults the spleen. Once the spleen is insulted, then abdominal pain occurs. *Bái sháo* (Radix Paeoniae Alba) can nourish blood and soften the liver, can retain liver yin by astringing, and can relax liver qi. If the liver is softened, then the spleen will not be insulted and pain can be eliminated. Hence, there is a saying that *bái sháo* (Radix Paeoniae Alba) restrains liver wood. *Chái hú* (Radix Bupleuri) has the function to disperse liver qi, soothe the liver and release stagnation. Any liver qi that is constrained or stagnated induces hypochondriac pain; *chái hú* (Radix Bupleuri) is the most suitable for this. Of the two herbs combined together, one is astringent and the other is soothing; one is moving and the other is static. Soothing and softening are mutually enhanced, the moving and static are combined; the physical and the function are both taken care of. In addition, *chái hú* (Radix Bupleuri) is bitter and acrid. The aromatic disperses and sedates. Its properties are raising and dispersing. Our ancestors had a saying that the acrid and dispersing injure yin. If combined with *bái sháo* (Radix Paeoniae Alba) which retains yin by astringing and softens the liver, this can prevent this side effect. They are really the best compatibility to soothe the liver, release stagnation, soften the liver and alleviate pain. *Sì Nì Sǎn* (Counterflow Cold Powder, 四逆散) and *Dà Chái Hú Tāng* (Major Bupleurum Decoction, 大柴胡汤) in Zhong-jing's *Discussion on Cold Damage* as well as *Xiāo Yáo Sǎn* (Free Wanderer Powder, 逍遥散) in *Formulary of the Bureau of Medicines of the Taiping Era*, all have the *bái sháo* (Radix Paeoniae Alba) and *chái hú* (Radix Bupleuri) compatibility.·

Case 1

Cholecystitis.

Initial Visit

A male patient, 16 years old, January 13th, 1988.

The patient had right hypochondriac pain extending to include back pain which was accompanied by stomach and abdominal distention and fullness, which were worse after eating. He had already had it for a couple of months. He had taken reducing food and strengthening stomach medicine but they did not work and so he came to visit. On palpation the patient's pulse was wiry and forceful; the tongue was red with a white coating; the right hypochondriac gallbladder area had a pressing pain, the pain was worse on percussion. A B-ultrasound check showed: The gallbladder was slightly enlarged, the gallbladder wall thickness was 0.5cm and not smooth and he was diagnosed with chronic cholecystitis. The treatment plan was to soothe the liver and eliminate stagnation, benefit the gallbladder and alleviate pain. A modified *Dà Chái Hú Tāng* (Major Bupleurum Decoction) was used to treat him.

Prescription

柴胡	*chái hú*	15g	Radix Bupleuri
白芍	*bái sháo*	15g	Radix Paeoniae Alba
黄芩	*huáng qín*	9g	Radix Scutellariae
枳实	*zhǐ shí*	10g	Fructus Aurantii Immaturus
川军	*chuān jūn*	6g	Radix et Rhizoma Rhei
蒲公英	*pú gōng yīng*	15g	Herba Taraxaci
半夏	*bàn xià*	9g	Rhizoma Pinelliae
延胡索	*yán hú suǒ*	10g	Rhizoma Corydalis
郁金	*yù jīn*	9g	Radix Curcumae
金钱草	*jīn qián cǎo*	24g	Herba Lysimachiae
甘草	*gān cǎo*	6g	Radix Glycyrrhizae

Water was used to decoct and one pack per day was taken.

Second Visit

January 19th: After taking six packs of this formula, the hypochondriac pain stopped, and the abdominal distention also remarkably reduced. Pressing on the gallbladder area still produced slight pain and discomfort. He continued on with the same formula.

Third Visit

January 28th: The hypochondriac pain had already been eliminated; his appetite increased, and after eating there was no stomach and abdominal distention. He was then prescribed a herbal tablet *Xiāo Yán Lì Dǎn Piàn* (Anti-Inflammatory and Gallbladder-Disinhibiting Tablet) instead to complete the treatment.

Case 2

Left side chest and hypochondriac pain.

Initial Visit

A female patient, 52 years old, August 20th, 1995.

The patient, due to having left side chest and hypochondriac pain for a couple of days without recovery, came to visit. The patient complained that one week previously after getting angry over some domestic issues and overexertion later, she felt left chest and hypochondriac pain. When breathing out or coughing, the pain was worse. On pressing down on her left chest and hypochondria, there was no mass or tender point. Her EKG was normal. Her pulse was deep, wiry and forceful. Her tongue coating was thin and white. The diagnosis was liver qi stagnation inducing hypochondriac pain. The treatment plan was to use modified *Xiāo Yáo Sǎn* (Free Wanderer Powder) to treat it.

Prescription

柴胡	*chái hú*	18g	Radix Bupleuri
白芍	*bái sháo*	20g	Radix Paeoniae Alba
当归	*dāng guī*	15g	Radix Angelicae Sinensis
薄荷	*bò he*	9g	Herba Menthae
茯苓	*fú líng*	15g	Poria
枳壳	*zhǐ qiào*	12g	Fructus Aurantii
延胡索	*yán hú suǒ*	15g	Rhizoma Corydalis
白术	*bái zhú*	10g	Rhizoma Atractylodis Macrocephalae
红花	*hóng huā*	9g	Flos Carthami
丹参	*dān shēn*	20g	Radix et Rhizoma Salviae Miltiorrhizae
甘草	*gān cǎo*	6g	Radix Glycyrrhizae

Water was used to decoct and one pack per day was taken.

Second Visit

December 28th: After taking four packs of the above formula, the pain disappeared and there was no reoccurrence for a couple of months. However, in recent days due to getting angry again, she had left chest and hypochondriac pain as before, so she came to visit again. Her pulse and symptoms were diagnosed basically the same as the last time.

The same formula was used to treat her.

Case 3

Hepatitis and hypochondriac pain.

Initial Visit

A male patient, 20 years old, April 30th, 1993.

The patient was a hepatitis B virus carrier. Recently due to overwork he was tired and felt a whole body lassitude, which was especially obvious on his lower limbs. His stomach and abdomen were slightly distended, his appetite, urination and bowel movement were all normal except for his liver area pain. He had a check up at a local county hospital for his liver function: Except for the ALT being 160u/L, the rest were all normal, and the hepatitis B surface antigen was positive. On observation, the patient had good physical development. His liver area had percussion tenderness. The inferior edge of the liver could be palpated and it was soft. His pulse was wiry, thready and forceful. His tongue coating was thin and white and his tongue was red and slightly dark. The treatment plan was to soothe the liver, transform stasis, soften the liver and alleviate pain.

Prescription

白花蛇舌草	bái huā shé shé cǎo	30g	Herba Hedyotis Diffusae
蒲公英	pú gōng yīng	15g	Herba Taraxaci
白芍	bái sháo	20g	Radix Paeoniae Alba
柴胡	chái hú	15g	Radix Bupleuri
丹参	dān shēn	20g	Radix et Rhizoma Salviae Miltiorrhizae
泽兰	zé lán	15g	Herba Lycopi
虎杖	hǔ zhàng	15g	Rhizoma et Radix Polygoni Cuspidati
枳壳	zhǐ qiào	10g	Fructus Aurantii
白术	bái zhú	10g	Rhizoma Atractylodis Macrocephalae
五味子	wǔ wèi zǐ	9g	Fructus Schisandrae Chinensis
土茯苓	tǔ fú líng	20g	Rhizoma Smilacis Glabrae
丹皮	dān pí	9g	Cortex Moutan

Water was used to decoct, then it was divided into two servings, one pack per day.

Second Visit

May 7th: After taking seven packs of the above formula, all the symptoms had reduced. The liver area pain also reduced a lot. The same formula was continued with, for treatment.

Third Visit

May 27th: After taking more than twenty packs more of the above formula, all symptoms disappeared, and the liver function ALT tested normal.

4. *Bái Sháo* (Radix Paeoniae Alba) ——
Bái Zhú (白术, Rhizoma Atractylodis Macrocephalae)

Function

To harmonise the liver and spleen, alleviate pain and stop diarrhoea.

Application

This is used for liver and spleen disharmony inducing abdominal pain, diarrhoea and borborygmus. Modern clinics use it for chronic colitis.

Compatibility Analysis

Bái sháo (Radix Paeoniae Alba) is bitter, sour and slightly cold. It functions to nourish blood, retain yin by astringing, soften the liver and alleviate pain. The liver is a strong *zang* organ. Its qi easily rebels horizontally to insult the spleen. When the spleen is insulted then there is abdominal pain. *Bái sháo* (Radix Paeoniae Alba) can relax liver qi and soften the liver. If the liver has been softened, then the spleen will not be insulted and pain can be eliminated. Hence, there is a saying that *bái sháo* (Radix Paeoniae Alba) restrains liver wood. *Bái sháo* (Radix Paeoniae Alba) is bitter, sweet and warm. Bitterness and warmth can overcome dampness. The sweet and the aromatic can strengthen the spleen and relax the middle *jiao*. Huang Gong-xiu said *bái sháo* (Radix Paeoniae Alba) is *"the first important herb to tonify spleen qi"*. The *bái sháo* (Radix Paeoniae Alba) and *bái zhú* (Rhizoma Atractylodis Macrocephalae) combination functions to restrain wood and cultivate earth, alleviate pain and stop diarrhoea. Wu He-ben said: *"For diarrhoea one should blame the spleen. For pain one should blame the liver. The liver is blamed for excess, the spleen is blamed for deficiency. If there is spleen deficiency and liver excess, this produces pain and diarrhoea."* The above two herbs in combination are suitable to regulate the liver and spleen as well as to stop pain and diarrhoea. This compatibility comes from *Bái Zhú Sháo Yào Sǎn* (White Atractylodes and Peony Powder, 白术芍药散), a formula of Liu Cao-chuang's. It is composed of four herbs *bái zhú* (Rhizoma Atractylodis Macrocephalae), *bái sháo* (Radix Paeoniae Alba), *fáng fēng* (Radix Saposhnikoviae) and *chén pí* (Pericarpium Citri Reticulatae). This formula is good for treating abdominal pain and diarrhoea, it is, therefore, also named *Tòng Xiè Yào Fāng* (Important Formula for Painful Diarrhea, 痛泻要方). In clinic the treatment affect is good for any abdominal pain followed by diarrhoea as the main symptom of enteritis and dysentery. If the stool is watery, *chē qián zǐ* (Semen Plantaginis), *fú líng* (Poria) and *gān jiāng* (Rhizoma Zingiberis) can be added to warm the middle *jiao* and help to separate the turbid and the clear. If with pus and blood, add *bái tóu wēng* (白头翁, Radix Pulsatillae), *huáng qín* (Radix Scutellariae) and *huáng lián* (Rhizoma Coptidis) to clear the heat and stop dysentery. For tenesmus patients, add *mù xiāng* (木香, Radix Aucklandiae) and *bīng láng piàn* (槟榔片, Sliced Semen Arecae) to promote qi movement and guide the stagnation. For severe abdominal pain, double the dosage of *bái sháo* (Radix Paeoniae Alba) and add *xiāng fù* (Rhizoma Cyperi) to soothe the liver and calm the wood. For prolonged diarrhoea in deficiency cold patients, add *wú zhū yú* (吴茱萸, Fructus Evodiae) and *gān jiāng* (Rhizoma Zingiberis) to warm the middle *jiao* and alleviate pain. In patients with prolonged diarrhoea that is sometimes severe, and sometimes better, or seasonal diarrhoea patients, most of them have blood stasis and so *dān shēn* (Radix et Rhizoma Salviae Miltiorrhizae), *dāng guī* (Radix Angelicae Sinensis) and *hóng huā* (Flos Carthami), etc. can be added to invigorate blood and transform stasis.

Case Study

Initial Visit

A male patient, 38 years old, July 5th, 1997.

The patient had had abdominal pain and diarrhoea for a year. He used berberine and norfloxacin, etc. to treat it but without effect. It was sometimes better and sometimes severe and, particularly in the morning when he woke up, the abdominal pain and diarrhoea were pronounced. He recently had watery diarrhoea two to three times a day. A stool culture showed up (-). His tongue coating was white and turbid. His pulse was wiry and his right pulse tended toward slow. A colonoscopy showed: The colon mucosal membrane had hyperoedema, it was congested with blood and so nothing abnormal was observed. He was diagnosed with colitis. The syndrome was liver excess and spleen deficiency, with the spleen having lost its transforming and transporting ability. The treatment plan was to regulate and harmonise the liver and spleen and assist with transforming stasis and eliminating dampness.

Prescription

白芍	*bái sháo*	30g	Radix Paeoniae Alba
白术	*bái zhú*	15g	Rhizoma Atractylodis Macrocephalae
陈皮	*chén pí*	9g	Pericarpium Citri Reticulatae
防风	*fáng fēng*	10g	Radix Saposhnikoviae
丹参	*dān shēn*	24g	Radix et Rhizoma Salviae Miltiorrhizae
红花	*hóng huā*	10g	Flos Carthami
茯苓	*fú líng*	24g	Poria
泽泻	*zé xiè*	15g	Rhizoma Alismatis
山药	*shān yào*	15g	Rhizoma Dioscoreae
生苡米	*shēng yì mǐ*	24g	Semen Coicis (raw)
桔梗	*jié gěng*	8g	Radix Platycodonis
柴胡	*chái hú*	12g	Radix Bupleuri
枳壳	*zhǐ qiào*	12g	Fructus Aurantii
川牛膝	*chuān niú xī*	12g	Radix Cyathulae
甘草	*gān cǎo*	6g	Radix Glycyrrhizae

Water was used to decoct and it was then taken.

Second Visit

June 3rd: After taking five packs of the above formula, the abdominal pain and diarrhoea symptoms all had reduced. As it was so effective, the prescription was not changed and he continued to take it.

Third Visit

June 8th: After another five packs, the abdominal pain was eliminated, the bowel movements were once or twice a day and slightly watery, the pulse turned slower, and the tongue coating was white but slightly turbid. The formula matched the syndrome, so he continuously used the original formula's modification for treatment.

Fourth Visit

June 20th: After taking more than twenty packs of the above formula, all his symptoms disappeared and he recovered.

Bái zhǔ's (白术 , Rhizoma Atractylodis Macrocephalae) taste and property are sweet, bitter and warm. It enters the spleen channel of the foot *taiyin* and the stomach channel of the foot *yangming* meridians. Its functions can tonify the spleen and qi, transform dampness and promote urination. Huang Gong-xiu said: *"Why is bái zhú (Rhizoma Atractylodis Macrocephalae) especially tonifying for spleen qi? It is because the spleen suffers dampness, one should immediately eat bitter to dry it. When the spleen needs relaxation, one should immediately eat sweet to relax it. Bái zhú (Rhizoma Atractylodis Macrocephalae) tastes bitter and sweet. It not only can dry dampness to strengthen the spleen but also can relax the spleen to generate fluid. If taken it can strengthen the digestive function to eliminate food. Also, its property is the warmest. It is the best herb to tonify spleen qi."* In addition, the *Divine Husbandman's Classic of the Materia Medica* said it mainly treats *"wind, cold, damp painful obstruction......."* It is also an important herb to treat wind-dampness arthralgia syndrome. Its application scope can be very broad if combined compatibily.

1. *Bái Zhú* (Rhizoma Atractylodis Macrocephalae) ——
Shēng Yì Rén (生薏仁, raw Semen Coicis)

Function

To eliminate wind-damp, and to treat painful obstruction.

Application

This is used for wind-dampness arthralgia syndrome.

Compatibility Analysis

Bái zhú (Rhizoma Atractylodis Macrocephalae) is sweet, bitter and warm. It can strengthen the spleen, eliminate dampness, tonify qi and secure the exterior and eliminate wind-damp. *Shēng yì rén* (raw Semen Coicis) is sweet, bland and cold. It can nourish the tendons, relax or relieve spasm pain, strengthen the spleen and eliminate dampness. Its characteristic is to enter the liver channel of the foot *jueyin* meridian in order to eliminate dampness pathogens in the tendons and meridians. The two herbs combined can increase the effectiveness of eliminating dampness and alleviating pain. Moreover, the pharmacological property is gentle. *Yì rén* (Semen Coicis) has the function to nourish the tendons. *Bái zhú* (Rhizoma Atractylodis Macrocephalae) has effects to secure the exterior and strengthen the spleen. Hence, the two herbs eliminate the dampness pathogen but do not injure the upright qi. Professor Fu Xian-fang of Tianjin University of Traditional Chinese Medicine, liked to use *bái zhú* (Rhizoma Atractylodis Macrocephalae) combined with *shēng yì rén* (raw Semen Coicis) to treat impediment syndrome, each time he used it he got a good result. Professor Fu believed that: *"Though impediment syndrome has wind, cold, dampness and heat differentiations, the dampness pathogen obstructing and stagnating the tendon and meridians is the main aetiology. They only differ in the amount of wind, cold or heat, therefore, impediment syndrome is difficult to cure. In clinic it is essential to treat dampness first in impediment syndrome".*

Case Study

Initial Visit

A female patient, 16 years old, November 21st, 1989.

One week previously, both the patient's feet, knees and ankle joints suddenly became red, swollen and painful. When touched, it got worse. On palpation, it was extremely hot. She had been diagnosed, by a hospital, with rheumatic fever; ESR 50mm/h was accompanied by irritability, dry mouth, brownish urine, a wiry and fast pulse, a red tongue and a thin yellow coating. The syndrome was damp-heat pouring downward, with stagnation obstructing the meridians causing heat painful obstruction. The treatment plan was to clear heat, eliminate dampness, and extinguish wind.

Prescription

生薏仁	*shēng yì rén*	30g	Semen Coicis (raw)
白术	*bái zhú*	15g	Rhizoma Atractylodis Macrocephalae
金银藤	*jīn yín téng*	30g	Lonicerae Caulis
防己	*fáng jǐ*	15g	Radix Stephaniae Tetrandrae
川牛膝	*chuān niú xī*	9g	Radix Cyathulae
木瓜	*mù guā*	9g	Fructus Chaenomelis
防风	*fáng fēng*	9g	Radix Saposhnikoviae
木通	*mù tōng*	5g	Caulis Akebiae
竹叶	*zhú yè*	6g	Herba Lophatheri
甘草梢	*gān cǎo shāo*	9g	Radix Glycyrrhizae Uralensis

Water was used to decoct, and then it was divided into two servings.

Second Visit

November 24th: After taking three packs of the above formula, the redness, swelling and pain had all reduced, and her urine had become clear. She was then prescribed a modified version of the above formula with *mù tōng* (Caulis Akebiae) and *zhú yè* (竹叶 , Herba Lophatheri) eliminated, and *huáng bǎi* (黄柏 , Cortex Phellodendri Chinensis) 9g added. She continued to take it.

Third Visit

November 28th: The redness and swelling were basically eliminated; the pain had reduced a lot. She took the same formula again.

Fourth Visit

November 8th: After continuously taking more than ten packs of the above formula, the pain and swelling disappeared, ESR 15mm/h. She stopped using the formula and recovered.

2. *Bái Zhú* (Rhizoma Atractylodis Macrocephalae) —— *Cāng Zhú* (苍术, Rhizoma Atractylodis)

Function

To strengthen the spleen, dry dampness, eliminate swelling, and promote movement of the internal and external dampness pathogen.

This can be used to treat conditions due to spleen deficiency not transforming and transporting inducing the dampness pathogen to stay and stagnate; or syndromes due to the dampness pathogen, which can induce all kinds of sicknesses, such as oedema due to nephritis, excessive leukorrhoea in women, viral hepatitis with ALT not reducing and patients with dampness encumbered spleen.

Compatibility Analysis

Bái zhú (Rhizoma Atractylodis Macrocephalae) is sweet, bitter and warm. It can strengthen the spleen and eliminate dampness, tonify qi and the middle *jiao*. *Cāng zhú* (苍术, Rhizoma Atractylodis) is sweet, acrid and warm. It can dry dampness and strengthen the spleen. It also can eliminate wind and expel the turbid. The two herbs' functions are similar. However, one tends to move externally and the other tends to move internally. Li Shi-chai stated of *cāng zhú* (Rhizoma Atractylodis) that: "*It soothes the chest and promotes sweating. Its function is more superior than bái zhú (Rhizoma Atractylodis Macrocephalae), however, its power is less than bái zhú (Rhizoma Atractylodis Macrocephalae) in tonifying the middle jiao and eliminating dampness. Generally speaking, for spleen deficiency we should use bái zhú (Rhizoma Atractylodis Macrocephalae) and for pathogens obstructing the middle jiao we should use cāng zhú (Rhizoma Atractylodis).*" Zhang Yin-an even thought that: "*The property of bái zhú (Rhizoma Atractylodis Macrocephalae) is superior to that of cāng zhú (Rhizoma Atractylodis). If tonifying the spleen is required, then use bái zhú (Rhizoma Atractylodis Macrocephalae), but if promoting the spleen's transforming and transporting function is required, then use cāng zhú (Rhizoma Atractylodis). Hence, both are required if one wants to tonify while promoting the transforming and transporting function. If more tonification is needed than the transforming and transporting, then more bái zhú (Rhizoma Atractylodis Macrocephalae) and less cāng zhú (Rhizoma Atractylodis) should be used. If more transforming and transporting and less tonification is needed, then use more cāng zhú (Rhizoma Atractylodis) but less bái zhú (Rhizoma Atractylodis Macrocephalae).*" Though the two *zhús* functions are different, their strengthening the spleen and drying dampness functions are the same. However, *bái zhú* (Rhizoma Atractylodis Macrocephalae) has stronger power to strengthen the spleen and *cāng zhú* (Rhizoma Atractylodis) has stronger power to dry dampness. If the two *zhús* are used together, they can both tonify as well as transform and transport, and treat the exterior and interior at the same time. Hence, it is effective for both external and internal dampness inducing nephritic oedema. Modern pharmacological research discovered that *cāng zhú* (Rhizoma Atractylodis) has the function to protect the renal tubular epithelial cells, alleviate the kidneys' internal impediment and enhance the kidneys' function. It also has the function to enhance the human body's immunity function. In addition, it is also effective for spleen deficiency and cold dampness sinking downward inducing leukorrhoea. An example of this compatibility of the two *zhús* is Fu Qing-zhu's *Wán Dài Tāng* (Discharge-Ceasing Decoction, 完带汤).

Case 1

Leukorrhoea syndrome.

Initial Visit

A female patient, 46 years old, April 7th, 1991.

The patient had had low back pain and excessive leukorrhoea for more than a year. In recent years she had lower back and abdominal pain. When the day was cloudy or she encountered cold, the pain got worse, her lower abdomen felt cold, and she had a distenting and dragging down sensation, with a lot of leukorrhoea similar to white nasal discharge. Her menstruation was basically normal but had a scanty flow, when menstruating she had abdominal pain, accompanied by poor appetite, stomach distention, a pale tongue, with a thin white coating and a deep pulse. The diagnosis was both spleen and kidney deficiency, with cold and damp pouring downward syndrome. Spleen deficiency meant she could not transform and transport and so the dampness pathogen obstructed internally. Kidney deficiency meant she could not warmly transform, and so the water and dampness stayed internally. With both spleen and kidney deficiency the dampness stagnated below and induced leukorrhoea. The treatment was to strengthen the spleen and warm the kidney, to transform dampness and stop leukorrhoea. She was treated with modified *Wán Dài Tāng* (Discharge-Ceasing Decoction).

Prescription

白术	*bái zhú*	20g	Rhizoma Atractylodis Macrocephalae
苍术	*cāng zhú*	15g	Rhizoma Atractylodis
川断	*chuān duàn*	15g	Radix Dipsaci
杜仲	*dù zhòng*	15g	Cortex Eucommiae
附子	*fù zǐ*	6g	Radix Aconiti Lateralis Praeparata (decoct first)
山药	*shān yào*	30g	Rhizoma Dioscoreae
荆芥穗	*jīng jiè suì*	8g	Spica Schizonepetae (fried)
车前子	*chē qián zǐ*	15g	Semen Plantaginis
党参	*dǎng shēn*	15g	Radix Codonopsis
柴胡	*chái hú*	9g	Radix Bupleuri
甘草	*gān cǎo*	6g	Radix Glycyrrhizae

Water was used to decoct, then it was divided into two servings, one pack per day.

Second Visit

April 15th: After taking seven packs of the above formula, the leukorrhoea had reduced, the coldness in the lower abdomen and the dripping sensation, etc. was also lighter. However, there were still low back pain, a poor appetite and stomach distention. To the above formula *shā rén* (Fructus Amomi) 8g and *lù jiǎo shuāng* (鹿角霜 , Cornu Cervi Degelatinatum) 15g were added. She continued taking it.

Third Visit

April 22nd: After taking another seven packs, all her symptoms basically disappeared. The leukorrhea had stopped, and her appetite increased. Again, she used the same formula for three more packs and recovered.

Case 2

Acute nephritic oedema.

Initial Visit

A male patient, 21 years old, October 5th, 1984.

Ten days previously, the patient had had flu, fever, headache, aversion to cold, and a sore and swollen throat. It was followed by oedema of the face, eyes and whole body for two days and so he came to visit. On observation of the patient's face, his eyes had oedema, his eyelids were swollen and he could not open his eyes, his lower limbs were also slightly swollen. He said his urine was scanty and red, he had aversion to cold, body tiredness, a poor appetite and was eating less, and he had a slight cough with thin phlegm. His pulse was superficial. His tongue coating was thin and white. A routine urine test showed: protein ++, occult blood +++, occasionally tube type particles were visible, blood pressure 150/100mm Hg. It was diagnosed as acute glomerulonephritis that is wind water syndrome in TCM, the pathogenesis is that an exterior pathogen had attacked the exterior, and the lung qi lost its dispersing function. The lung is the upper resource of water and governs the whole body's superficies. With the lung qi not dispersing, it cannot regulate the water pathway and deliver down to the urinary bladder thus inducing wind restriction and water obstruction. Wind and water mutually fight; they overflow to the skin and form oedema. The treatment plan was to extinguish wind, clear heat, disperse the lung and promote urination. A modified *Yuè Bì Jiā Zhú Tāng* (Spleen-Effusing Decoction Plus White Atractylodes, 越婢加术汤) was used for treatment.

Prescription

麻黄	*má huáng*	8g	Herba Ephedrae
苍术	*cāng zhú*	20g	Rhizoma Atractylodis
白茅根	*bái máo gēn*	30g	Rhizoma Imperatae
白术	*bái zhú*	20g	Rhizoma Atractylodis Macrocephalae
益母草	*yì mǔ cǎo*	20g	Herba Leonuri
茯苓	*fú líng*	15g	Poria
蝉衣	*chán yī*	10g	Periostracum Cicadae
金银花	*jīn yín huā*	30g	Flos Lonicerae Japonicae
车前子	*chē qián zǐ*	20g	Semen Plantaginis (wrapped up)
甘草	*gān cǎo*	6g	Radix Glycyrrhizae

Water was used to decoct, then it was divided into two servings, one pack per day.

Second Visit

October 12th: After taking seven packs of the above formula, the oedema disappeared, the routine urine test showed: Protein +, occult blood ++. The body heat and aversion to cold were also eliminated, and the amount of urination increased. To the original formula *xiān hè cǎo* (Herba Agrimoniae) 30g, and *hàn lián cǎo* (旱莲草 , Herba Ecliptae) 30g, were added. He continued to take it.

Third Visit

October 19th: After taking another seven packs, all his symptoms basically disappeared, the routine urine test was normal, and his blood pressure was 130/85mm Hg. He continued to take seven more packs to complete the treatment.

Fourth Visit

October 27th: The routine urine test was, again, normal and so the formula was stopped and he was observed. Later, at a follow up visit no reoccurrence was seen.

3. *Bái Zhú* (Rhizoma Atractylodis Macrocephalae) —— *Zhǐ Shí* (枳实, Fructus Aurantii Immaturus)

Function

To strengthen the spleen, promote movement of qi, expel nodules, eliminate fullness, and free bowel movement.

Application

This can be used for water retention or food accumulation collecting under the heart (epigastric), infecting the spleen, making it lose its transforming and transporting function inducing a focal stuffiness. It also can be used for spleen deficiency not transforming and transporting and so inducing stomach deficiency, fullness and distention. In addition, it also has treatment results for stomach prolapse, stomach and intestinal functional disorder, poor digestion and habitual constipation.

Compatibility Analysis

Bái zhú (Rhizoma Atractylodis Macrocephalae) tonifies qi, strengthens the spleen and eliminates dampness. *Zhǐ shí* (Fructus Aurantii Immaturus) promotes movement of qi, expels nodules and eliminates retention. Two herbs, one is tonifying and the other is eliminating. *Bái zhú's* (Rhizoma Atractylodis Macrocephalae) tonifying is for spleen deficiency not transforming and transporting thus inducing qi stagnation. If the spleen is strengthened, then qi stagnation can be eliminated. This is elimination while tonifying. *Zhǐ shí's* (Fructus Aurantii Immaturus) elimination is due to water retention and food stagnation in the middle *jiao* affecting the spleen's healthy transforming and transporting. If retention is gone and food eliminated, then the spleen's transforming and transporting will be healthy. This is tonifying while eliminating. The two herbs combined together can both tonify and eliminate. They can mutually enhance each other. Zhong-jing's *Zhǐ Zhú Tāng* (Unripe Bitter Orange and Atractylodes Decoction, 枳术汤) uses this compatibility, but *zhǐ shí's* (Fructus Aurantii Immaturus) dosage is twice that of *bái zhú* (Rhizoma Atractylodis Macrocephalae). This is elimination more than tonifying i.e. tonifying within elimination. Zhang Jie-gu's *Zhǐ Zhú Wán* (Unripe Bitter Orange and Atractylodes Pill, 枳术丸) also uses this compatibility, however, *bái zhú's* (Rhizoma Atractylodis Macrocephalae) dosage is twice that of *zhǐ shí's* (Fructus Aurantii Immaturus). This is tonifying more than elimination i.e. elimination within tonification. Therefore, using this compatibility in clinic, one should analyze whether it is due to spleen deficiency inducing fullness, or due to stagnation inducing deficiency; then decide the dosages of *zhǐ shí* (Fructus Aurantii Immaturus) and *bái zhú* (Rhizoma Atractylodis Macrocephalae). In ancient times, *zhǐ shí* (Fructus Aurantii Immaturus) and *zhǐ qiào* (Fructus Aurantii) were not differentiated. In current clinics, *zhǐ qiào* (Fructus Aurantii) is more often used with *bái zhú* (Rhizoma Atractylodis Macrocephalae) for patients with spleen deficiency inducing qi stagnation and the middle *jiao* fullness. *Zhǐ qiào's* (Fructus Aurantii) function is the same as *zhǐ shí* (Fructus Aurantii Immaturus), but its effect is slow and gentle. It does not consume qi or injure the middle *jiao*. Based on modern research, *zhǐ shí* (Fructus Aurantii Immaturus) and *zhǐ qiào* (Fructus Aurantii) both have the function to excite the stomach and intestinal smooth muscles, to encourage the stomach and intestines' peristalsis contraction rhythm to increase and be more forceful. *Bái zhú* (Rhizoma Atractylodis Macrocephalae) contains a volatile oil; this has the function to promote secretions in the stomach and intestines. Therefore, the two herbs combined have effective results for treatment of stomach prolapse, stomach and intestinal functional disorder, poor digestion and habitual constipation.

Case 1

Stomach prolapse.

Initial Visit

A female patient, 30 years old, May 18th, 1988.

The patient had a chronic stomach distention and fullness disease. Recently, the stomach fullness and distention had got worse and was accompanied by umbilical and abdominal pain, anorexia, lassitude, low back soreness, no strength in the knees, palpitations and insomnia, etc. A barium radiography report showed: The gastric mucosa was rough. stomach peristalsis and evacuation was slow. The bottom of the stomach was lower than the sacroiliac by 2cm. She was diagnosed with gastritis and stomach prolapse. Her pulse was deep and thready. Her tongue coating was thin and yellow. This was spleen and stomach deficiency, middle *jiao* qi deficiency and stagnation. The treatment plan was to tonify qi, strengthen the spleen, harmonise the stomach and eliminate fullness.

Prescription

枳实	*zhǐ shí*	20g	Fructus Aurantii Immaturus
白术	*bái zhú*	15g	Rhizoma Atractylodis Macrocephalae
百合	*bǎi hé*	24g	Bulbus Lilii
乌药	*wū yào*	10g	Radix Linderae
檀香	*tán xiāng*	8g	Lignum Santali Albi
砂仁	*shā rén*	4g	Fructus Amomi

丹参	*dān shēn*	20g	Radix et Rhizoma Salviae Miltiorrhizae
香附	*xiāng fù*	12g	Rhizoma Cyperi
黄芪	*huáng qí*	24g	Radix Astragali
知母	*zhī mǔ*	9g	Rhizoma Anemarrhenae
杭白芍	*háng bái sháo*	15g	Radix Paeoniae Lactiflorae
甘草	*gān cǎo*	6g	Radix Glycyrrhizae

Water was used to decoct, then it was divided into two servings, to be taken in the morning and at night, one pack per day.

Second Visit

May 27th: After continuously taking eight packs of the above formula, the stomach distention and fullness improved, her appetite increased, but her stomach, umbilicus and abdomen still had slight pain, she also sometimes had head heaviness or dizziness, and insomnia. A modification of the above formula was used.

Prescription

枳实	*zhǐ shí*	20g	Fructus Aurantii Immaturus
白术	*bái zhú*	15g	Rhizoma Atractylodis Macrocephalae
百合	*bǎi hé*	24g	Bulbus Lilii
乌药	*wū yào*	10g	Radix Linderae
檀香	*tán xiāng*	8g	Lignum Santali Albi
砂仁	*shā rén*	4g	Fructus Amomi
丹参	*dān shēn*	20g	Radix et Rhizoma Salviae Miltiorrhizae
香附	*xiāng fù*	12g	Rhizoma Cyperi
枸杞	*gǒu qǐ*	15g	Fructus Lycii
川芎	*chuān xiōng*	15g	Rhizoma Chuanxiong
夜交藤	*yè jiāo téng*	30g	Caulis Polygoni Multiflori
甘草	*gān cǎo*	6g	Radix Glycyrrhizae

Third Visit

June 1st: All her symptoms had reduced except she still had shortness of breath, a heavy sensation in her back and her pulse was still deep. To the above formula *shēng huáng qí* (raw Radix Astragali) 30g was added and she continued to take it.

Fourth Visit

June 6th: Her appetite was back to normal, and all her symptoms had reduced a lot. The patient requested to take ready made (patent) Chinese medicine instead, so she was prescribed *Rén Shēn Guī Pí Wán* (Ginseng Spleen-Returning Pill, 人参归脾丸) and *Jiā Wèi Zhǐ Zhú Wán* (Supplemented Unripe Bitter Orange and Atractylodes Pill, 加味枳术丸).

Case 2

Poor digestion, food accumulation.

Initial Visit

A female patient, 1 year old, October 11th, 1995.

The patient was a full year old. Her body was emaciated, she had no appetite, every day she only suckled her mother's milk, her hair was thin and without lustre, the index finger vein was not clearly visible, on pressing her abdomen, distention and fullness were palpable. Her mother said that the child had had good develpoment six or seven months ago and could eat some chicken's eggs, and milk, etc. but recently the child had no appetite. This was food accumulation and poor digestion syndrome. When asked about bowel movement, her mother said she went two to three times a day and it was always undigested food. The treatment plan was to strengthen the spleen, reduce food and harmonise the stomach.

Prescription

炒枳实	chǎo zhǐ shí	15g	Fructus Aurantii Immaturus (fried)
焦白术	jiāo bái zhú	15g	Rhizoma Atractylodis Macrocephalae Praeparata
焦槟榔片	jiāo bīng láng piàn	10g	Semen Arecae Praeparata (sliced)
神曲	shén qū	10g	Massa Medicata Fermentata (fried)
炒二丑	chǎo èr chǒu	10g	Semen Pharbitidis (fried)

The above formula was ground together into a fine powder, 300g of flour, two eggs, and 50g of white sugar were added with a little bit of water to mix. It was baked as a cookie shaped thin pizza and was stored in a plastic bag for the child to eat as she wished.

After one pack of the above formula, the patient recovered.

This is the author's empirical formula. In clinic, this is very effective for treating small children's food accumulation or poor digestion.

Case 3

Habitual constipation.

Initial Visit

A female patient, 35 years old, July 30th, 1999.

The patient had had habitual constipation for four to five years. There were a couple of days between bowel movements. The constipation was accompanied by slight abdominal distention. For every bowel movement she needed to take phenolphthalein for it to occur. In the beginning, it was effective, but later, after taking four tablets, she still had no bowel movement and it was very painful. The patient had a strong body, was well developed physically, and had a wiry and forceful pulse, and a white tongue coating. The treatment plan was to moisten the intestines and free the bowel movement.

Prescription

| 生白术 | shēng bái zhú | 50g | Rhizoma Atractylodis Macrocephalae (raw) |
| 枳壳 | zhǐ qiào | 15g | Fructus Aurantii |

芦荟	*lú huì*	3g	Aloe
当归	*dāng guī*	15g	Radix Angelicae Sinensis
何首乌	*hé shǒu wū*	12g	Radix Polygoni Multiflori
桃仁	*táo rén*	9g	Semen Persicae
甘草	*gān cǎo*	6g	Radix Glycyrrhizae

Water was used to decoct, then it was divided into two servings, to be taken in the morning and at night on an empty stomach, one pack per day.

Second Visit

August 6th: After two packs, the bowel movement had freed, after taking seven packs, she had one bowel movement per day accompanied by slight abdominal pain. To the original formula *háng bái sháo* (Radix Paeoniae Lactiflorae) 15g, was added to be taken again.

Third Visit

August 13th: The abdominal pain was eliminated, and her bowel movement was free, once a day. For reinforcing the effect of the treatment, the original formula dosage was increased by five times, made into powder, and honey was used to pack it as pills, each pill was 9g. She was instructed to take one in the morning, and one again in the evening.

Follow up visit:

After taking the formula, her bowel movements occurred every one to two days. She had constipation occasionally. She followed the same formula for two more packs after which, the constipation was eliminated.

Bái zhǐ's (白芷 , Radix Angelicae Dahuricae) taste and property are acrid and warm. It enters the lung channel of the foot *taiyin* and the stomach channel of the foot *yangming* meridians. It has the function to release the exterior, expel wind, dry dampness, eliminate swelling, alleviate pain and drain pus. Clinically it is more often used for a flu headache, stuffy nose, sinusitis inducing headache, white or red leukorrhoea in women, and external abscesses, swellings and sores, etc. The *Divine Husbandman's Classic of the Materia Medica* said: "*It manages red or white vaginal discharge in women, blood stagnation inducing swelling in the genital area, cold or fever, wind in the head, invasion of the eyes with tears, it generates flesh, brightens colour, and can be used as a facial make up.*" So far, in clinic, this product is more often used because its acridity can expel wind; its warmth can eliminate dampness; and its aromatic property can open orifices. This is an herb to treat wind and alleviate pain. Meanwhile, it has the function to reduce swelling and drain pus. It is also a commonly used external herb. In the *Divine Husbandman's Classic of the Materia Medica* it mainly treated red or white vaginal discharge and blood stagnation in women inducing genital area swelling syndromes that belong to cold-damp. However, it is usually not used as a main herb. In addition, because *bái zhǐ* (Radix Angelicae Dahuricae) is aromatic and can unblock the nine orifices, if used in external medicine it can increase the permeability of the orifices and improve the treatment effect.

1. *Bái Zhǐ* (Radix Angelicae Dahuricae) —— *Shí Chāng Pú* (石菖蒲, Rhizoma Acori Tatarinowii)

Function

It is aromatic and so it can transform dampness, open orifices and alleviate pain.

Application

It is used for chronic rhinitis, sinusitis, stuffy nose, runny nose and external invasion of wind-damp, a heavy head sensation, supraobital bone pain, etc.

Compatibility Analysis

Bái zhǐ (Radix Angelicae Dahuricae) is acrid and warm. It enters the foot *yangming* meridian. It is a commonly used herb for sinusitis. The *Grand Materia Medica* said *bái zhǐ* (Radix Angelicae Dahuricae) "*treats sinusitis, epistaxis, toothache, supraobital bone pain......*" *Shí chāng pú* (石菖蒲 , Rhizoma Acori Tatarinowii) is acrid, bitter and warm. Its qi is aromatic. It is also good at opening the nine orifices. It can transform phlegm and eliminate dampness. The two herbs combined together have a stronger power to permeate and open orifices. *Bái zhǐ* (Radix Angelicae Dahuricae) also can drain pus, reduce swelling and alleviate pain. *Shí chāng pú* (Rhizoma Acori Tatarinowii) also can expel dampness, transform the turbid and invigorate the brain. Pus is a damp pathogen; when pus is eliminated then dampness is eliminated; when dampness transforms then pus is eliminated. Hence, the two herbs combined can be used for sinusitis, turbid nasal discharge inducing headache with good results.

Case 1

Chronic rhinitis.

Initial Visit

A male patient, 43 years old, March, 20th, 1994.

The patient had chronic rhinitis for a couple of years. It got worse each spring and autumn season. It was better in hot weather. During an episode, there were stuffy nose, watery nasal discharge and insensitivity to smell. He had it checked at the Otolaryngology Department of a hospital and was diagnosed with chronic rhinitis. On observation, the patient had good physical development, his nasal cavity mucosa was pale with a lot of secretions. His tongue coating was white and his tongue was red with a wiry pulse. The patient complained that in recent days he had a headache, slight aversion to cold and intermittent sneezing. This was external wind-cold invasion, with the lung orifices stagnated and closed. The treatment plan was to expel wind-cold, open the orifices and disperse the lung.

Prescription

白芷	*bái zhǐ*	10g	Radix Angelicae Dahuricae
石菖蒲	*shí chāng pú*	10g	Rhizoma Acori Tatarinowii
防风	*fáng fēng*	12g	Radix Saposhnikoviae
荆芥	*jīng jiè*	10g	Herba Schizonepetae
苍耳子	*cāng ěr zǐ*	15g	Fructus Xanthii (fried)
金银花	*jīn yín huā*	20g	Flos Lonicerae Japonicae
辛夷	*xīn yí*	10g	Flos Magnoliae
藿香	*huò xiāng*	9g	Herba Agastachis
甘草	*gān cǎo*	6g	Radix Glycyrrhizae

Water was used to decoct, then it was divided into two servings, to be taken in the morning and at night, one pack per day.

Second Visit

March 26th: After taking five packs of the above formula, all the symptoms of headache, aversion to cold, etc. were already eliminated, and the stuffy nose also basically disappeared. As the patient had had the condition for many years, it was decided to regulate the whole body's immunization power as a main goal of treatment and slowly build up its resistance. The following formula was therefore prescribed:

生黄芪	*shēng huáng qí*	100g	Radix Astragali (raw)
石菖蒲	*shí chāng pú*	80g	Rhizoma Acori Tatarinowii
白芷	*bái zhǐ*	80g	Radix Angelicae Dahuricae
辛夷	*xīn yí*	60g	Flos Magnoliae
金钱草	*jīn qián cǎo*	100g	Herba Lysimachiae
党参	*dǎng shēn*	80g	Radix Codonopsis
苍耳子	*cāng ěr zǐ*	80g	Fructus Xanthii (fried)
乌梅	*wū méi*	50g	Fructus Mume
甘草	*gān cǎo*	6g	Radix Glycyrrhizae

The above herbs were all ground together into a fine powder, 6g to be taken each time, twice a day.

Note

After about a total of two months on the formula all symptoms had disappeared. The herbs were stopped and there was no reoccurrence.

Case 2

Sinusitis.

Initial Visit

A male patient, 32 years old, July 18th, 1989.

The patient had a history of headache for a year; the pain was more pronounced on his right supraobital bone and was worse in the afternoon, better in the morning, accompanied by a turbid nasal discharge and ugly fishy smell. An Otolaryngology Department diagnosed it as sinusitis. On observation the the patient had a red tongue, with a white but turbid tongue coating, a wiry, slippery and forceful pulse. His right supraobital bone was noticeably painful on pressure; his nasal cavity had a yellowish purulent fluid. Chinese medicine differentiated it as sinusitis. It was due to damp-heat stagnating and obstructing the nasal orifices. The treatment plan was to clear heat, transform dampness, aromatically open the orifices, drain pus and alleviate pain.

Prescription

金银花	jīn yín huā	30g	Flos Lonicerae Japonicae
连翘	lián qiáo	20g	Fructus Forsythiae
石菖蒲	shí chāng pú	12g	Rhizoma Acori Tatarinowii
白芷	bái zhǐ	12g	Radix Angelicae Dahuricae
生苡仁	shēng yǐ rén	30g	Semen Coicis (raw)
川芎	chuān xiōng	15g	Rhizoma Chuanxiong
蒲公英	pú gōng yīng	15g	Herba Taraxaci
地丁	dì dīng	15g	Herba Violae
苍耳子	cāng ěr zǐ	15g	Fructus Xanthii (fried)
薄荷	bò hé	9g	Herba Menthae
甘草	gān cǎo	9g	Radix Glycyrrhizae

Water was used to decoct, then it was divided into two servings, to be taken in the morning and night, one pack per day.

Second Visit

July 25th: After taking seven packs of the above formula, the headache had reduced a lot, the nasal secretion had also reduced, and his sense of smell returned. As it was effective the prescription remained unchanged and he continued taking it.

Third Visit

August 2nd: After taking another seven packs, all symptoms improved a lot. He was instructed to take a further seven packs to complete the treatment.

2. *Bái Zhǐ* (Radix Angelicae Dahuricae) ——
Dīng Xiāng (丁香, Flos Caryophylli),
Wú Zhū Yú (吴茱萸, Fructus Evodiae), *Ròu Guì* (肉桂, Cortex Cinnamomi)

Function

Warm the middle *jiao*, expel cold, regulate qi, alleviate pain, strengthen the spleen, tonify the kidney and stop diarrhoea.

Application

Used for poor digestion, abdominal pain, diarrhoea, anorexia, emaciation, etc., in young children.

Compatibility Analysis

This compatibility is for external use. Equal amounts of each are ground into a fine powder to cover the umbilicus. In the formula: *Dīng xiāng* (丁 香 , Flos Caryophylli) is acrid and warm. It enters the spleen channel of the foot *taiyin*, the stomach channel of the foot *yangming*, the lung channel of the hand *taiyin* and the kidney channel of the foot *shaoyin* meridians. It is a commonly used herb to warm the middle *jiao*, expel cold and alleviate pain. The *Materia Medica of the Kaibao Era* (开宝本草 , *Kāi Bǎo Běn Cǎo*) said: "*It warms the spleen, stops cholera......*" and the *Grand Materia Medica* said: "*It treats deficiency inducing nausea and vomiting, vomiting and diarrhoea in young children.......*" *Wú zhū yú* (Fructus Evodiae) is also a kind of acrid and warm herb. It enters the spleen channel of the foot *taiyin*, the stomach channel of the foot *yangming*, the liver channel of the foot *jue yin* and the kidney channel of the foot *shaoyin* meridians. It can warm the middle *jiao*, eliminate stagnation and alleviate pain. The *Divine Husbandman's Classic of the Materia Medica* said: "*It governs warming the middle jiao, descending qi, alleviating pain.......*" *Ròu guì* (Cortex Cinnamomi) is sweet, acrid and very hot. It enters the liver channel of the foot *jueyin* and the kidney channel of the foot *shaoyin* meridians. It functions to warm the kidney and tonify fire, expel cold and alleviate pain. It is a chief herb to tonify the lower *jiao mingmen* (life-gate) fire. Wang Hao-gu said that it: "*tonifies mingmen (life-gate) deficiency, and tonifies fire to eliminate yin pathogens*". The above three herbs all function to warm the middle *jiao* to expel cold and can promote the spleen and stomach qi mechanism. Meanwhile, *ròu guì* (Cortex Cinnamomi) can also warm the lower *jiao* to enhance the middle *jiao* qi's transforming and fermenting functions of the spleen and stomach. Hence, any spleen or kidney deficiency could induce abdominal pain, diarrhoea, anorexia, poor digestion, etc., so all can use this compatibility with effective results. As all three herbs are acrid and warm, they easily injure yin and consume qi, so they are recommended for external use only, especially in young children, and so it was changed to a medicinal for external use. Aromatic herbs that open orifices such as *bái zhǐ* (Radix Angelicae Dahuricae) are added to increase the other herbs permeability, in order to assist the herbs to penetrate skin more easily for a more effective treatment. Meanwhile, *bái zhǐ* (Radix Angelicae Dahuricae) is also acrid and warm. It functions to dry dampness and expel cold. For treating young children's diarrhoea and abdominal pain it can also have a certain effect. All of the above herbs put together achieve functions to warm the middle *jiao*, expel cold, regulate qi, alleviate pain, strengthen the spleen, tonify the kidney and stop diarrhoea. It is more easily accepted by the sick children and their family members, because it is for external use. Moreover, there are no side effects.

Case 1

Diarrhoea.

Initial Visit

A male patient, 1 year old, September 20th, 1987.

The sick child had abdominal distention and diarrhoea for a week and so came to visit. His mother said he had diarrhoea for a week, with a watery stool, abdominal distention and fullness, and no appetite. He had taken polyzyme tablets and lactasin, etc. but they did not work. More recently he had started vomiting food and milk. On observation the the sick child appeared dispirited, his abdomen was distended and full like a drum, a routine stool test did not seem abnormal, occasionally a lipid drop could be seen, a routine blood test was normal. His tongue was pale, his tongue coating was white, and his index finger vein was dark. This was the syndrome of spleen and stomach deficiency cold, with disorder in transforming and transporting. The treatment plan was to tonify the kidney, strengthen the spleen, warm the middle *jiao* and stop diarrhoea. It was considered inconvenient to encourage a young child to take a decoction orally, and so external medicine was used for treatment instead.

Prescription

白芷	*bái zhǐ*	20g	Radix Angelicae Dahuricae
丁香	*dīng xiāng*	20g	Flos Caryophylli
吴茱萸	*wú zhū yú*	20g	Fructus Evodiae
肉桂	*ròu guì*	20g	Cortex Cinnamomi

The above herbs were all ground together into a fine powder, each time 10g were administered. Warm water was used to make it into a paste, which was used to cover the child's umbilicus area (RN8, *shén què*), with gauze or cotton tape put over the top. The dressing was changed once a day.

Second Visit

September 25th: The above formula was applied for three days, after which the diarrhoea stopped, the abdominal distention disappeared and he could eat. Five days later, all the symptoms disappeared. The family asked if continuous treatment would be required. They were instructed to take two more days formula then stop it and observe. Later it was discovered that the child had recovered.

Case 2

Aversion to food.

Initial Visit

A male patient, 3 years old April 15th, 1990.

The patient was emaciated and had an aversion to food for half a year. His mother complained that he had no appetite every day, if he saw food, he would try to avoid eating it. At meal times she forced him to eat, but he only ate a couple of bits. On observation the the child was emaciated, his spirit was fine, he had a red tongue, with a white slightly turbid tongue coating. This was the children's aversion to food syndrome. From syndrome differentiation: Young children on an irregular diet, injure the spleen and stomach. When the middle *jiao* spleen and stomach are injured, then they cannot transform and transport water and food and so induce food stagnation internally. The treatment plan was to strengthen the spleen, reduce food intake, regulate qi and break stagnation.

Internal use formula:

炒枳实	*chǎo zhǐ shí*	20g	Fructus Aurantii Immaturus (fried)
焦白术	*jiāo bái zhú*	20g	Rhizoma Atractylodis Macrocephalae Praeparata
焦槟榔片	*jiāo bīng láng piàn*	15g	Semen Arecae Praeparata (sliced)
神曲	*shén qū*	16g	Massa Medicata Fermentata
炒二丑	*chǎo èr chǒu*	10g	Semen Pharbitidis (fried)

The above herbs were all ground together into a fine powder, flour 400g, two eggs, white sugar 50g, were added with a little amount of water, it was mixed and baked as thin biscuit to let him eat it as he wished. (For this formula explanation, please refer to *bái zhú* (Rhizoma Atractylodis Macrocephalae).

External use formula:

白芷	*bái zhǐ*	50g	Radix Angelicae Dahuricae
丁香	*dīng xiāng*	50g	Flos Caryophylli
吴茱萸	*wú zhū yú*	50g	Fructus Evodiae
肉桂	*ròu guì*	50g	Cortex Cinnamomi

The above herbs were all ground together into a fine powder, each time 10g were used mixed with warm water to form a glue, then applied to the umbilicus, and dressed with gauze to cover it, and cotton tape to hold it in place. Every night, before sleep, the dressing was placed on the umbilicus. The next morning, it was removed after he had woken up.

Second Visit

May 6th: After dressing the umbilicus externally for twenty days and taking one pack of the cookies internally his appetite increased, and his tongue coating had already turned thin. He was instructed to continue with the same formula for treatment again.

Note

After taking two packs, all symptoms were eliminated, after stopping the herbs, his appetite was normal. An alternative method that could have been used is to put the powdered medicine into a cloth bag and to put it on the umbilicus and use gauze cloth to wrap and fix it in place, also using 10g each time. The treatment will be just as effective. This method is better than the previous one as it does not contaminate clothes and bedding.

Bái Jiè Zǐ (白芥子 , Sinapis Semen)

Bái jiè zǐ's (白芥子 , Sinapis Semen) property and taste are acrid and warm. It enters the lung channel of the hand *taiyin* meridian. It functions to warm the lung, expel phlegm, reduce swelling and alleviate pain. It can be used for cold phlegm inducing cough or asthma, a lot of phlegm and rapid breathing, as well as external ulcers flowing toward becoming abscesses. Zhu Dan-xi said: *"If phlegm is in the hypochondria and between skin and the outside membrane, only bái jiè zǐ (Sinapis Semen) can reach it."* Miu Zhong-chun said: *"It can search and expel internal and external phlegm nodules. It is especially effective for patients with chest and diaphragm cold phlegm, and cold saliva stagnating and obstructing."* In addition, *bái jiè zǐ* (Sinapis Semen) is a foaming agent. When used externally it can cause redness in the local skin, a burning heat and induce bubbles, therefore it can sterilise and alleviate pain.

1. *Bái Jiè Zǐ* (Sinapis Semen) ——
Dì Lóng (地龙, Pheretima)

Function

To invigorate blood and unblock collaterals, to clear phlegm and promote speaking.

Application

This is used for post stroke sequelae with influent talk.

Compatibility Analysis

Bái jiè zǐ (Sinapis Semen) is acrid and warm. It can clear and expel internal and external phlegm nodules. The *Essentials of the Materia Medica* said that *bái jiè zǐ* (Sinapis Semen) can *"promote free movement in the meridians and the collaterals"*. Hence, for the phlegm nodules in the meridians or collaterals, *bái jiè zǐ* (Sinapis Semen) has the function to clear and expel them. The taste and property of *dì lóng* (Pheretima) are salty and cold. It is an important herb to invigorate the collaterals and unblock the meridians. Its property is freeing and good at reaching all of the limbs and bones. Wang Qing-ren used *dì lóng* (Pheretima) with *huáng qí* (Radix Astragali) to treat paralysis after stroke. It uses its characteristic of freeing and unblocking collaterals. *Bái jiè zǐ* (Sinapis Semen) combined with *dì lóng* (Pheretima), the power to free and unblock the collaterals is even stronger. *Bái jiè zǐ* (Sinapis Semen) relies on the unblocking the collaterals' property of *dì lóng* (Pheretima) to break phlegm nodules in the meridians and collaterals. *Dì lóng* (Pheretima) relies on the ability to dredge and expel phlegm of *bái jiè zǐ* (Sinapis Semen) to enhance its unblocking meridians and invigorating collaterals power. Moreover, the acridity and warmth of *bái jiè zǐ* (Sinapis Semen) can reduce the effect of the salty coldness of *dì lóng* (Pheretima) inducing stagnation. The coldness of *dì lóng* (Pheretima) also can restrict the acridity and heat of *bái jiè zǐ* (Sinapis Semen) to make the compatibility balanced and the effectiveness more remarkable. The author uses it to treat the sequelae of post stroke. It is an effective treatment for phlegm and blood stasis obstructing and stagnating the orifices and collaterals inducing influent speaking.

Case Study

Initial Visit

A male patient, 67 years old, March 5th, 2000.

The patient suddenly had a stroke after the previous autumn and was left with hemiplasia. He was investigated at the local city hospital: A CT scan showed mild brain atrophy and multiple-infarct dementia. After treatment, his situation improved. He could cope with his life but his influent speaking did not improve and so he came to visit. On observation, the patient was physically well developed, his mind was clear, walking and activities were fine except for reduced strength with a slight slow movement in his right upper and lower limbs, his tongue was stiff causing the influent speaking. His pulse was wiry and forceful. His tongue was red and tongue coating was thin and white. The patient complained of dizziness and a numb sensation in his limbs; his blood pressure was 200/140mm Hg. This was the sequelae of stroke, phlegm and blood stasis obstructing and stagnating in the orifices and collaterals inducing influent speaking, also with liver yang rising inducing dizziness. The treatment plan was to use a formula to calm the liver and subdue the yang, invigorate blood and unblock the collaterals, clear phlegm and promote speaking.

Prescription

地龙	*dì lóng*	20g	Pheretima
天冬	*tiān dōng*	15g	Radix Asparagi
杭白芍	*háng bái sháo*	30g	Radix Paeoniae Lactiflorae
川牛膝	*chuān niú xī*	20g	Radix Cyathulae
白芥子	*bái jiè zǐ*	10g	Sinapis Semen
代赭石	*dài zhě shí*	20g	Haematitum
全蝎	*quán xiē*	6g	Scorpio
天麻	*tiān má*	15g	Rhizoma Gastrodiae
川军	*chuān jūn*	8g	Radix et Rhizoma Rhei
桑枝	*sāng zhī*	30g	Ramulus Mori
鸡血藤	*jī xuè téng*	30g	Caulis Spatholobi
甘草	*gān cǎo*	6g	Radix Glycyrrhizae

Water was used to decoct, then it was divided into two servings, one pack per day.

Second Visit

March 14th: After taking fourteen packs of the above formula, his symptoms had reduced and his blood pressure was 180/100mm Hg. To the above formula *chì sháo* (Radix Paeoniae Rubra) 15g, and *guì zhī* (Ramulus Cinnamomi) 10g, were added, and *zhě shí* (Haematitum) was eliminated and he continued to take it.

Third Visit

April 11th: His speech was remarkably better. He, himself, also felt his speech was not as difficult as before, his headache, dizziness, and limb numbness also reduced a lot.

In consideration of the limitations of farm life, the formula was changed to a powdered medicine to be taken continuously.

Prescription

丹参	*dān shēn*	100g	Radix et Rhizoma Salviae Miltiorrhizae
葛根	*gé gēn*	100g	Radix Puerariae Lobatae
赤芍	*chì sháo*	80g	Radix Paeoniae Rubra
地龙	*dì lóng*	80g	Pheretima
白芥子	*bái jiè zǐ*	40g	Sinapis Semen
桃仁	*táo rén*	60g	Semen Persicae
红花	*hóng huā*	60g	Flos Carthami
全蝎	*quán xiē*	30g	Scorpio
天麻	*tiān má*	40g	Rhizoma Gastrodiae
杭白芍	*háng bái sháo*	100g	Radix Paeoniae Lactiflorae
川牛膝	*chuān niú xī*	100g	Radix Cyathulae
甘草	*gān cǎo*	30g	Radix Glycyrrhizae
三七粉	*sān qī fěn*	40g	Radix et Rhizoma Notoginseng Powder

The above herbs were all ground together into a fine powder, each time 5g were taken, twice a day.

At a follow up visit half a year later, the patient was stable, with no reoccurrence of dizziness and limb numbness, and his speaking was basically normal.

Bái huā shé shé cǎo's (白花蛇舌草 , Herba Hedyotis Diffusa) taste and property are sweet, bland and slightly cold. It enters the liver channel of the foot *jueyin* and the spleen channel of the foot *taiyin* meridians. It has the function to clear heat and detoxify toxins, expel stasis and reduce swellings. Clinically it is more often used for upper respiratory tract infections inducing lung heat cough and asthma, tonsillitis, pharyngolaryngitis, appendicitis, enteritis, adnexitis, pelvic inflammatory disease, etc. Based on clinical reports, this herb also has an anti-viral and anti-cancer function. It is also used for viral hepatitis and all kinds of cancers.

1. *Bái Huā Shé Shé Cǎo* (Herba Hedyotis Diffusae) —— *Pú Gōng Yīng* (蒲公英, Herba Taraxaci)

Function

To clear heat, detoxify toxins, protect the liver and benefit the gallbladder.

Application

This is used for all kinds of viral hepatitis, and carriers of the hepatitis B virus. It is also used for cholecystitis and upper respiratory tract infections.

Compatibility Analysis

Bái huā shé shé cǎo (白花蛇舌草 , Herba Hedyotis Diffusae) is sweet, bland and slightly cold. Though it is a herb that clears heat and detoxifies toxins, its property is gentle. Within its clearing and detoxifying, it also can invigorate blood and expel stasis. Modern medical research discovered that *bái huā shé shé cǎo* (Herba Hedyotis Diffusae) can stimulate lattice endothelial cell proliferation, enhance the power of phagocytes to ingest cells, and increase an organism's non-specificity immunity function. In addition, it functions well to protect liver cells. *Pú gōng yīng* (Herba Taraxaci) is sweet, bitter and cold. It is also a herb that clears heat and detoxifies toxins. Based on reports, it functions very well to protect the liver and benefit the gallbladder. *Pú gōng yīng* (Herba Taraxaci) decoction or injection for treating acute jaundice hepatitis and non jaundice hepatitis has a remarkable promotional effect on the liver function and on the jaundice index restoration. The two herbs' medicinal properties are gentle and the taste is sweet, hence, they do not injure the upright qi when clearing heat and detoxifying toxins. The two herbs both function to invigorate blood and expel stasis, therefore, they are suitable for all kinds of viral hepatitis, and also for patients with chronic hepatitis with cholecystitis.

Case Study

Initial Visit

A male patient, 39 years old. May 28th, 1991.

The patient was admitted to a county hospital due to abnormal liver function three months previously. He was transferred to another infection hospital as there were no great treatment results from the first one. Later, he was referred by his friend to come to our hospital for treatments. Immediate assessments showed: Liver function: Jaundice index 20u, serum thymol turbidity test 10.1u, ALT 225 u/L, Hbs Ag (+). The patient had a history of gastric and duodenal ulcer along with chronic cholecystitis. Currently he had symptoms of lassitude, anorexia, distention after eating, liver area discomfort, etc. His pulse was deep, thready and wiry. His tongue was dark and his tongue coating was slightly turbid. The treatment plan was to invigorate blood, transform stasis, soothe and tonify the liver, detoxify toxins and reduce transaminase. A modified *Gān Níng Tāng* (Liver-Quietening Decoction, 肝宁汤), one of the author's own formulae was used for treatment.

Prescription

白花蛇舌草	*bái huā shé shé cǎo*	30g	Herba Hedyotis Diffusae
蒲公英	*pú gōng yīng*	15g	Herba Taraxaci
丹参	*dān shēn*	20g	Radix et Rhizoma Salviae Miltiorrhizae
柴胡	*chái hú*	15g	Radix Bupleuri
泽兰	*zé lán*	15g	Herba Lycopi
虎杖	*hǔ zhàng*	15g	Rhizoma et Radix Polygoni Cuspidati
枳壳	*zhǐ qiào*	12g	Fructus Aurantii
白术	*bái zhú*	12g	Rhizoma Atractylodis Macrocephalae
太子参	*tài zǐ shēn*	15g	Radix Pseudostellariae
生苡仁	*shēng yǐ rén*	24g	Semen Coicis (raw)
茵陈	*yīn chén*	15g	Herba Artemisiae Scopariae
茯苓	*fú líng*	15g	Poria
金钱草	*jīn qián cǎo*	20g	Herba Lysimachiae
甘草	*gān cǎo*	6g	Radix Glycyrrhizae

Water was used to decoct, then it was divided into two servings, one pack per day.

Second Visit

June 11th: After taking fourteen packs of the above formula, his symptoms reduced a little bit, he still had distention after eating, anorexia, a constipated bowel movement, and yellowish urine. *Dà huáng* (Radix et Rhizoma Rhei) 8g was added and was taken continuously.

Third Visit

June 20th: After taking the above formula, one pack per day, all symptoms had reduced a lot, his appetite increased, his bowel movement was free and smooth, his urine turned clear. The original formula was modified and he continued treatment.

Prescription

白花蛇舌草	*bái huā shé shé cǎo*	30g	Herba Hedyotis Diffusae
蒲公英·	*pú gōng yīng*	15g	Herba Taraxaci
丹参	*dān shēn*	20g	Radix et Rhizoma Salviae Miltiorrhizae
柴胡	*chái hú*	15g	Radix Bupleuri
泽兰	*zé lán*	15g	Herba Lycopi

虎杖	*hǔ zhàng*	15g	Rhizoma et Radix Polygoni Cuspidati
五味子	*wǔ wèi zǐ*	8g	Fructus Schisandrae Chinensis
枳壳	*zhǐ qiào*	12g	Fructus Aurantii
白术	*bái zhú*	12g	Rhizoma Atractylodis Macrocephalae
砂仁	*shā rén*	6g	Fructus Amomi
金钱草	*jīn qián cǎo*	20g	Herba Lysimachiae
甘草	*gān cǎo*	6g	Radix Glycyrrhizae

Water was used to decoct to be taken as before.

Fourth Visit

July 3rd: The liver function tested normal, ALT 36 u/L, and all the clinical symptoms basically disappeared. The herbal formula was changed to be taken every other day and continuously for a month, then return for a check up.

Fifth Visit

August 2nd: The liver function tested normal, Hbs Ag (−).

Note

For composition, formula explanation and clinical modification of *Gān Níng Tāng* (Liver-Quietening Decoction), please refer to the *zé lán* (Herba Lycopi) and *hǔ zhàng* (虎杖, Rhizoma et Radix Polygoni Cuspidati) compatibility.

Bái qián (白前 , Rhizoma et Radix Cynanchi Stauntonii) is acrid, bitter and slightly warm. It enters the lung channel of the hand *taiyin* meridian. It functions to expel phlegm, descend qi and stop cough. In clinic, it can be used for lung qi stagnation and excess, increased phlegm with cough discomfort, and rebellious qi rapidly inducing asthma. It is a commonly used herb to stop cough, expel phlegm, disperse the lung and descend qi. Modern research data has shown that *bái qián* (Rhizoma et Radix Cynanchi Stauntonii) contains saponin and has the function to expel phlegm.

1. *Bái Qián* (Rhizoma et Radix Cynanchi Stauntonii) —— *Qián Hú* (前胡, Radix Peucedani)

Function

To descend qi, disperse the lung, expel phlegm and stop cough.

Application

This is used for all syndromes regardless of whether it is due to cold, heat, deficiency or excess, if it is a patient with cough with a lot of phlegm, stuffy chest and rebellious qi it can be applied.

Compatibility Analysis

Bái qián (Rhizoma et Radix Cynanchi Stauntonii) is acrid, bitter and slightly warm. It stops cough, descends qi, and favours treating cold phlegm. *Qián Hú* (前胡 , Radix Peucedani) is acrid, bitter and slightly cold. It also has functions to stop cough, transform phlegm and descend qi but favours expelling hot phlegm. The two herbs' functions are similar, however, one of them is warm and the other is cold; their properties are opposites. The two herbs combined together can neutrally regulate cold and heat as well as enhance their functions to stop cough, transform phlegm, disperse the lung and descend qi, therefore regardless of whether it is cold or heat inducing the cough with phlegm both can be treated. Yue Mei-zhong thought: *"Bái qián (Rhizoma et Radix Cynanchi Stauntonii) can clear the phlegm from the small bronchi."* and *"Qián Hú (Radix Peucedani) can dredge out the phlegm in the large bronchi."* Hence, both herbs used together can treat many kinds of cough with phlegm symptoms.

Case Study

Initial Visit

A male patient, 32 years old, driver, September 11th, 1995.

The patient caught the flu, and had a cough and fever more than twenty days ago. After treatment, the fever had gone but the cough persisted. He had taken anti-inflammatory medicine and patent Chinese medicine for stopping cough and expelling phlegm but they had no effect. Now, he had cough, asthma, stuffiness in the chest, a lot of white and slightly sticky phlegm, at night time the cough got worse. His tongue coating was white and turbid with a dark tongue. His pulse was wiry and slippery. This was phlegm and dampness obstructing internally, with lung qi not dispersing. The treatment plan was to stop cough, expel phlegm, and disperse the lung and calm asthma.

Prescription

百部	*bǎi bù*	15g	Radix Stemonae
桔梗	*jié gěng*	10g	Radix Platycodonis
陈皮	*chén pí*	9g	Pericarpium Citri Reticulatae
白前	*bái qián*	12g	Rhizoma et Radix Cynanchi Stauntonii
前胡	*qián hú*	12g	Radix Peucedani
川贝	*chuān bèi*	10g	Bulbua Fritillariae Cirrhosae
杏仁	*xìng rén*	10g	Semen Armeniacae Dulcis
当归	*dāng guī*	15g	Radix Angelicae Sinensis
茯苓	*fú líng*	20g	Poria
半夏	*bàn xià*	9g	Rhizoma Pinelliae
蝉衣	*chán yī*	10g	Periostracum Cicadae
甘草	*gān cǎo*	6g	Radix Glycyrrhizae

Water was used to decoct, then it was divided into two servings, served warm, one pack per day.

Second Visit

September 15th: After continuously taking four packs, the cough and phlegm improved, but the asthma was still heavy. To the original formula *zhì má huáng* (processed Herba Ephedrae) 9g and *bái guǒ* (白果 , Semen Ginkgo) 9g, were added to be taken continuously.

Third Visit

September 20th: After taking a further five packs, the cough and asthma basically calmed, the phlegm also reduced, and the tongue coating turned a thin white. Three packs of the original formula were used to reinforce the results.

Bái Máo Gēn (白茅根 , Rhizoma Imperatae)

Bái Máo Gēn's (白茅根 , Rhizoma Imperatae) taste and property are sweet and cold. It enters the heart channel of the hand *shaoyin*, the lung channel of the hand *taiyin* and the stomach channel of the foot *yangming* meridians. It has functions to sedate fire and generate fluid, cool blood and stop bleeding. It is a gentle and commonly used herb. It can treat syndromes of lung heat inducing cough, epistaxis, haemoptysis, haematuria, oedema, etc.

1. *Bái Máo Gēn* (Rhizoma Imperatae) ——
Cè Bǎi Yè (侧柏叶, Cacumen Platycladi)

Function
To clear the lung and stop cough.

Application
It is mainly used to treat lung heat induced cough.

Compatibility Analysis
Bái máo gēn (白茅根 , Rhizoma Imperatae) is sweet and cold. It functions to sedate fire, clear heat and nourish yin. The *Grand Materia Medica* said that *bái máo gēn* (Rhizoma Imperatae) treats *"lung heat inducing rapid asthma"*. As it functions to nourish yin and generate fluid, it is the best suited to treat young children's lung heat induced cough. *Cè bǎi yè* (Cacumen Platycladi) is bitter, and slightly cold. It functions to cool blood and stop bleeding. The author in clinic uses it to treat lung heat inducing cough or persisitent chronic cough. *Cè bǎi yè* (Cacumen Platycladi) treats cough and asthma in a Korean folk empirical formula. As its property is astringent, it functions to retain the lung and stop coughing. The *Essentials of the Materia Medica* said *cè bǎi yè* (Cacumen Platycladi) has a tonifying lung function. For clinical practice, it is peaceful and effective. It is especially used with *bái máo gēn* (Rhizoma Imperatae) as they have the function to clear the lung, nourish yin, retain the lung and stop coughing. It is especially good for children's upper respiratory tract infections inducing fever and cough. For lung heat inducing haemoptysis, it is also good.

Case 1
External invasion inducing lung heat and cough.

Initial Visit
A male patient, 3 years old, September 7th, 1985.

The patient had an external invasion inducing lung heat and cough for a week. After treatment, the fever reduced but the cough did not; he had a dry cough without phlegm which was worse at night time. The breathing sounds from both lungs were rough. The index finger vein was floating with a red purplish colour. His tongue coating was thin and white. This was wind-heat attacking the lung, with lung heat stagnating inside and losing its descending function causing the cough. The treatment plan was to clear heat, moisten the lung and stop the cough.

Prescription

白茅根	*bái máo gēn*	15g	Rhizoma Imperatae
侧柏叶	*cè bǎi yè*	9g	Cacumen Platycladi
桑叶	*sāng yè*	10g	Folium Mori
杏仁	*xìng rén*	6g	Semen Armeniacae Dulcis
川贝	*chuān bèi*	6g	Bulbua Fritillariae Cirrhosae
桔梗	*jié gěng*	4g	Radix Platycodonis
陈皮	*chén pí*	6g	Pericarpium Citri Reticulatae
百部	*bǎi bù*	6g	Radix Stemonae
甘草	*gān cǎo*	3g	Radix Glycyrrhizae

Water was used to decoct and reduced to 150ml, which was divided into three servings.

Second Visit

September 11th: After taking three packs of the above formula, the cough stopped, he took another two packs and no cough reoccurred and he recovered.

Case 2

Bronchiectasis and haemoptysis.

Initial Visit

A male patient, 54 years old, March 26th, 1991.

The patient had a history of bronchiectasis and haemoptysis. Recently, due to flu, he coughed persistently for half a month. The day before his visit he had coughed out phlegm with blood. The colour of the blood was a fresh red; accompanied by symptoms of palpitations, shortness of breath, etc. A routine blood test was normal. A chest x-ray showed: Both lungs' interstitial markings had increased and were thicker, but signs of tuberculosis were not seen. His tongue was red, and his tongue coating was white. His pulse was wiry and fast. This was the aftermath of an external invasion; lung heat had stagnated inside, the stagnated heat pushed blood to flow restlessly and induced haemoptysis. The treatment plan was to clear heat, stop cough, retain the lung and stop bleeding.

Prescription

百部	*bǎi bù*	15g	Radix Stemonae
侧柏叶	*cè bǎi yè*	15g	Cacumen Platycladi
白茅根	*bái máo gēn*	24g	Rhizoma Imperatae
紫菀	*zǐ wǎn*	10g	Radix et Rhizoma Asteris
桔梗	*jié gěng*	15g	Radix Platycodonis
陈皮	*chén pí*	9g	Pericarpium Citri Reticulatae
大贝	*dà bèi*	10g	Bulbus Frittilariae Thunbergii

生龙骨	*shēng lóng gǔ*	15g	Os Draconis (raw)
杏仁	*xìng rén*	9g	Semen Armeniacae Dulcis
丹皮	*dān pí*	10g	Cortex Moutan
五味子	*wǔ wèi zǐ*	6g	Fructus Schisandrae Chinensis
麦冬	*mài dōng*	10g	Radix Ophiopogonis
地榆	*dì yú*	20g	Radix Sanguisorbae
甘草	*gān cǎo*	10g	Radix Glycyrrhizae

Water was used to decoct, one pack was taken per day.

Second Visit

April 1st: After taking six packs of the above formula, the cough had reduced a lot, the haemoptysis was also better, although he occasionally expectorated phlegm with pink silk like blood. He continued taking the original formula.

Third Visit

April 6th: After a further five packs, the haemoptysis stopped and all symptoms were eliminated. The patient was instructed to take two more packs of the same formula to strengthen the effectiveness of treatment.

2. *Bái Máo Gēn* (Rhizoma Imperatae) —— *Yì Mǔ Cǎo* (益母草, Herba Leonuri)

Function

To promote urination, to reduce swelling.

Application

This is used for oedema due to acute or chronic nephritis, haematuria, proteinuria and hypertension.

Compatibility Analysis

Bái máo gēn (Rhizoma Imperatae) is sweet and cold. It can clear heat, promote urination and reduce swelling. It is the only herb that, alone, is effective for oedema due to acute nephritis. Based on modern pharmacological research, this herb contains abundant sucrose, glucose and sylvite, therefore, it has a strong diuretic function. Although it can promote urination and reduce swelling, it does not injure yin. *Yì mǔ cǎo* (Herba Leonuri) is also called *kūn cǎo* (坤草). It is acrid, bitter and slightly cold. It is a good herb for gynaecology to regulate menstruation and invigorate blood. It also can promote urination and reduce swelling. In recent years, it has been used to treat oedema due to acute or chronic nephritis. It has the effect to increase the amount of bowel and urine excretion, and quickly reduces oedema. It also can reduce blood pressure. Meanwhile, *yì mǔ cǎo* (Herba Leonuri) invigorates blood and transforms stasis. It also can improve the kidney organ's blood circulation and enhance recovery from nephritis. The two herbs combined achieve a mutually enhancing purpose, they can prompt elimination of oedema and improve the kidney's function, meanwhile, they can reduce swelling and promote urination without injury to yin.

Case 1

Initial Visit

A male patient, 16 years old, March 12th, 1987.

The patient suddenly had facial oedema ten days ago and his eyelids were the most swollen. His urine colour was red, like strong tea. His throat was red and painful. A local doctor treated it as acute nephritis and gave him penicillin, but did not see a remarkable effect and so came to visit. A routine urine test reported: Protein ++++, WBC +, occult blood +++, an haematological investigation showed: RBC ++, tube type +, blood pressure 150/95mm Hg. His tongue was red and tongue coating was thin white and slightly yellowish. His pulse was superficial, slippery and wiry. Diagnosis: Acute renal glomerulonephritis. Syndrome differentiation in TCM: Damp-heat stagnation and obstruction, water and fluid could not transform and transport, so overflowed outward to the muscle and skin, hence, the oedema and red urine. The treatment plan was to clear heat, resolve dampness and reduce swelling.

Prescription

白茅根	*bái máo gēn*	50g	Rhizoma Imperatae
益母草	*yì mǔ cǎo*	30g	Herba Leonuri
泽泻	*zé xiè*	20g	Rhizoma Alismatis
茯苓	*fú líng*	20g	Poria
车前子	*chē qián zǐ*	20g	Semen Plantaginis (wrapped up)
猪苓	*zhū líng*	20g	Polyporus
蒲公英	*pú gōng yīng*	15g	Herba Taraxaci
竹叶	*zhú yè*	9g	Herba Lophatheri
大腹皮	*dà fù pí*	15g	Pericarpium Arecae
苏叶	*sū yè*	10g	Folium Perillae
甘草	*gān cǎo*	6g	Radix Glycyrrhizae
蝉蜕	*chán tuì*	9g	Periostracum Cicadae

Water was used to decoct, then it was divided into two servings, one pack per day.

Second Visit

March 20th: After taking seven packs of the above formula, the swelling visibly reduced and the routine urine test showed: Protein ++, WBC +, occult blood ++, haematological investigation showed: WBC ++. To the above formula *chuān niú xī* (Radix Cyathulae) 15g, was added, he continued to take the formula.

Third Visit

March 27th: After taking another seven packs, the swelling was eliminated, his blood pressure was 125/70mm Hg. The routine urine test showed: Protein +, occult blood +, WBC +; his throat was still red, swollen and painful. To the last formula *bǎn lán gēn* (Radix Isatidis) 20g, was added.

Fourth Visit

April 9th: After taking twelve packs of the above formula, all his symptoms basically disappeared. A routine urine test showed: Protein (-), an haematological investigation showed: Few RBCs. He was instructed to rest as often as possible. He was then prescribed the following formula to regulate the effects.

Prescription

白茅根	bái máo gēn	50g	Rhizoma Imperatae
益母草	yì mǔ cǎo	30g	Herba Leonuri
茯苓	fú líng	15g	Poria
泽泻	zé xiè	15g	Rhizoma Alismatis
车前子	chē qián zǐ	15g	Semen Plantaginis (wrapped up)
地榆	dì yú	10g	Radix Sanguisorbae
枸杞	gǒu qǐ	12g	Fructus Lycii
蒲公英	pú gōng yīng	20g	Herba Taraxaci
蝉蜕	chán tuì	9g	Periostracum Cicadae
白术	bái zhú	9g	Rhizoma Atractylodis Macrocephalae
甘草	gān cǎo	6g	Radix Glycyrrhizae

Water was used to decoct. After taking more than twenty packs of the above formula, all tests were normal.

Case 2

Pyelonephritis.

Initial Visit

A female patient, 3 years old, August 5th, 1987.

The patient had a urinary system infection for a couple of months. A routine urine test showed: Protein (+), RBC (+), a few pus cells, and a urine culture found an E. coli infection. After treatment with Chinese medicine and biomedicine it improved, except that urine protein (+), RBC (+), and she still had frequent urination. Her tongue coating was thin and white, and her tongue was a pale red. From syndrome differentiation: Small children are little yang bodies. It is easy to become deficient and easy to become excessive. Prolonged infection consumed the upright qi. The kidney is the root of pre-heaven and the spleen is the root of post-heaven qi, therefore the spleen and kidney must be blamed first. The kidney governs storing essence; and the spleen governs containing essence. If both the spleen and kidney are deficient, the storing and containing functions are in disorder, hence her urine protein took a long time to reduce. The treatment plan was to use a formula to tonify the kidney, strengthen the spleen and promote urination.

Prescription

白茅根	bái máo gēn	40g	Rhizoma Imperatae
益母草	yì mǔ cǎo	20g	Herba Leonuri
山药	shān yào	12g	Rhizoma Dioscoreae

茯苓	*fú líng*	12g	Poria
生地	*shēng dì*	12g	Radix Rehmanniae
山萸肉	*shān yú ròu*	9g	Fructus Corni
泽泻	*zé xiè*	9g	Rhizoma Alismatis
丹皮	*dān pí*	9g	Cortex Moutan
附片	*fù piàn*	4g	Radix Aconiti Lateralis Praeparata (sliced)
瞿麦	*qú mài*	9g	Herba Dianthi
甘草	*gān cǎo*	6g	Radix Glycyrrhizae

Water was used to decoct, then it was divided into two servings, to be taken in the morning and at night, one pack per day.

Second Visit

August 11th: After taking six packs of the above formula, the symptoms were as before, she still had frequent urination, and there was no change in her routine urine test. It was decided that the power to strengthen the spleen and tonify qi was weak in the formula, so it was changed to the following:

Prescription

生黄芪	*shēng huáng qí*	15g	Radix Astragali (raw)
知母	*zhī mǔ*	6g	Rhizoma Anemarrhenae
山药	*shān yào*	15g	Rhizoma Dioscoreae
茯苓	*fú líng*	12g	Poria
白茅根	*bái máo gēn*	24g	Rhizoma Imperatae
益母草	*yì mǔ cǎo*	9g	Herba Leonuri
蝉蜕	*chán tuì*	9g	Periostracum Cicadae
苏叶	*sū yè*	9g	Folium Perillae
益智仁	*yì zhì rén*	9g	Fructus Alpiniae Oxyphyllae
甘草	*gān cǎo*	6g	Radix Glycyrrhizae

Third Visit

August 15th: After taking three packs of the above formula, the frequent urination was eliminated and the routine urine test showed: Protein (-), with occasional RBCs. To the above formula *sū yè* (Folium Perillae) was eliminated, and *dì yú* (Radix Sanguisorbae) 6g was added.

Fourth Visit

August 19th: The routine urine test did not seem abnormal. She was instructed to follow the same formula for three packs more in order to strengthen the results.

Note

Three months after stopping the formula the routine urine test was checked again, and it still did not seem abnormal.

Gān cǎo's (甘草 , Radix Glycyrrhizae) taste and property are sweet and neutral. It enters the meridians of the heart channel of the hand *shaoyin*, the lung channel of the hand *taiyin*, the spleen channel of the foot *taiyin*, the stomach channel of the foot *yangming*. It functions to tonify the middle *jiao* and qi, clear heat and detoxify toxins, stop cough and expel phlegm, relax urgency and alleviate pain, and it can harmonise all kinds of herbs. It is the most broadly used herb in clinical practice. If used raw, its property is neutral. It favours tonifying the spleen and stomach, clearing heat and detoxifying toxins, sedating heart fire, expelling phlegm and stopping cough. If used processed, its property is warm, tending toward tonifying the qi of the middle *jiao* and treating many kinds of qi deficiency problems. If properly combined, its application is very broad. Based on pharmacological research, *gān cǎo* (Radix Glycyrrhizae) contains glycyrrhizin. The potassium and the calcium salt in the glycyrrhizin are the main ingredients for the sweet taste of liquorice. In addition, there are liquorice flavanoids, such as licoflavone, the liquorice aglycone, as well as asparagine, mannitol, etc. The glycyrrhizin or its calcium salt can detoxify toxins, from the bacteriotoxin, to snake venom, to globefish poison (tetrodotoxion, TTX) as well as food toxins and body metabolite poison, etc. The glycyrrhetinic acid has a function similar to the adrenocorticotropic hormone (ACTH), it can promote the water and sodium retention inside the body and discharge potassium ions outside, it has an anti-diuretic function. Liquorice extract can alleviate smooth muscle spasm in the stomach and intestines, and inhibit histamine induced stomach acid secretion. When using for ulcerative conditions, it has a protective function over the surface of the ulcer.

1. *Gān Cǎo* (Radix Glycyrrhizae) —— *Kǔ Shēn* (苦参, Radix Sophorae Flavescentis)

Function

To tonify heart qi, regulate the heart rhythm.

Application

This is used for patients with irregular heart rhythm, knotted and regularly intermittent pulse, chest tightness and shortness of breath.

Compatibility Analysis

Gān cǎo (Radix Glycyrrhizae) tonifies the heart qi and functions to recover the pulse and regulate the heart rhythm. Formulae such as *Zhì Gān Cǎo Tāng* (Honey-Fried Liquorice Decoction, 炙甘草汤) in the *Discussion on Cold Damage* treat a knotted or regularly intermittent pulse with heart palpitations, use *zhì gān cǎo* (Radix et Rhizoma Glycyrrhizae Praeparata cum Melle) as its king herb. The taste and property of *kǔ shēn* (苦参 , Radix Sophorae Flavescentis) are bitter and cold. Our ancestors used it to clear heat, dry dampness and kill worms. Modern clinical reports have stated that it functions to regulate the heart rhythm and treat premature ventricular beat. However, *kǔ shēn* (Radix Sophorae Flavescentis) tastes extremely bitter, if taken it will more easily injure the heart yang qi and the spleen and stomach in the middle *jiao*. It is not pertinent to use it for patients with body deficiency and the middle *jiao* deficiency cold. When combined with *zhì gān cǎo* (Radix et Rhizoma Glycyrrhizae Praeparata cum Melle) that tastes sweet, is warm in property and tonifies the middle *jiao*, it can relax its bitter and cold property and prevent injury to the middle *jiao* and heart yang. Meanwhile, the two herbs can achieve a synergistic actionwhich can increase the regulating the heart rhythm function.

Case 1

Initial Visit

A male patient, 52 years old, September 8th, 1986.

The patient had a history of high blood pressure. Recently he suddenly had chest stuffiness, a suffocating sensation, palpitations and shortness of breath. He was suspected of having coronary heart disease and came to the hospital clinic for treatment. An EKG reported: A frequent premature ventricular beat. His pulse was wiry and knotted. His tongue coating was thin and white. The treatment plan was to use a formula to regulate the heart rhythm and tonify heart qi.

Prescription

苦参	*kǔ shēn*	20g	Radix Sophorae Flavescentis
炙甘草	*zhì gān cǎo*	15g	Radix et Rhizoma Glycyrrhizae Praeparata cum Melle
党参	*dǎng shēn*	12g	Radix Codonopsis
麦冬	*mài dōng*	15g	Radix Ophiopogonis
五味子	*wǔ wèi zǐ*	6g	Fructus Schisandrae Chinensis
山楂	*shān zhā*	20g	Fructus Crataegi
益母草	*yì mǔ cǎo*	15g	Herba Leonuri

Second Visit

November 14th: The above formula was modified and he took a total of more than sixty packs; all his symptoms were eliminated. Everyday, once or twice a premature beat occurred but without remarkable discomfort. However, after stopping the formula when fatigued, he still had shortness of breath and chest stuffiness. The above formula was changed and the decoction was prepared as a powder to be taken often.

Prescription

苦参	*kǔ shēn*	100g	Radix Sophorae Flavescentis
炙甘草	*zhì gān cǎo*	100g	Radix et Rhizoma Glycyrrhizae Praeparata cum Melle
西洋参	*xī yáng shēn*	30g	Radix Panacis Quinquefolii
麦冬	*mài dōng*	50g	Radix Ophiopogonis
五味子	*wǔ wèi zǐ*	50g	Fructus Schisandrae Chinensis
丹参	*dān shēn*	100g	Radix et Rhizoma Salviae Miltiorrhizae
三七粉	*sān qī fěn*	30g	Radix et Rhizoma Notoginseng Powder
益母草	*yì mǔ cǎo*	100g	Herba Leonuri

The above herbs were all ground together into a fine powder, each time 6g were taken with warm water, three times a day.

Case 2

Initial Visit

A male patient, 73 years old, April 18th, 1990.

The patient had a heart rhythm disorder for more than a year, accompanied by symptoms of palpitations, chest stuffiness and shortness of breath, etc. An EKG reported: A premature ventricular beat, and insufficient blood supply to the cardiac muscle. He had taken the medicines isosorbide dinitrate, *Dān Shēn Piàn* (Compound Salvia Tablet, 丹参片) and propafenone etc., but with no good effect. When his condition was severe he needed hospitalization for treatment to relax. Later, he invited the author to diagnose and treat. His pulse was deficienct, wiry, and forceless accompanied by a knotted and regularly intermittent pulse. His tongue coating was thin and white. The patient was doing long term leadership work, it overloaded his heart and consumed his spirit, inducing both qi and yin deficiency of the heart vessels. With both qi and yin deficiency the pulse could not be continuous, hence he had the palpitations as well as the knotted and regularly intermittent pulse. The treatment plan was to use modified *Zhì Gān Cǎo Tāng* (Honey-Fried Liquorice Decoction, 炙甘草汤) to tonify qi, nourish yin and regulate the pulse.

Prescription

苦参	kǔ shēn	12g	Radix Sophorae Flavescentis
炙甘草	zhì gān cǎo	10g	Radix et Rhizoma Glycyrrhizae Praeparata cum Melle
百合	bǎi hé	20g	Bulbus Lilii
丹参	dān shēn	20g	Radix et Rhizoma Salviae Miltiorrhizae
葛根	gé gēn	20g	Radix Puerariae Lobatae
桂枝	guì zhī	9g	Ramulus Cinnamomi
西洋参	xī yáng shēn	9g	Radix Panacis Quinquefolii
麦冬	mài dōng	15g	Radix Ophiopogonis
五味子	wǔ wèi zǐ	9g	Fructus Schisandrae Chinensis
枣仁	zǎo rén	15g	Semen Ziziphi Spinosae
三七粉	sān qī fěn	4g	Radix et Rhizoma Notoginseng Powder (dissolved)

Water was used to decoct, then it was divided into two servings, to be taken in the morning and at night, one pack per day.

Second Visit

April 25th: After taking seven packs of the above formula, the chest stuffiness, shortness of breath and palpitations, etc. symptoms reduced a lot, the premature beat was not obvious. He continued to take the original formula again.

Third Visit

May 3rd: The clinical symptoms basically disappeared, the premature beat also never reoccurred. He was instructed to take the original formula every other day continuously.

The above formula modification was used for treatment for two months. After he had taken a total of more than sixty packs, he stopped the medicine. He was followed up for a couple years, and until now, he is still alive and healthy.

2. *Gān Cǎo* (Radix Glycyrrhizae) ——
Bái Sháo (白芍, Radix Paeoniae Alba)

Function

To soften the liver and relax urgency, to relieve spasm and alleviate pain.

Application

This compatibility is also named *Sháo Yào Gān Cǎo Tāng* (Peony Liquorice Decoction, 芍药甘草汤). It was originally recorded in the *Discussion on Cold Damage*. It mainly treats fluid injury due to mistreated sweating inducing foot cramps. Modern uses of this formula fall into three categories: 1. Blood deficiency and fluid injury, or cold invasion inducing hands and foot spasm and pain; 2. All kinds of stomach and abdominal pain, such as stomach spasm inducing stomachache, and abdominal pain, hypochondriac pain and painful menstruation; 3. Stomach and spleen disharmony inducing diarrhoea.

Compatibility Analysis

Gān cǎo (Radix Glycyrrhizae) tastes sweet and can relax tension and alleviate pain. *Bái sháo* (Radix Paeoniae Alba) tastes sour and can soften the liver and relax tension, resolve tetany and alleviate pain. The two herbs combined together, sweet and sour can transform yin pathogens as well as have an even stronger power to nourish the blood, retain yin by astringing, soften the liver and alleviate pain. Cheng Zhong-ling admired it and said that: "*It has a miraculous effect on alleviating pain*". It is also equivalent to the meaning of the the *Basic Questions - Methods of Treating Visceral Qi in Accordance with the Seasons* (素问·藏气法时论, *Sù Wèn - Zàng Qì Fǎ Shí Lùn*) statement: "*If the liver is suffering tension, immediately eat sweet things to relax it and use sour to sedate it.*" If *fù zǐ* (Radix Aconiti Lateralis Praeparata) is added it is called *Sháo Yào Gān Cǎo Fù Zǐ Tāng* (Peony, Liquorice and Aconite Decoction, 芍药甘草附子汤) and can effectively treat cold feet. If *quán xiē* (Scorpio) and *wú gōng* (Scolopendra) are added it is called *Sháo Gān Zhǐ Jìng Tāng* (Peony and Liquorice Convulsion Stopping Decoction, 芍甘止痉汤) and can effectively treat encephalitis patients after reducing the fever, or hands and feet hypertonicity, or neck and back ankylosis.

Case 1

Hiccough.

Initial Visit

A male patient, 55 years old, May 5th, 1990.

The patient suddenly started hiccoughing persistently after eating shortly after a bout of anger and this meant water and food could not be ingested. It stopped during sleep, but after he woke up, he continued to hiccough for two days. He had treated it with chlorpromazine without effect. It was a syndrome of a qi mechanism disorder, rebellious qi overshooting upward inducing the hiccough. It is classified as a diaphragm spasm in biomedicine. The treatment plan was to use a method to regulate the qi mechanism and resolve spasm.

Prescription

白芍	bái sháo	40g	Radix Paeoniae Alba
甘草	gān cǎo	20g	Radix Glycyrrhizae
川芎	chuān xiōng	15g	Rhizoma Chuanxiong
当归	dāng guī	15g	Radix Angelicae Sinensis
桔梗	jié gěng	6g	Radix Platycodonis
柴胡	chái hú	9g	Radix Bupleuri
枳壳	zhǐ qiào	10g	Fructus Aurantii
桃仁	táo rén	9g	Semen Persicae
红花	hóng huā	9g	Flos Carthami
川牛膝	chuān niú xī	12g	Radix Cyathulae
生地	shēng dì	12g	Radix Rehmanniae

Water was used to decoct and it was taken. After one pack, the hiccough reduced a lot. After three packs, the hiccough stopped and he recovered.

Note

This hiccough syndrome case treatment used a combination of two formulae: *Sháo Yào Gān Cǎo Tāng* (Peony Liquorice Decoction) and *Xuè Fǔ Zhú Yū Tāng* (House of Blood Stasis-Expelling Decoction). In clinic, this method for treating hiccoughs is better for the excess syndrome patients. It is not appropriate for deficient patients. Hiccough, as a single symptom of an excess syndrome, is more often due to a qi mechanism disorder, and so *Xuè Fǔ Zhú Yū Tāng* (House of Blood Stasis-Expelling Decoction) was used to transform stasis and regulate the qi mechanism (please refer to the previous *chuān niú xī* (Radix Cyathulae), *jié gěng* (Radix Platycodonis) and *chái hú* (Radix Bupleuri compatibility). Also as hiccough was induced by a diaphragm spasm, *Sháo Yào Gān Cǎo Tāng* (Peony Liquorice Decoction) was used to resolve it.

Case 2

Initial Visit

A female patient, 6 years old, August 4th, 1987.

The sick child had pain in both legs for more than a year, worse at night time. Every night at bed-time or during sleep, the two legs were painful even bothering her sleep. The leg pain was accompanied by ankylosis in the four limbs. Her appetite, bowel movement and urination were normal. Her pulse was wiry and thready. Her tongue coating was white and slightly greasy at the root. This was prolonged sickness inducing more blood stasis. Blood stasis caused the tendons and blood vessels to lose nourishment, hence the pain and its being worse at night time. The tendon lost moisture and nourishment, and this caused the ankylosis. The treatment plan was to nourish blood, soften the liver, transform stasis and alleviate pain. *Sháo Yào Gān Cǎo Tāng* (Peony Liquorice Decoction) with modified *Huó Luò Xiào Líng Dān* (Network-Quickening Miraculous Effect Elixir, 活络效灵丹) was used for treatment.

Prescription

白芍	*bái sháo*	30g	Radix Paeoniae Alba
生薏米	*shēng yì mǐ*	30g	Semen Coicis (raw)
川牛膝	*chuān niú xī*	15g	Radix Cyathulae
当归	*dāng guī*	10g	Radix Angelicae Sinensis
丹参	*dān shēn*	20g	Radix et Rhizoma Salviae Miltiorrhizae
甘草	*gān cǎo*	9g	Radix Glycyrrhizae

Water was used to decoct, and then it was divided into two servings

Second Visit

September 17th: After taking four packs of the above formula, there was no leg pain, but there was still ankylosis in the four limbs. The original formula was used with *shēng yì mǐ* (raw Semen Coicis) eliminated, and *jī xuè téng* (Caulis Spatholobi) 20g, and *huò xiāng* (Herba Agastachis) 6g, added. She continued to take it again.

Third Visit

September 22nd: After taking another five packs, all symptoms basically recovered. Again she used five packs of the same formula to complete the treatment.

3. *Gān Cǎo* (Radix Glycyrrhizae) —— *Jié Gěng* (桔梗, Radix Platycodonis)

Function

To clear heat and detoxify toxins, diffuse the lung and benefit the throat, drain pus and expel phlegm.

Application

This is used for pathogenic heat lodging in the *shaoyin* inducing swollen and painful throat. It is also used for lung abscess and coughing or spitting up blood with pus, etc. The author uses these two herbs more often in clinic if anyone presents with an upper respiratory tract infection inducing cough, using them to clear heat, detoxify toxins, diffuse the lung and expel phlegm. Moreover for the patients with chronic pharyngitis inducing a dry itchy and painful pharynx and larynx, the results are also good.

Compatibility Analysis

Gān cǎo (Radix Glycyrrhizae) clears heat and detoxifies toxins. *Jié gěng* (Radix Platycodonis) is bitter, acrid and neutral. It enters the lung channel of the hand *taiyin* meridian. It functions to open and diffuse lung qi, benefit the throat, expel phlegm and drain pus. The two herbs combined together have an even stronger power to clear and resolve the throat, diffuse the lung and expel phlegm. *Jié Gěng Tāng* (Platycodon Decoction, 桔梗汤) in the *Discussion on Cold Damage* uses this compatibility to treat sore throat and in the *Essentials from the Golden Cabinet* it is used to treat lung abscesses.

Case Study

Initial Visit

A female patient, 37 years old, October 21st, 1995.

The patient had chronic pharyngitis. She always had a sore and itchy throat like there was something there, and a dry and itchy cough. In recent days, it had got worse. After ineffective treatment with medicines like *Xī Guā Shuāng* (Mirabilitum Praeparatum, 西 瓜 霜), *Cǎo Shān Hú* (Sarcandrae Ramulus et Folium), etc., she came to our clinic for treatments. Her pulse was diagnosed as wiry, thready and fast; her tongue was red, her tongue coating was thin and white, her throat was red with follicles on the pharynx and larynx walls. This was lung qi stagnating and clumping without dispersing, with the prolonged stagnation transforming into heat and injuring the yin. The lung qi stagnated and clumped and the throat felt like something was caught in it and she had a cough. The transformed heat injured yin and then this produced the dry, sore throat, thready and fast pulse. The treatment plan was to use a formula to clear heat, nourish yin, diffuse the lung and benefit the throat.

Prescription

桔梗	*jié gěng*	10g	Radix Platycodonis
甘草	*gān cǎo*	9g	Radix Glycyrrhizae
麦冬	*mài dōng*	15g	Radix Ophiopogonis
生地	*shēng dì*	10g	Radix Rehmanniae
元参	*yuán shēn*	12g	Radix Scrophulariae
金果榄	*jīn guǒ lǎn*	10g	Radix Tinosporae (cracked apart)

Water was used to decoct, to be taken often, drinking it like tea, one pack per day.

Second Visit

October 28th: After taking seven packs continuously, the symptoms of sore and itchy throat and dry cough, etc., all reduced a lot. The patient was instructed to continuously take this formula.

Third Visit

November 4th: All symptoms improved, the throat follicles disappeared. The pulse was still wiry and fast. She was given the original formula to take again.

After continuously taking thirty packs of the formula, all symptoms disappeared.

Shí gāo's (石膏 , Gypsum Fibrosum) tastes and property are acrid, sweet and very cold. It enters the lung channel of the hand *taiyin* and the stomach channel of the foot *yangming* meridians. It is an important herb to clear heat and sedate fire, and is especially good at clearing lung and stomach fire. It is appropriately used for exuberant heat with sweating, vexation, thirst, preference for drinking cold drinks and loss of consciousness due to high fever. The coldness of *shí gāo* (Gypsum Fibrosum) is essential to resolve the heat of a heat pathogen in the *yangming* qi level. It is often combined with *zhī mǔ* (Rhizoma Anemarrhenae) as in *Bái Hǔ Tāng* (White Tiger Decoction, 白虎汤). It may clear and sedate lung heat in lung heat cough and asthma patients. It is often combined with *má huáng* (Herba Ephedrae), as in *Má Xìng Shí Gān Tāng* (Ephedra, Apricot Kernel, Gypsum, and Liquorice Decoction, 麻杏石甘汤). *Shí gāo* (Gypsum Fibrosum) enters the stomach channel of the foot *yangming* meridian to clear stomach heat. It is a common herb to treat stomachache, toothache and headache. Laboratory tests have shown that when there is fever, *shēng shí gāo* (raw Gypsum Fibrosum) can suppress the overexcited thermotaxis centre, as well as has a strong and quick reducing heat function, but it does not last for a long time. As *shí gāo* (Gypsum Fibrosum) can also suppress sweat gland secretion, it can therefore reduce fever, without perspiration.

1. *Shí Gāo* (Gypsum Fibrosum) ——
Quán Xiē (全蝎, Scorpio)

Function

To clear the stomach, sedate fire, and alleviate pain.

Application

This is used for exuberant stomach fire inducing swollen and painful gums, toothache and headache, etc.

Compatibility Analysis

Shēng shí gāo (raw Gypsum Fibrosum) enters the stomach channel of the foot *yangming* meridian and is an important herb to clear its excess fire. *Quán xiē* (Scorpio) is acrid, neutral and enters the liver channel of the foot *jueyin* meridian. It can extinguish wind, unblock collaterals and alleviate pain. Regardless of whether it is a headache, toothache or nerve pain, all have better results at resolving tetany and alleviating pain with *Quán xiē* (Scorpio). *Shí gāo* (Gypsum Fibrosum) combined with *quán xiē* (*Scorpio*) together, can sedate stomach fire and alleviate pain. Use *shí gāo* (Gypsum Fibrosum) to clear the stomach and sedate fire to treat the root. Use *quán xiē* (*Scorpio*) to extinguish wind, resolve tetany and alleviate pain to treat the manifestations. This combination has good results for toothache or headache, regardless of whether it is the stomach channel of the foot *yangming* or the liver channel of the foot *jueyin* meridian's exuberant fire inducing them.

Case 1

Migraine.

Initial Visit

A male patient, 32 years old, April 6th, 1987.

The patient had right side migraine for a couple of years. It was sometimes better and sometimes worse, but especially noticeable around noon, it was accompanied by insomnia. He had been diagnosed in a hospital as having trigeminal neuralgia. His pulse was wiry, big and forceful. His tongue was red and his tongue coating was white. This was liver and stomach fire exuberance. When there is liver and stomach fire exuberance, upward clear yang overshoots and induces headache and a wiry pulse. The *Inner Classic* said: "*At noon, yang qi is prosperous*" hence, the symptoms of the exuberant fire syndrome will become worse around noon time. Prolonged exuberant heat must injure yin. With yin injured then fire will inflame, hence it will be prolonged without recovery. The treatment plan was to nourish yin, clear fire, descend rebellion and alleviate pain.

Prescription

生石膏	*shēng shí gāo*	20g	Gypsum Fibrosum (raw)
川牛膝	*chuān niú xī*	20g	Radix Cyathulae
生地	*shēng dì*	15g	Radix Rehmanniae
麦冬	*mài dōng*	20g	Radix Ophiopogonis
全蝎	*quán xiē*	6g	Scorpio
川芎	*chuān xiōng*	20g	Rhizoma Chuanxiong
枣仁	*zǎo rén*	15g	Semen Ziziphi Spinosae
甘草	*gān cǎo*	15g	Radix Glycyrrhizae

Water was used to decoct, then it was divided into two servings, to be taken in the morning and at night, one pack per day.

Second Visit

April 11th: After taking four packs of the medicine, the headache reduced, but the insomnia with a lot of dreams was the same as before. The original formula was modified and taken again.

Prescription

生石膏	*shēng shí gāo*	20g	Gypsum Fibrosum (raw)
川牛膝	*chuān niú xī*	20g	Radix Cyathulae
川芎	*chuān xiōng*	20g	Rhizoma Chuanxiong
龟甲	*guī jiǎ*	10g	Carapax et Plastrum Testudinis
全蝎	*quán xiē*	6g	Scorpio
知母	*zhī mǔ*	10g	Rhizoma Anemarrhenae
远志	*yuǎn zhì*	10g	Radix Polygalae
川连	*chuān lián*	6g	Rhizoma Coptidis
肉桂	*ròu guì*	3g	Cortex Cinnamomi
麦冬	*mài dōng*	15g	Radix Ophiopogonis
代赭石	*dài zhě shí*	15g	Haematitum
甘草	*gān cǎo*	6g	Radix Glycyrrhizae

Third Visit

April 20th: After taking eight packs of the above formula continuously, the headache did not reoccur, he could sleep, but sometimes had a lot of dreams. The same formula was used with *shēng lóng gǔ* (raw Os Draconis) 20g, and *yè jiāo téng* (Caulis Polygoni Multiflori) 20g, added. He continued taking it.

Fourth Visit

May 18th: After taking more then ten packs of the above formula, he never had another headache, and his sleep was also good. He was instructed to stop the formula and observe.

Case 2

Toothache.

Initial Visit

A male patient, 50 years old, April 20th, 1972.

The patient had a toothache for one week. It was worse at night and he could not sleep. He always used cold dressings on the painful area and felt more comfortable as a result. He took pain killers like Metamizole Sodium, etc., but they did not work and so he invited me to visit. The patient's right side upper gum was swollen and painful, but there was no cavity. His pulse was wiry, big, thready and fast. His tongue was red and his tongue coating was thin and yellow. This was stomach fire inducing toothache. The treatment plan was to use *Yù Nǚ Jiān* (Jade Lady Brew, 玉女煎) with *quán xiē* (Scorpio) added.

Prescription

生石膏	*shēng shí gāo*	40g	Gypsum Fibrosum (raw)
知母	*zhī mǔ*	12g	Rhizoma Anemarrhenae
川牛膝	*chuān niú xī*	15g	Radix Cyathulae
生地	*shēng dì*	15g	Radix Rehmanniae
麦门冬	*mài mén dōng*	15g	Radix Ophiopogonis
全蝎	*quán xiē*	8g	Scorpio
细辛	*xì xīn*	4g	Radix et Rhizoma Asari
甘草	*gān cǎo*	6g	Radix Glycyrrhizae

Water was used to decoct, then it was divided into three servings to be taken, one pack per day.

Note

The patient in this case took one pack of the above formula which stopped the pain; and after three packs he recovered. He had no toothache for the following ten years. This can be called an effective formula for stopping toothache. The author in clinic uses this formula to treat many toothache cases with very good results. This formula is best for treating stomach fire induced toothache. If it's a kidney deficiency induced toothache, then it is not appropriate to use it.

2. *Shí Gāo* (Gypsum Fibrosum) ——
Guì Zhī (桂枝, Ramulus Cinnamomi)

Function

To unblock vessels, promote moving for painful obstruction, to clear heat and alleviate the pain.

Application

This is used for wind-damp inducing heat painful obstruction, red, swollen, hot and painful joints. For modern clinical wind-damp type arthritis or rheumatoid arthritis that has signs of heat, it has functions to reduce swelling and alleviate the pain.

Compatibility Analysis

Guì zhī (Ramulus Cinnamomi) is acrid, warm and enters the blood level. It can warmly unblock meridians and blood vessels, promote moving for painful obstruction, alleviate pain and promote blood circulation. *Shí gāo* (Gypsum Fibrosum) is acrid and cold. The acridity can expel and the coldness can clear. It can clear and expel heat pathogens. With *guì zhī* (Ramulus Cinnamomi) and *shí gāo* (Gypsum Fibrosum) combined, *guì zhī* (Ramulus Cinnamomi) relies on the coldness of *shí gāo* (Gypsum Fibrosum) to clear stagnated heat in the blood vessels; *shí gāo* (Gypsum Fibrosum) relies on the ability of *guì zhì* (Ramulus Cinnamomi) to warmly unblock the meridians and blood vessels to expel heat stagnation. The two herbs mutually supplement each other and can unblock blood vessels and promote moving for painful obstruction, clear stagnated heat and alleviate painful obstruction. There are good results for heat painful obstructions.

Case Study

Initial Visit

A female patient, 12 years old, February 23rd, 1987.

The patient's joints in two fingers were red, swollen, hot and painful for more than twenty days. There was a limited range of motion when flexing or extending the joints. The pain got worse when she encountered cold or cold water. Her ESR was 30mm/h, her pulse was wiry and fast, her tongue was red and her tongue coating was thin and white. This was damp-heat stagnating and obstructing the meridians and blood vessels inducing heat painful obstruction. The treatment plan was to unblock the blood vessels, invigorate blood, clear heat and alleviate pain.

Prescription

Modified *Bái Hǔ Jiā Guì Zhī Tāng* (White Tiger Decoction Plus Cinnamon Twig, 白虎加桂枝汤).

桂枝	*guì zhī*	10g	Ramulus Cinnamomi
白芍	*bái sháo*	12g	Radix Paeoniae Alba
当归	*dāng guī*	10g	Radix Angelicae Sinensis
生石膏	*shēng shí gāo*	24g	Gypsum Fibrosum (raw)
知母	*zhī mǔ*	10g	Rhizoma Anemarrhenae
细辛	*xì xīn*	3g	Radix et Rhizoma Asari
丹参	*dān shēn*	20g	Radix et Rhizoma Salviae Miltiorrhizae
通草	*tōng cǎo*	4g	Medulla Tetrapanacis
甘草	*gān cǎo*	6g	Radix Glycyrrhizae

Water was used to decoct, then it was divided into two servings, served warm, one pack per day.

Second Visit

February 28th: After taking four packs of the above formula the heat and pain reduced; the redness and swelling also reduced by more than half. She continued taking another four packs of the original formula.

Third Visit

March 6th: On this visit the redness, swelling and pain were basically eliminated with the exception that her finger joints were slightly thicker than normal and still not very flexible. The treatment method was changed to invigorate blood, unblock meridians, eliminate dampness and expel stasis.

Prescription

桂枝	*guì zhī*	9g	Ramulus Cinnamomi
白芍	*bái sháo*	12g	Radix Paeoniae Alba
生薏仁	*shēng yì rén*	20g	Semen Coicis (raw)
当归	*dāng guī*	10g	Radix Angelicae Sinensis
丹参	*dān shēn*	15g	Radix et Rhizoma Salviae Miltiorrhizae
桑枝	*sāng zhī*	15g	Ramulus Mori
甘草	*gān cǎo*	6g	Radix Glycyrrhizae

Fourth Visit

March 16th: After taking ten packs of the above formula, the finger joints basically recovered to normal and she stopped taking the medicine. She was instructed to keep a check on it periodically.

3. *Shí Gāo* (Gypsum Fibrosum) ——
Fáng Fēng (防风, Radix Saposhnikoviae)

Function

To clear and sedate spleen heat.

Application

This is mainly used to treat symptoms of the middle *jiao* spleen and stomach hidden fire, mouth and tongue ulceration, dry mouth, sweet taste inside the mouth, etc.

Compatibility Analysis

The spleen opens its orifice at the mouth. If the spleen channel of the foot *taiyin* meridian's hidden fire steams upward, it can manifest as mouth and tongue ulceration, dry mouth and bad breath. The spleen channel of the foot *taiyin* meridian has stagnated heat, it steams fluid to overflow upward, therefore producing a sweet taste. *Shí gāo* (Gypsum Fibrosum) is acrid, sweet and cold. It is an important herb to clear and resolve heat pathogens. *Fáng fēng* (Radix Saposhnikoviae) is acrid, sweet and warm. It enters the spleen channel of the foot *taiyin* meridian. Its property is raising and dispersing. The *Inner Classic* said: "*If fire is stagnated treat it by effusion.*" Therefore, the spleen channel of the foot *taiyin* meridian's hidden fire cannot be dispersed and resolved without *fáng fēng* (Radix Saposhnikoviae). The two herbs combined together, *fáng fēng* (Radix Saposhnikoviae) gets assistance from *shí gāo* (Gypsum Fibrosum), the power to clear and resolve hidden accumulated fire is even stronger. *Shí gāo* (Gypsum Fibrosum) gets assistance from the power of *fáng fēng* (Radix Saposhnikoviae), its power to disperse and expel stagnated heat is even greater. The *Xiè Huáng Sǎn* (Yellow Draining Powder, 泻黄散) in the *Craft of Medicines and Patterns for Children* (小儿药证直诀 , *Xiǎo Ér Yào Zhèng Zhí Jué*) uses this as the main compatibility to treat spleen and stomach hidden fire inducing mouth and tongue ulcers, dry mouth and dry lips, and also to treat spleen heat in small children causing them to frequently stick their tongue out. In clinic, the author always uses this formula to treat spleen heat inducing mouth sweetness syndrome with mostly good results.

Case Study

Initial Visit

A female patient. 49 years old, September 5th, 1987.

The patient was usually strong and healthy. A couple of months previously she suddenly developed a sweet taste in her mouth. In the beginning, the sweet taste was light, but then day by day it gradually got worse. When dining or drinking, she felt like sugar had been added in. She was afraid she might have diabetes so she went to a city hospital to have it checked. Her fasting blood and urine glucose were normal, hence, she came to clinic for treatments. On observation the patient had good physical development. Except for the sweetness in her mouth, she had no other complaints. Her pulse tended toward forceful and big. Her tongue coating was thin and white. This belonged to the spleen channel of the foot *taiyin* meridian heat stagnation syndrome. As the spleen opens into the mouth and, of the five tastes, it governs sweetness, so the spleen channel of the foot *taiyin* meridian heat stagnation steamed body fluid to overflow upward, and therefore produced the sweetness in the mouth. The treatment plan was to clear spleen heat, and expel hidden fire. A modified *Xiè Huáng Sǎn* (Yellow Draining Powder) was chosen for treatment.

Prescription

生石膏	*shēng shí gāo*	30g	Gypsum Fibrosum (raw)
防风	*fáng fēng*	12g	Radix Saposhnikoviae
栀子	*zhī zǐ*	9g	Fructus Gardeniae
藿香	*huò xiāng*	9g	Herba Agastachis
荷叶	*hé yè*	10g	Folium Nelumbinis
甘草	*gān cǎo*	6g	Radix Glycyrrhizae

Water was used to decoct, to be taken one pack per day.

Note

She continuously took five packs of the above formula, the sweetness in her mouth was eliminated; follow up visits over a couple of years showed no reoccurrence. This formula was used to treat ten patients with the symptoms of a sweet taste in the mouth and fire in the spleen, all had good results.

Xiān hè cǎo's (仙鹤草 , Herba Agrimoniae) taste and property are bitter, astringent and slightly warm. It enters the heart channel of the hand *shaoyin*, the lung channel of the hand *taiyin*, the spleen channel of the foot *taiyin*, the stomach channel of the foot *yangming*, the large intestine channel of the hand *yangming* and the liver channel of the foot *jueyin* channel meridians. It functions to stop bleeding and tonify deficiency. It is a commonly used herb to stop bleeding in clinical practice. It can be used for all kinds of bleeding syndromes such as haemoptysis, epistaxis, haematochezia, haematuria, uterine bleeding, etc. Aside from stopping bleeding, it can tonify yin and treat lost strength due to overexertion injury and treat chronic fatigue syndrome (please refer to *dà zǎo* - Fructus Jujubae with *xiān hè cǎo* - Herba Agrimoniae).

1. *Xiān Hè Cǎo* (Herba Agrimoniae) ——
Ē Jiāo (阿胶, Colla Corii Asini)

Function

To tonify blood and stop bleeding.

Application

This is used for thrombocytopaenia syndrome. It is also used for all kinds of bleeding syndromes.

Compatibility Analysis

Ē jiāo (Colla Corii Asini) is sweet and neutral. It is a product of flesh and blood with emotion and an important herb to tonify yin and blood. Cheng Wu-ji said: "*For yin deficient patients, tonify with taste. Ē jiāo (Colla Corii Asini) tastes sweet; it can tonify yin and blood.*" *Xiān hè cǎo* (Herba Agrimoniae) is an herb for stopping bleeding. It also can tonify the middle *jiao* qi and treat deficiency. Tangible blood is generated from intangible qi. If you tonify qi then you can generate blood, therefore, with the two herbs combined together, their power to tonify blood and stop bleeding is even stronger. They are effective for the blood deficiency induced bleeding syndrome and ITP (idiopathic thrombocytopaenic purpura, also known as immune thrombocytopaenic purpura) syndrome.

Case Study

Initial Visit

A female patient, 45 years old, August 3rd, 1998.

The patient had lassitude and anorexia for more than a month. Recently, bleeding spots appeared on her lower limbs. It started off small and later defused into pieces. She had been investigated in a hospital. Her blood platelet count was $56 \times 10^9/L$ and she was diagnosed with thrombocytopaenia. She took medicine but it did not improve and so she came to visit a Chinese medicine doctor. The patient had good physical development; her chronic condition was obvious from her face. Her sleep was not good because she was stressed about her disease. Her pulse was deep and choppy. Her tongue coating was thin and white and tongue was pale. This was both heart and spleen deficiency. The deficient spleen could not control the blood, hence the blood overflowed outside. The treatment plan was to tonify qi and strengthen the spleen, nourish blood and stop bleeding.

Prescription

仙鹤草	*xiān hè cǎo*	30g	Herba Agrimoniae
生黄芪	*shēng huáng qí*	20g	Radix Astragali (raw)
当归	*dāng guī*	15g	Radix Angelicae Sinensis
丹参	*dān shēn*	24g	Radix et Rhizoma Salviae Miltiorrhizae
山萸肉	*shān yú ròu*	15g	Fructus Corni
太子参	*tài zǐ shēn*	20g	Radix Pseudostellariae
茜草	*qiàn cǎo*	10g	Radix et Rhizoma Rubiae
丹皮	*dān pí*	10g	Cortex Moutan
生地	*shēng dì*	15g	Radix Rehmanniae
黄精	*huáng jīng*	15g	Rhizoma Polygonati
夜交藤	*yè jiāo téng*	30g	Caulis Polygoni Multiflori
枣仁	*zǎo rén*	15g	Semen Ziziphi Spinosae
甘草	*gān cǎo*	6g	Radix Glycyrrhizae
阿胶	*ē jiāo*	15g	Colla Corii Asini (melted)

Water was used to decoct, then it was divided into two servings, one pack per day.

Second Visit

August 19th: After taking a total of fourteen packs of the above formula, the purplish spots on her lower limbs were visibly reduced, her appetite had increased, but she still had a feeling of lassitude, and her blood platelet count was 66×10^9/L. 10 pieces of *dà zǎo* (Fructus Jujubae) were added to the above formula; and she continued to take it again.

Third Visit

September 16th: The purplish spots on her lower limbs had disappeared, and all her other symptoms also basically disappeared, her blood platelet count increased to 85×10^9/L. She continued to take the above formula.

Fourth Visit

December 18th: After taking a total of more than a hundred packs, all her symptoms disappeared. Her blood platelet count was 130×10^9/L. She stopped taking the formula and periodically re-checked for signs of reoccurrence.

She had follow up visits for a year and no reoccurrence was noted.

Shí chāng pú (石菖蒲 , Rhizoma Acori Tatarinowii) is acrid and warm. It enters the heart channel of the hand *shaoyin*, the liver channel of the foot *jueyin* and the kidney channel of the foot *shaoyin* meridians. It functions to open orifices and eliminate phlegm, arouse the brain and calm the spirit, harmonise the stomach and transform dampness. The aromatic clear, light and fresh qi of *shí chāng pú* (Rhizoma Acori Tatarinowii), repels the turbid pathogens and can excite clear yang, open orifices and make ear and eyes alert, transform the turbid and eliminate toxins, break up blockage and arouse the spirit. Only aromatic herbs can unblock the orifices to treat turbid phlegm misting and closing the orifices and unconsciousness. If there is tinnitus and deafness, the head and eyes are not clear, without these clear, light and fresh properties, they cannot be unblocked. The *Divine Husbandman's Classic of the Materia Medica* said that it: *"Opens the heart orifices, tonifies the five zang organs, unblocks the nine orifices, brightens the ears and eyes, makes voice come out, and governs tinnitus."* *Shí chāng pú*'s (Rhizoma Acori Tatarinowii) taste is aromatic and it can eliminate dampness and increase the appetite. It is an aromatic herb and can strengthen stomach herbs. Based on modern pharmological research, this herb can stimulate stomach fluid secretion and inhibit abnormal fermentation in the stomach and intestines; hence, it is effective for treating poor digestion inducing abdominal distention.

1. *Shí Chāng Pú* (Rhizoma Acori Tatarinowii) ——
Lù Lù Tōng (路路通, Fructus Liquidambaris)

Function

To invigorate stasis and unblock the orifices.

Application

This is used for deafness, tinnitus or a blocked nose that cannot smell and belongs to blood stasis or phlegm stagnation patients.

Compatibility Analysis

Shí chāng pú (Rhizoma Acori Tatarinowii) is acrid, warm and aromatic. It is effective for unblocking and benefiting the nine orifices, expelling turbid phlegm and breaking up obstructions. *Lù lù tōng* (路路通 , Fructus Liquidambaris) is acrid and neutral. It enters the liver channel of the foot *jueyin* meridian. It unblocks channels, promotes urination and is good at eliminating stasis and stagnation in the meridians and collaterals. A folk empirical formula uses *lù lù tōng* (Fructus Liquidambaris) alone, as a single herb, combined with water to decoct for treating tinnitus. Meanwhile it can also unblock nasal orifices. Modern clinic reports have stated that *lù lù tōng* (Fructus Liquidambaris) has an anti-allergy function and is effective for allergic rhinitis. *Shí chāng pú* (Rhizoma Acori Tatarinowii) and *lù lù tōng* (Fructus Liquidambaris) combined together have a mutually supplementing function. *Lù lù tōng* (Fructus Liquidambaris) can unblock collaterals and can help *shí chāng pú* (Rhizoma Acori Tatarinowii) to open orifices and unblock the closed stagnation. *Shí chāng pú* (Rhizoma Acori Tatarinowii) is acrid and can unblock which also helps *lù lù tōng* (Fructus Liquidambaris) to eliminate stasis and stagnation. The two herbs are more effective used together than either one used alone.

Case 1

Initial Visit

A female patient, 60 years old, August 6ᵗʰ, 1987.

The patient had ringing like waves in her left ear for two years. For the previous week, her left ear felt stuffy, blocked and she gradually could not hear. Her blood pressure was 135/80mm Hg, her pulse was big with both *chi* pulses tending toward deficiency, her tongue coating was thin and white accompanied by low back soreness and lower limb lassitude and flaccidity. This was kidney deficiency, the kidney essence could not rise up to nourish the clear orifices so the clear orifices lost nourishment, turbid phlegm then stagnated and caused blockage due to deficiency.

Prescription

石菖蒲	*shí chāng pú*	15g	Rhizoma Acori Tatarinowii
路路通	*lù lù tōng*	10g	Fructus Liquidambaris
生地	*shēng dì*	20g	Radix Rehmanniae
山萸肉	*shān zhū yú*	10g	Fructus Corni
山药	*shān yào*	15g	Rhizoma Dioscoreae
泽泻	*zé xiè*	9g	Rhizoma Alismatis
丹皮	*dān pí*	9g	Cortex Moutan
茯苓	*fú líng*	10g	Poria
甘草	*gān cǎo*	6g	Radix Glycyrrhizae

Water was used to decoct, then it was divided into two servings, to be taken in the morning and at night, one pack per day.

Second Visit

July 18ᵗʰ: After continuously taking ten packs of the above formula, her symptoms had reduced, although she occasionally had a dizziness sensation. To the above formula *wēi líng xiān* (威灵仙, Radix et Rhizoma Clematidis) 9g, and *hé yè* (Folium Nelumbinis) 10g, were added. She continued to take it.

Third Visit

July 27ᵗʰ: Her symptoms had reduced a lot, her left ear could hear regular talking voices but her dizziness was the same as before. Again, *dān shēn* (Radix et Rhizoma Salviae Miltiorrhizae) 20g, and *gé gēn* (Radix Puerariae Lobatae) 24g, were added to enhance the power to transform stasis and raise up the clear. She continued to take it.

Fourth Visit

August 7ᵗʰ: After taking ten packs of the above formula, all her symptoms basically disappeared. Her prescription was changed to *Liù Wèi Dì Huáng Wán* (Six-Ingredient Rehmannia Pill, 六味地黄丸), one pill each in the morning and evening to complete the treatment and she recovered.

Case 2

Initial Visit

A female patient, 57 yeas old, January 4th, 1996.

The patient had flu and fever a couple of months previously. After treatment and recovery, both ears felt blocked and she had a reduced ability to hear sounds. She could not hear people talking in a low voice. Her nose was blocked and she could not smell perfume or foul smell. Her pulse was wiry and thready and her tongue coating was thin and white. This resulted after external invasion because the residual pathogen was not totally cleared, and so stagnated and blocked the clear orifices and induced the stuffy nose.

Prescription

路路通	lù lù tōng	20g	Fructus Liquidambaris
石菖蒲	shí chāng pú	15g	Rhizoma Acori Tatarinowii
柴胡	chái hú	15g	Radix Bupleuri
辛夷	xīn yí	10g	Flos Magnoliae
丹参	dān shēn	20g	Radix et Rhizoma Salviae Miltiorrhizae
赤芍	chì sháo	15g	Radix Paeoniae Rubra
钩藤	gōu téng	20g	Ramulus Uncariae Cum Uncis
泽泻	zé xiè	15g	Rhizoma Alismatis
茯苓	fú líng	20g	Poria
生地	shēng dì	15g	Radix Rehmanniae
甘草	gān cǎo	6g	Radix Glycyrrhizae

Water was used to decoct, then it was divided into two servings, to be taken warm, in the morning and at night, one pack per day.

Second Visit

January 10th: After taking five packs, her symptoms had reduced. She could hear clearly for the regular talking voice but it was still difficult, and she could smell again. She took the same formula again.

Third Visit

January 16th: All her symptoms had improved a lot. She followed the original formula for seven packs and recovered.

2. *Shí Chāng Pú* (Rhizoma Acori Tatarinowii) —— *Guī Jiǎ* (龟甲, Carapax et Plastrum Testudinis)

Function

To tonify the heart and kidney, to nourish the brain.

Application

This is used for syndromes of memory loss, forgetfulness, palpitations and insomnia, etc.

Compatibility Analysis

Shí chāng pú (Rhizoma Acori Tatarinowii) is acrid and can open the orifices, arouse the brain and calm the spirit. It is an important herb to excite clear yang. *Guī jiǎ* (龟甲, Carapax et Plastrum Testudinis) is cold and salty. It enters the heart channel of the hand *shaoyin*, the liver channel of the foot *jueyin*, the spleen channel of the foot *taiyin* and the kidney channel of the foot *shaoyin* meridians. It is an extreme yin product; hence, Li Shizhen said it: *"Tonifies heart, kidney and blood. All are nourishing yin."* In clinic, it is more often used for deficiency inducing bone steaming, blood drying and spirit tiredness. *Shí chāng pú* (Rhizoma Acori Tatarinowii) and *guī jiǎ* (Carapax et Plastrum Testudinis) used together, one is active and the other is calm, it is a combination of active and calm. *Guī jiǎ* (Carapax et Plastrum Testudinis) is extreme yin, it uses the acrid, warm and aromatic property of *shí chāng pú* (Rhizoma Acori Tatarinowii) to make yin to pursue yang, as well as yin and yang to mutually supplement each other. *Shí chāng pú* (Rhizoma Acori Tatarinowii) arouses the brain and opens the orifices. It is assisted by *guī jiǎ* (Carapax et Plastrum Testudinis), to help in tonifying the heart, tonifying the kidney and nourishing yin; it can use yang to pursue yin thus infinitely generating and transforming. In clinic one can see disharmony of the heart and kidney inducing palpitations, insomnia, forgetfulness and kidney deficiency in the elderly, and insufficiency of the sea of the marrow inducing memory loss.

Case 1

Initial Visit

A male patient, 59 years old, May 29th, 1987.

From 1973 he started having high blood pressure with insomnia. In recent years it gradually became severe. Recently he had severe dizziness and insomnia, memory loss, tinnitus and spirit lassitude. His blood pressure was 160/95mm Hg. His tongue was red and his tongue coating was thin and white. His pulse was wiry and forceful. This was the sign of liver and kidney deficiency with deficient yang floating. With liver and kidney deficiency, the essence could not go upward to nourish, and so, the sea of marrow was insufficient. The clear orifices lost nourishment, and caused the spirit tiredness, insomnia and memory loss. With yin deficiency, deficient yang floated up to cover the clear upper orifice, and caused the dizziness and tinnitus. The treatment plan was to tonify the liver and kidney, wake up the brain and calm the spirit.

Prescription

石菖蒲	*shí chāng pú*	20g	Rhizoma Acori Tatarinowii
龟甲	*guī jiǎ*	20g	Carapax et Plastrum Testudinis
钩藤	*gōu téng*	30g	Ramulus Uncariae Cum Uncis (decoct later)
天麻	*tiān má*	15g	Rhizoma Gastrodiae
代赭石	*dài zhě shí*	20g	Haematitum
杭白芍	*háng bái sháo*	20g	Radix Paeoniae Lactiflorae
远志	*yuǎn zhì*	10g	Radix Polygalae

生地	*shēng dì*	15g	Radix Rehmanniae
甘草	*gān cǎo*	6g	Radix Glycyrrhizae

Water was used to decoct, then it was divided into two servings, one pack per day.

Second Visit

June 6th: After taking seven packs, the dizziness and insomnia improved, the spirit felt clearer and happier than before. He continued to take the same formula.

Third Visit

June 13th: The symptoms improved, his blood pressure was 150/90mm Hg, and he could fall asleep. For strengthening the effectiveness, the decoction was switched to a powder.

Prescription

石菖蒲	*shí chāng pú*	30g	Rhizoma Acori Tatarinowii
龟甲	*guī jiǎ*	30g	Carapax et Plastrum Testudinis
远志	*yuǎn zhì*	30g	Radix Polygalae
生龙骨	*shēng lóng gǔ*	50g	Os Draconis (raw)
代赭石	*dài zhě shí*	50g	Haematitum
天麻	*tiān má*	30g	Rhizoma Gastrodiae
川连	*chuān lián*	15g	Rhizoma Coptidis
肉桂	*ròu guì*	5g	Cortex Cinnamomi

All the above herbs were ground together into a fine powder, each time 3g were taken, twice a day, to be taken with warm water.

Later at a follow up visit, he had taken the above formula and all symptoms were basically gone; and his blood pressure was also more stable.

Case 2

Initial Visit

A female patient, 50 years old, September 10th, 1989.

For the previous few months, the patient felt her spirit was confused and she felt depressed and had memory loss, therefore, her job was affected and she could not work. On examining the patient, she was physically well developed, but physically obese, and had a wiry, slow but forceful pulse, a dark tongue and a white, slightly turbid tongue coating. Her appetite, bowel movement and urination were normal. This was the kidney deficiency inducing phlegm and dampness syndrome. If her kidney was deficient then the sea of marrow would be insufficient, therefore, she had memory loss and was dispirited. Her body obesity and turbid tongue coating were indications of phlegm and dampness. Phlegm and dampness going upward, mists the clear orifices, and therefore, her spirit was confused. The treatment plan was to tonify the kidney, transform phlegm, eliminate dampness and arouse the brain.

Prescription

石菖蒲	*shí chāng pú*	20g	Rhizoma Acori Tatarinowii
龟甲	*guī jiǎ*	10g	Carapax et Plastrum Testudinis
茯苓	*fú líng*	20g	Poria
陈皮	*chén pí*	9g	Pericarpium Citri Reticulatae
远志	*yuǎn zhì*	10g	Radix Polygalae
半夏	*bàn xià*	9g	Rhizoma Pinelliae
首乌	*shǒu wū*	15g	Radix Polygoni Multiflori
藿香	*huò xiāng*	9g	Herba Agastachis
甘草	*gān cǎo*	6g	Radix Glycyrrhizae

This was decocted with water and divided to be taken twice, once each in the morning and the evening, one pack per day.

Second Visit

September 20th: After taking five packs of the above formula, she felt her spirit was improving. After another five packs, she felt her brain was clear and her memory was improving. Her tongue coating had turned white and thin. Using the original formula, *tù sī zǐ* (Semen Cuscutae) 15g, and *gé gēn* (Radix Puerariae Lobatae) 20g, were added. She continued taking it.

After taking yet another ten packs, all of her symptoms basically disappeared, and she could return to work.

3. *Shí Chāng Pú* (Rhizoma Acori Tatarinowii) ——
Huò Xiāng (藿香, Herba Agastachis)

Function

To strengthen the stomach, transform dampness and eliminate fullness, using its aromatic property.

Application

This is used for poor digestion inducing stomach distention and fullness, nausea, vomiting and diarrhoea. It also can be used for stomachache or cold-damp stagnation due to eating or drinking raw or cold products inducing stomach and abdominal pain and distention, etc.

Compatibility Analysis

Shí chāng pú (Rhizoma Acori Tatarinowii) is acrid, warm and aromatic. It can transform dampness and harmonise the stomach. *Huò xiāng* (Herba Agastachis) is acrid, sweet and slightly warm. It is also an aromatic herb that transforms dampness, strengthens the stomach and arouses the spleen. Zhang Shan-lei said that: "*Huò xiāng (Herba Agastachis) is aromatic but not harsh. It gently warms but does not lead to dryness and heat. It can eliminate hazy dampness pathogen and assist the normal qi of the spleen and the stomach. It is the fastest herb to release dampness trapped in the spleen yang, causing fatigue and lassitude, and a turbid tongue coating"*. Based on modern pharmacological research, both *shí chāng pú* (Rhizoma Acori Tatarinowii) and *huò xiāng* (Herba Agastachis) can stimulate the stomach fluid secretions and inhibit abnormal fermentation of the stomach and intestines. Therefore, these two herbs combined together can increase the effectiveness of transforming dampness and harmonising the stomach. It is irrelevant whether it is cold-damp or food stagnation patients inducing the stomach or the abdominal distention and fullness, nausea and vomiting, diarrhoea or stomachache, abdominal pain, thick and greasy tongue coating, good results have been found with all of them.

Case 1

Initial Visit

A female patient, 46 years old, March 15th, 1986.

The patient had a naturally weak spleen and stomach. One week previously, due to occasionally eating raw and cold food, she had stomachache, distention and fullness, nausea, diarrhoea, and anorexia. Her pulse was deep and slightly wiry. Her tongue coating was thin, greasy and white. This was cold-damp stagnating in the stomach and the intestines syndrome. The treatment plan was to treat it by warming the middle *jiao*, regulating qi, transforming dampness and harmonising the stomach.

Prescription

菖蒲	*chāng pú*	20g	Rhizoma Acori Tatarinowii
藿香	*huò xiāng*	10g	Herba Agastachis
半夏	*bàn xià*	9g	Rhizoma Pinelliae
陈皮	*chén pí*	20g	Pericarpium Citri Reticulatae
白术	*bái zhú*	12g	Rhizoma Atractylodis Macrocephalae
枳壳	*zhǐ qiào*	10g	Fructus Aurantii
厚朴	*hòu pò*	9g	Cortex Magnoliae Officinalis
茯苓	*fú líng*	25g	Poria
良姜	*liáng jiāng*	6g	Rhizoma Alpiniae Officinarum
香附	*xiāng fù*	12g	Rhizoma Cyperi
神曲	*shén qū*	10g	Massa Medicata Fermentata
甘草	*gān cǎo*	6g	Radix Glycyrrhizae

This was decocted with water and divided to be taken three times, once each in the morning, at noon and in the evening, one pack per day.

Second Visit

March 18th: After taking three packs of the above formula, her stomachache stopped. Her symptoms of abdominal distention, nausea, diarrhoea, etc., were also eliminated, except for the anorexia which was still the same. She was instructed to take the same formula continuously for three more packs after which she recovered.

Case 2

Initial Visit

A female patient, 67 years old, March 22nd, 1988.

The patient had a medical history of stomachache. Recently due to cold, her stomachache happened again accompanied by stomach and abdominal distention and fullness, nausea and vomiting of clear water. Her pulse was wiry and slightly slippery. Her tongue coating was white and greasy. This was a cold pathogen injuring the middle *jiao*, this meant that the stomach and the spleen could not transform and transport and so induced dampness generating internally. Cold and dampness coagulated and gathered in the middle *jiao*, therefore, she had the stomach distention and fullness, nausea and vomiting of clear water. She was immediately treated by transforming dampness, warming the middle *jiao*, harmonising the stomach and stopping nausea.

Prescription

石菖蒲	*shí chāng pú*	15g	Rhizoma Acori Tatarinowii
藿香	*huò xiāng*	10g	Herba Agastachis
茯苓	*fú líng*	24g	Poria
白术	*bái zhú*	12g	Rhizoma Atractylodis Macrocephalae
半夏	*bàn xià*	9g	Rhizoma Pinelliae
陈皮	*chén pí*	10g	Pericarpium Citri Reticulatae
枳壳	*zhǐ qiào*	12g	Fructus Aurantii
砂仁	*shā rén*	4g	Fructus Amomi
乌药	*wū yào*	10g	Radix Linderae
檀香	*tán xiāng*	8g	Lignum Santali Albi
桂枝	*guì zhī*	9g	Ramulus Cinnamomi
甘草	*gān cǎo*	6g	Radix Glycyrrhizae

This was decocted with water and divided to be taken twice, once each in the morning and evening, one pack per day.

Second Visit

March 29th: After taking five packs of the above formula, her stomachache reduced a lot, and her nausea and vomiting of clear water stopped. She still had slight abdominal distention. The above formula was modified by adding *xiāng fù* (Rhizoma Cyperi) 12g and she continued the treatment.

Third Visit

April 3rd: After taking four more packs, all of her symptoms disappeared. She was instructed not to eat raw or cold food. She used three packs more of the same formula to complete the treatment and strengthen the results.

Gōng láo yě's (功劳叶 , Folium Ilex) taste and property are slightly bitter and cold. It enters the lung channel of the hand *taiyin* and the kidney channel of the foot *shaoyin* meridians. Its functions can nourish yin, moisten the lung, and tonify the liver and the kidney. It is an important herb to treat pulmonary tuberculosis inducing cough, by reducing heat and eliminating steaming. Modern clinics mainly use it for pulmonary tuberculosis inducing hot flushes and cough.

1. *Gōng Láo Yè* (Folium Ilex) ——
Bǎi Bù (百部, Radix Stemonae)

Function

To clear heat, moisten the lungs, nourish yin and stop cough, to kill parasites.

Application

This is used for pulmonary tuberculosis inducing cough, fatigue, lassitude, afternoon hot flushes and spontaneous sweating, etc.

Compatibility Analysis

Gōng láo yè (Folium Ilex) can tonify the lung and the kidney, nourish yin, reduce deficiency heat, kill parasites and stop cough. It is a commonly used herb to treat pulmonary tuberculosis. In the *Grand Materia Medica* it states: "*It can moisten the lung and treat the lung heat induced cough.*" *Bǎi bù* (百部 , Radix Stemonae) contains tuberostemonine. Modern clinical research has proved that it has anti-tuberculosis functions and can stop cough. These two herbs combined together can nourish yin, moisten the lung, stop cough, reduce deficiency heat, eliminate bone steaming and kill parasites. This is a main compatibility to treat pulmonary tuberculosis.

Case Study

Initial Visit

A female patient, 32 years old, November 10th, 1985.

In May 1980, due to cough and afternoon hot flushes, the patient was investigated at a local farmers' hospital and diagnosed with tuberculosis in the right upper lobe of her lung. She was treated with anti-tuberculosis medicine and became better, up until the winter of 1984. Due to overexertion, her health became worse. At a check up in an hospital, her chest x-ray showed tuberculosis in the right upper lobe of her lung with cavities.spreading. She was treated with anti-tuberculosis medicine but without good results. Her current diagnosis: The chronic sickness was evident from the patient's face. She was emaciated. She had cough with little phlegm, suffocation, anorexia, shortness of breath, amenorrhoea for two years, spontaneous sweating at night, a thready and fast pulse, a red tongue and thin and white tongue coating. This was both qi and yin deficiency of the lung channel of the hand *taiyin* meridian inducing pulmonary tuberculosis. The treatment plan was to nourish yin, moisten the lung, stop cough and kill the parasites. A modified *Gōng Láo Bǎi Bù Tāng* (Folium Ilicis and Stemona Root Decoction, 功劳百部汤), one of the author's own personal prescriptions, was used.

Prescription

功劳叶	*gōng láo yè*	30g	Folium Ilex
百部	*bǎi bù*	15g	Radix Stemonae
白及	*bái jí*	15g	Rhizoma Bletillae
桔梗	*jié gěng*	10g	Radix Platycodonis
川贝	*chuān bèi*	10g	Bulbus Fritillariae Cirrhosae
杏仁	*xìng rén*	9g	Semen Armeniacae Dulcis
紫菀	*zǐ wǎn*	10g	Radix et Rhizoma Asteris
地榆	*dì yú*	15g	Radix Sanguisorbae
地骨皮	*dì gǔ pí*	30g	Cortex Lycii
乌梅	*wū méi*	9g	Fructus Mume
太子参	*tài zǐ shēn*	15g	Radix Pseudostellariae
甘草	*gān cǎo*	6g	Radix Glycyrrhizae

This was decocted with water and divided to be taken twice, one pack per day.

She also used *Kàng Jié Hé Wán* (Anti Tuberculosis Pill, 抗结核丸) externally to rub on her back, once a week. (Please refer to the compatibility of *xióng huáng* - Realgar, for further detail.)

Second Visit

November 17th: After taking seven packs of the formula, her cough became better, but her lassitude, spontaneous sweating, anorexia, etc. were the same as before. Using the above formula, *shā rén* (Fructus Amomi) 8g, *shēng huáng qí* (raw Radix Astragali) 24g, and *sāng yè* (Folium Mori) 15g, were added, and she continued taking it.

Third Visit

February 2nd: After four months treatment, her clinical symptoms were basically eliminated. Her appetite increased and her body weight increased by 5kg. Her chest x-ray reported: The right upper lung had dense shadow spots. She was diagnosed with pulmonary tuberculosis fibrosis calcification.

Note

Gōng Láo Bǎi Bù Tāng (Folium and Stemona Root Decoction) is one of the author's own personal formulations.

Prescription

功劳叶	*gōng láo yè*	20~30g	Folium Ilex
百部	*bǎi bù*	15g	Radix Stemonae
地榆	*dì yú*	15g	Radix Sanguisorbae
白及	*bái jí*	12g	Rhizoma Bletillae

桔梗	*jié gěng*	10g	Radix Platycodonis
太子参	*tài zǐ shēn*	15g	Radix Pseudostellariae
紫菀	*zǐ wǎn*	10g	Radix et Rhizoma Asteris
甘草	*gān cǎo*	6g	Radix Glycyrrhizae

Method: First, immerse the above herbs in clear water for half an hour, then decoct for thirty minutes. Reserve the decoction, and add water to the strained herbs and decoct for another twenty minutes. Mix the two decoctions together and divide in two, to be taken twice, while warm.

Function

To moisten the lung, nourish yin, stop cough, and to kill parasites.

Application

Pulmonary tuberculosis.

Modifications

1. For patients with fever, and hot flushes, add *qīng hāo* (Herba Artemisiae Annuae), *biē jiǎ* (Carapax Trionycis) and *chái hú* (Radix Bupleuri), etc.

2. For patients with spontaneous night sweating, add *huáng qí* (Radix Astragali), *sāng yè* (Folium Mori) and *wū méi* (Fructus Mume), etc.

3. For patients with cough and phlegm, add *chuān bèi* (Bulbus Fritillariae Cirrhosae), *xìng rén* (Semen Armeniacae Dulcis) and *chén pí* (Pericarpium Citri Reticulatae), etc.

4. For patients with haemoptysis, add *sān qī* (Radix et Rhizoma Notoginseng), *huā ruǐ shí* (花蕊石 , Ophicalcitum), *shān yú ròu* (Fructus Corni) and *xuè yú tàn* (Crinis Carbonisatus), etc.

5. For patients with lassitude and anorexia, add *huáng qí* (Radix Astragali), *xiān hè cǎo* (Herba Agrimoniae) and *shā rén* (Fructus Amomi), etc.

Formula Explanation

Gōng láo yè (Folium Ilex) and *bǎi bù* (Radix Stemonae) nourish yin, moisten the lung, stop cough and kill parasites. *Dì yú* (Radix Sanguisorbae) and *bái jí* (Rhizoma Bletillae) promote binding, tonify the lung, kill parasites and they are suitable for tuberculosis with cavities. Modern pharmacological research indicates that *bái jí* (Rhizoma Bletillae) and *dì yú* (Radix Sanguisorbae) have an inhibitory function on human type tuberculosis. *Jié gěng* (Radix Platycodonis) and *zǐ wǎn* (Radix et Rhizoma Asteris) can disperse the lung, stop cough, moisten the lung and transform phlegm. *Zǐ wǎn* (Radix et Rhizoma Asteris) also has an inhibiting function on the Mycobacterium Tuberculosis. *Tài zǐ shēn* (Radix Pseudostellariae) and *gān cǎo* (Radix Glycyrrhizae) tonify qi and the middle *jiao* and harmonise all herbs. Tonifying the middle *jiao*, can tonify the lung, which is the same as cultivating the earth to generate metal.

Bàn xià's (半夏 , Rhizoma Pinelliae) taste and property are acrid, warm and toxic. It enters the spleen channel of the foot *taiyin*, the stomach channel of the foot *yangming* and the lung channel of the hand *taiyin* meridians. It can dry dampness and transform phlegm, harmonise the stomach and stop nausea, eliminate painful abdominal masses and expel nodules. The spleen is the source of generating phlegm. If the spleen has dampness it cannot transform and transport, so it gathers to form phlegm. *Bàn xià* (Rhizoma Pinelliae) enters the spleen channel of the foot *taiyin* and its properties are warm and dry therefore, it can dry dampness and transform phlegm. It is an important herb in clinical practice to treat phlegm and dampness. The stomach is a *zang* organ to store grains, if the stomach qi is in disharmony then it rebels upward and induces nausea and vomiting. *Bàn xià* (Rhizoma Pinelliae) also enters the stomach channel of the foot *yangming* meridian. Its functions specialise in harmonising the stomach and descending rebellions, therefore it is an important herb to treat rebellion induced nausea. The taste of *bàn xià* (Rhizoma Pinelliae) is acrid, acridity can expel nodules and eliminate stuffiness. For qi stagnation and phlegm clumping plum-stone qi (globus hystericus), phlegm and heat entangled and clumped in the chest, *bàn xià* (Rhizoma Pinelliae) is a necessary herb to treat it. The property of *bàn xià* (Rhizoma Pinelliae) is descending. In the *Miscellaneous Records of Famous Physicians* there is a saying that *bàn xià* (Rhizoma Pinelliae) can cause abortion. However, *Gān Jiāng Rén Shēn Bàn Xià Wán* (Dried Ginger, Ginseng and Pinellia Pill, 干姜人参半夏丸) in the *Essentials from the Golden Cabinet* treats morning sickness and vomiting but does not affect the pregnancy. This indicates: *"If with it has no injury, then, there is no injury."* In clinical practice, persistent pregnancy related nausea and vomiting, cannot be stopped without *bàn xià* (Rhizoma Pinelliae).

The clinical applications for *bàn xià* (Rhizoma Pinelliae) are broad. As it is toxic if used raw, except for external use, it should be processed for internal use. *Shēng jiāng* (Rhizoma Zingiberis Recens) can inhibit its toxin, therefore, *bàn xià* (Rhizoma Pinelliae) is more often processed with *shēng jiāng* (Rhizoma Zingiberis Recens).

1. *Bàn Xià* (Rhizoma Pinelliae) ——
Hòu Pò (厚朴, Cortex Magnoliae Officinalis)

Function

To descend rebellion, transform phlegm, regulate qi, break up stagnation and expel nodules.

Application

It is irrelevant whether it is due to phlegm and qi being mutually entangled inducing plum-stone qi (globus hystericus), or a throat neurosis syndrome or neurotic nausea and vomiting, etc., it can be used for all of these.

Compatibility Analysis

This compatibility is the main herb in Zhong-jing's *Bàn Xià Hòu Pò Tāng* (Pinellia and Official Magnolia Bark Decoction). This formula mainly treats phlegm and qi mutually entangled inducing plum-stone qi (globus hystericus). *Bàn xià* (Rhizoma Pinelliae) is acrid and warm. It can break stagnation, expel nodules and eliminate phlegm. *Hòu pò* (Cortex Magnoliae Officinalis) is acrid, bitter and warm. It can expel nodules, break stagnation and promote qi movement. If qi moves, then the dampness pathogen cannot gather and phlegm can be eliminated. When these two herbs are combined together, *bàn xià* (Rhizoma Pinelliae) gets *hòu pò* (Cortex Magnoliae Officinalis) to promote qi movement and this can increase the power to expel dampness and eliminate phlegm. *Hòu pò* (Cortex Magnoliae Officinalis) gets *bàn xià* (Rhizoma Pinelliae) to dry dampness and this can enhance the strength to break stagnation and expel nodules, therefore, it is suitable for treating symptoms due to entangled phlegm and qi .

Case Study

Initial Visit

A female patient, 27 years old, April 21st, 1989.

The patient felt a blockage in her throat for a couple of months. Recently, she got angry after an argument. She felt that her chest and diaphragm was full and stuffy, it felt like something jamming in her throat which she could not swallow down or spit out, it was accompanied by depression, bad quality sleep, a lot of dreams and being easily startled, dizziness, and anorexia. She was extremely afraid of getting oesophageal cancer, and so, came to our clinic for treatment. Her upper G.I. tract image after a barium radiograph showed that her oesophagus was clear without abnormality. On examination, her pulse was wiry, thready and deep. Her tongue coating was thin and white. This belongs to the qi mechanism stagnating and clumping. If qi is stagnated, then dampness gathers and becomes phlegm. Phlegm and qi become mutually entangled and induce this syndrome.

Prescription

半夏	*bàn xià*	10g	Rhizoma Pinelliae
厚朴	*hòu pò*	12g	Cortex Magnoliae Officinalis
茯苓	*fú líng*	15g	Poria
柴胡	*chái hú*	15g	Radix Bupleuri
杭白芍	*háng bái sháo*	15g	Radix Paeoniae Lactiflorae
薄荷	*bò he*	8g	Herba Menthae
苏梗	*sū gěng*	10g	Caulis Perillae
当归	*dāng guī*	12g	Radix Angelicae Sinensis
白术	*bái zhú*	10g	Rhizoma Atractylodis Macrocephalae
桔梗	*jié gěng*	9g	Radix Platycodonis
甘草	*gān cǎo*	6g	Radix Glycyrrhizae

This was decocted with water, to be taken one pack per day.

Second Visit

April 25th: After taking four packs of the above formula, her chest stuffiness and throat blockage were reduced by more than half. Her sleep still had a lot of dreams and she easily got over anxious. Modifying the original formula, *jié gěng* (Radix Platycodonis) and *bái zhú* (Rhizoma Atractylodis Macrocephalae) were eliminated, and *suān zǎo rén* (Semen Ziziphi Spinosae) 15g, *yè jiāo téng* (Caulis Polygoni Multiflori) 20g, *shēng lóng chǐ* (生龙齿 , raw Dens Dragonis) 20g, were added. She continued to take it.

Thrid Visit

April 30th: All of her symptoms were basically eliminated. Her sleep was also good except for occasional discomfort in her chest. The formula was changed to *Xiāo Yáo Săn* (Free Wanderer Pill, 逍遥丸) to complete the treatment.

2. *Bàn Xià* (Rhizoma Pinelliae) ——
Mài Mén Dōng (麦门冬, Radix Ophiopogonis)

Function

To nourish the stomach, generate fluid, regulate the penetrating vessel and descend rebellion.

Application

This is used for stomach yin deficiency, stomach qi rebelling upward inducing nausea, vomiting and hiccough; or lung or stomach yin deficiency, with body fluid consumed and injured inducing dry throat, cough or deficiency type asthma. Also used for women's irregularity in the penetrating vessel inducing the penetrating vessel rebelling upward syndrome.

Compatibility Analysis

The properties of *bàn xià* (Rhizoma Pinelliae) are slippery and descending. It is an important herb to harmonise the stomach and descend rebellions. *Mài mén dōng* (Radix Ophiopogonis) is sweet, bitter and slightly cold. It enters the stomach channel of the foot *yangming* meridian. It is an herb to nourish the stomach and generate fluid. It is a sweet, cold clear and moistening herb. If the stomach yin is consumed and injured, then the stomach fire will flare upward rebelliously. *Mài mén dōng* (Radix Ophiopogonis) can nourish the stomach yin and clear the stomach fire. The stomach resides in the middle *jiao* and is the sea of water and grains. Descending is the direction of flow for stomach qi. If the stomach qi descends then it can accept water and grains. The stomach likes moistness and dislikes dryness. If the stomach yin is sufficient, then fluid can be spread. These two herbs combined together can nourish the stomach yin and descend the stomach qi. *Mài mén dōng* (Radix Ophiopogonis) uses the descending property of *bàn xià* (Rhizoma Pinelliae) to increase its power to nourish the stomach, generate fluid and accept foods (to descend is to tonify the stomach). *Bàn xià* (Rhizoma Pinelliae) uses the cold property of *mài mén dōng* (Radix Ophiopogonis) to reduce its disadvantage of acridity and dryness as well as make the power to descend the stomach qi even stronger. This compatibility is from the *Mài Mén Dōng Tāng* (Ophiopogonis Decoction) in the *Essentials from the Golden Cabinet*. The original formula is mainly to treat the lung atrophy. This is cultivating the earth to generate the metal. The dosage proportion for *mài mén dōng* (Radix Ophiopogonis) and *bàn xià* (Rhizoma Pinelliae) in the original formula was 7:1. The author suggests that if it is stomach yin deficiency inducing the fire to rebel and qi travelling upward syndrome, the dosage of *mài mén dōng* (Radix Ophiopogonis) is better off a couple of times larger than that of *bàn xià* (Rhizoma Pinelliae) in order to prevent the acridity and dryness of *bàn xià* (Rhizoma Pinelliae) from injuring yin. If the stomach yin deficiency syndrome is not significant but it is only that the stomach has lost its harmony and descending function or the penetrating vessel is rebelling upward, then the ratio of *mài mén dōng* (Radix Ophiopogonis) and *bàn xià* (Rhizoma Pinelliae) can be smaller, between 2:1 to 3:1 is sufficient. This compatibility treating the penetrating vessel rebelling upward is based on our ancestors' saying that: *"The penetrating vessel belongs to the yangming."* This relates to the fact that descending is the route of the penetrating vessel and if descending the *yangming* then one is also descending the penetrating vessel. (Please refer to the compatibility of *chuān niú xī* (Radix Cyathulae)).

Case 1

Premenstrual phase mouth ulcer.

Initial Visit

A female patient, 27 years old, married, August, 5th, 1985.

Over the previous year, every month, one week before menstruation, mouth ulcers appeared. The patient tongue and lower lip had a couple of miliary ulcer spots. At meal times or when eating spicy food, it was painful and hard to bear. She had used some antibiotics (unknown) but to no effect. It kept reoccurring for more than a year. Her menstruation was basically normal but was accompanied by lower abdominal discomfort. Her pulse was wiry and thready. Her tongue was red and tongue coating was thin and white. This was caused by rebellious qi of the penetrating vessel accompanied by the heart and the spleen hidden fire attacking upward. The spleen opens its orifice at the mouth and the tongue is the heart's sprout. The heart and the spleen's hidden fire steamed upward, therefore, her mouth and tongue generated ulcers. The treatment plan was to use a formula to regulate the penetrating vessel, and descend rebellion, while clearing the heart and the spleen hidden fire.

Prescription

麦冬	*mài dōng*	24g	Radix Ophiopogonis
半夏	*bàn xià*	12g	Rhizoma Pinelliae
党参	*dǎng shēn*	15g	Radix Codonopsis
川牛膝	*chuān niú xī*	12g	Radix Cyathulae
荷叶	*hé yè*	9g	Folium Nelumbinis
防风	*fáng fēng*	4g	Radix Saposhnikoviae
栀子	*zhī zǐ*	6g	Fructus Gardeniae
益母草	*yì mǔ cǎo*	9g	Herba Leonuri
当归	*dāng guī*	10g	Radix Angelicae Sinensis
丹参	*dān shēn*	15g	Radix et Rhizoma Salviae Miltiorrhizae
代赭石	*dài zhě shí*	15g	Haematitum
甘草	*gān cǎo*	6g	Radix Glycyrrhizae

This was decocted with water and divided to be taken twice, once each in the morning and evening, one pack per day.

Second Visit

August 15th: After taking three packs of the above formula, before menstruation, there was no mouth ulcer. She was instructed to take the above formula again before her menstruation.

Later, after follow up visits for ten months, it did not happen again.

Case 2

Dizziness and vomiting.

Initial Visit

A female patient, 30 years old, June, 12th, 1986.

The patient had dizziness for a week which was accompanied by irritability, nausea and vomiting, anorexia and not wanting to eat, and her blood pressure was 90/60mm Hg. An otolaryngology examination did not show any abnormality. Her pulse was wiry, thready and fast. Her tongue was red with a scanty tongue coating. This was stomach yin deficiency and stomach qi rebelling upward syndrome. If the stomach yin was deficient then her stomach qi was in disharmony and rebelled upward, which manifested in her irritability, nausea and vomiting, anorexia, and not wanting to eat. If her yin was deficient then the fire would flare up. The deficiency fire flared upward and bothered the clear yang, which manifested in the dizziness in her head and eyes. Her pulse was wiry and thready. Her tongue was red with scanty tongue coating. These are yin deficiency signs. The treatment plan was to nourish the stomach, generate fluid, harmonise the stomach and descend rebellion.

Prescription

麦冬	*mài dōng*	30g	Radix Ophiopogonis
半夏	*bàn xià*	12g	Rhizoma Pinelliae
太子参	*tài zǐ shēn*	15g	Radix Pseudostellariae
菊花	*jú huā*	15g	Flos Chrysanthemi
葛根	*gé gēn*	20g	Radix Puerariae Lobatae
荷叶	*hé yè*	10g	Folium Nelumbinis
代赭石	*dài zhě shí*	15g	Haematitum
竹茹	*zhú rú*	9g	Caulis Bambusae in Taenia
五味子	*wǔ wèi zǐ*	5g	Fructus Schisandrae Chinensis
甘草	*gān cǎo*	6g	Radix Glycyrrhizae

This was decocted with water one pack to be taken per day.

Second Visit

June 15th: After taking three packs of the above formula, her dizziness in her head and irritability had reduced a lot. Her nausea and vomiting disappeared. Her appetite was normal. As it was effective, the prescription was not changed. She used three more packs of the same formula and she recovered.

3. *Bàn Xià* (Rhizoma Pinelliae) ——
Fú Líng (茯苓, Poria)

Function

To eliminate dampness and transform phlegm.

Application

It is useful for spleen deficiency that cannot transform and transport or for damp-phlegm causing symptoms such as cough with a lot of phlegm, chest and diaphragm fullness and a suffocating sensation, nausea and vomiting, dizziness and palpitations, etc..

Compatibility Analysis

Damp-phlegm in patients is due to the spleen yang not transforming and transporting, causing water and dampness to congeal and gather. There is a saying that *"the spleen is the source to generate phlegm."* The yang qi of the middle *jiao* does not transform and transport, dampness stagnates and phlegm is generated. Phlegm obstructs the qi mechanism, therefore, chest and diaphragm fullness and tightness manifest. Phlegm follows qi rising up, upward it attacks the lung, then there is cough with a lot phlegm. There is also a saying that *"the lung is a container to store phlegm."* Phlegm and dampness obstruct the middle *jiao*, the stomach loses its harmony and descending function and then nausea and vomiting manifest. Phlegm and dampness clumping under the heart when the heart yang is weak causes palpitations. Phlegm and dampness going upward bothering the clear yang causes dizziness. *Bàn xià* (Rhizoma Pinelliae) is acrid, warm and its property is dry. It has functions to dry dampness and eliminate phlegm. *Fú líng* (Poria) is sweet, neutral and bland. It has the power to leach out dampness and strengthen the spleen. The spleen likes dryness and dislikes dampness. *Bàn xià* (Rhizoma Pinelliae) can dry dampness and can help the spleen yang to healthily transform and transport. *Fú líng* (Poria) leaches out dampness and can help the spleen qi to move healthily. If the spleen qi moves healthily, then dampness cannot accumulate. If there is no dampness accumulation, then no phlegm will be generated. All symptoms will dissipate by themselves. This is a method to correct the primary problem and clear the source. *Èr Chén Tāng* (Two Matured Ingredients Decoction, 二陈汤) in the *Formulary of the Bureau of Medicines of the Taiping Era* uses this compatibility as the king herbs. It is a famous formula to treat phlegm syndromes.

Case 1

Initial Visit

A female patient, 41 years old, January 16th, 1987.

The patient had dizziness for half a year. It was sometimes better and sometimes severe. When severe, her head was dizzy and she could not open her eyes. She felt her surroundings were spinning and she could not stand. Meanwhile, she had nausea and vomiting. She was investigated in an hospital and diagnosed with meniere's disease.

She took medicine but with no effect and so came for Chinese medicine treatments. On examination, the patient's body was slightly overweight. Her pulse was wiry, thready and slightly slippery. Her tongue coating was white and turbid, and her tongue was dark. Our ancestors said that: "*If there is no phlegm, then, there is no dizziness.*" If the phlegm and dampness travel upward and bother the clear yang this induces dizziness.

Prescription

茯苓	*fú líng*	30g	Poria
半夏	*bàn xià*	15g	Rhizoma Pinelliae
陈皮	*chén pí*	12g	Pericarpium Citri Reticulatae
枳实	*zhǐ shí*	12g	Fructus Aurantii Immaturus
竹茹	*zhú rú*	10g	Caulis Bambusae in Taenia
丹参	*dān shēn*	30g	Radix et Rhizoma Salviae Miltiorrhizae
葛根	*gé gēn*	30g	Radix Puerariae Lobatae
钩藤	*gōu téng*	30g	Ramulus Uncariae Cum Uncis (decoct later)
泽泻	*zé xiè*	30g	Rhizoma Alismatis
荷叶	*hé yè*	10g	Folium Nelumbinis
甘草	*gān cǎo*	6g	Radix Glycyrrhizae

This was decocted with water, to be taken one pack per day.

Second Visit

January 19th: After taking three packs of the above formula, then her dizziness seemed to disappear. She took three more packs of the original formula.

She was seen half a year later and she said that after six packs of the formula, there was no reoccurrence.

Case 2

Initial Visit

A female patient, 44 years old, September 27th, 1985.

The patient had cough, chest tightness, with a lot of phlegm for more than half a month. The cough and phlegm became worse when she woke up in the morning. It was white phlegm, accompanied by loss of ability to taste food. Her pulse was wiry. Her tongue was red and her tongue coating was white and slightly turbid. This was phlegm and dampness obstructing internally, with the lung qi being unable to disperse. Phlegm and dampness obstructing internally, caused her chest stiffness, and produced a lot of phlegm, loss of taste for food; and the lung qi could not disperse, and therefore, she had the cough. The treatment plan was to disperse the lung, transform phlegm and stop cough.

Prescription

茯苓	*fú líng*	20g	Poria
半夏	*bàn xià*	10g	Rhizoma Pinelliae
陈皮	*chén pí*	9g	Pericarpium Citri Reticulatae

桔梗	jié gěng	9g	Radix Platycodonis
侧柏叶	cè bǎi yè	12g	Cacumen Platycladi
百部	bǎi bù	10g	Radix Stemonae
杏仁	xìng rén	9g	Semen Armeniacae Dulcis
川贝	chuān bèi	9g	Bulbus Fritillariae Cirrhosae
防风	fáng fēng	6g	Radix Saposhnikoviae
荆芥	jīng jiè	6g	Herba Schizonepetae
甘草	gān cǎo	6g	Radix Glycyrrhizae

This was decocted with water and divided to be taken twice, once each in the morning and evening, one pack per day.

Second Visit

October 3rd: After taking five packs of the above formula, her cough and phlegm were reduced by more than half. She was instructed to continue to take the original formula.

Third Visit

October 8th: After taking five more packs of the above formula, she had a slight cough and her phlegm was eliminated. She was instructed to continue to take another three packs of the original formula to complete the treatment.

4. *Bàn Xià* (Rhizoma Pinelliae) —— *Bái Jí* (白及, Rhizoma Bletillae), *Bǎi Hé* (百合, Bulbus Lilii)

Function

To harmonise the stomach, descend rebellion, transform blood stasis and alleviate pain.

Application

This is mainly used for the symptoms of reflux oesophagitis as well as oesophageal hiatus hernia inducing vomiting and pain behind the ribs, etc.

Compatibility Analysis

Regurgitation oesophagitis is a condition caused by reduced sphincter function at the inferior end of the oesophagus, it is an oesophageal peristalsis disorder, and regurgitation of gastric and intestinal contents inducing inflammation on the lower section mucosal membranes of the oesophagus. There are many aetiologies such as drinking, smoking, recurrent vomiting, food retention in the stomach, post operation on the oesophagus and the stomach, etc., all of these can reduce the oesophagus distal end sphincter function. In addition, oesophageal hiatus hernia, obesity, elevated abdominal pressure, etc., also can induce this disease. The treatment principle is mainly to reduce the reflux, avoid the stimulation from food regurgitation and improve the sphincter function. The Chinese medicine thought is that this disease is due to stomach qi disharmony, the stomach qi rebels upward, the stomach fluid follows the rebellious qi flowing upward to burn and injure the oesophagus. The treatment plan should mainly harmonise the stomach, descend rebellion, transform blood stasis and alleviate pain. Going down is the route of the stomach. If the stomach qi descends and the stomach fluid does not rebel upward, then all the symptoms will recover by themsleves.

Bàn xià (Rhizoma Pinelliae) is acrid and warm. Its property is slippery and beneficial for descending. It is an important herb to harmonise the stomach, descend the rebellion and stop vomiting. *Bǎi hé* (Bulbus Lilii) is sweet and neutral. It nourishes yin and tonifies the middle *jiao*. It is the king herb to tonify the spleen and stomach as well as harmonise the middle *jiao*. It, combined with *bàn xià* (Rhizoma Pinelliae), can harmonise the stomach and descend rebellion, as well as prevent the acrid and dry side effect of *bàn xià* (Rhizoma Pinelliae). *Bái jí* (Rhizoma Bletillae) is bitter and neutral. It can reduce swelling and generate flesh. Its property is extremely astringent and its quality is extremely sticky, therefore, it functions to protect the oesophageal mucosal membrane and increase the contracting function of the sphincter. In the *Essentials of the Materia Medica* it states that *bái jí* (Rhizoma Bletillae) can *"expel blood stasis and generate new blood"*, therefore, for oesophagitis, it can alleviate the pain and improve the condition. This is a compatibility to treat both the manifestations and the root problem.

Case 1

Reflux oesophagitis.

Initial Visit

A male patient, 26 years old, April 19th, 2000.

The patient had pain behind his ribs for more than a year. He had been investigated at a local hospital. His EKG was normal. His stomach endoscope test. showed: His oesophagus had hyperaemia and oedema and had some bleeding spots. He was diagnosed with reflux oesophagitis. A friend referred him to come to our clinic for treatment.

On examination, the patient was physically well developed. He stated that he had a history of alcoholism, a burning sensation pain behind his ribs, sometimes radiating to his scapulae, more often after meals or lying down on his side, and was accompanied by heartburn. His appetite was fine. His pulse was wiry and thready. His tongue was red and tongue coating was white. This was the stomach qi rebelling upward, with the stomach fluid burning and injuring the oesophagus. The treatment plan was to harmonise the stomach, descend rebellion, inhibit acid and alleviate pain.

Prescription

百合	*bǎi hé*	24g	Bulbus Lilii
半夏	*bàn xià*	10g	Rhizoma Pinelliae
白及	*bái jí*	15g	Rhizoma Bletillae
砂仁	*shā rén*	6g	Fructus Amomi
海螵蛸	*hǎi piāo xiāo*	24g	Endoconcha Sepiae
大贝母	*dà bèi mǔ*	10g	Bulbus Fritillariae Thunbergii
乌药	*wū yào*	12g	Radix Linderae

丹参	*dān shēn*	20g	Radix et Rhizoma Salviae Miltiorrhizae
藿香	*huò xiāng*	9g	Herba Agastachis
甘草	*gān cǎo*	6g	Radix Glycyrrhizae

This was decocted with water and divided to be taken twice, one pack per day.

Second Visit

April 27th: After taking seven packs of the above formula, the pain behind his ribs reduced a lot. His acid regurgitation and reflux were also reduced. The original formula was effective and so was continued with.

Third Visit

May 20th: After taking more than twenty packs of the above formula, all of his symptoms disappeared. The patient was instructed to drink less and to pay attention to his diet. His formula was stopped for observation.

Case 2

Oesophageal hiatus hernia.

Initial Visit

A male patient, 79 years old, March 22nd, 2000.

The patient had a history of heartburn and stomach discomfort. Recently, he suddenly had an episode of nausea and vomiting; he could not eat and had a burning pain sensation behind his ribs. He had been investigated in an hospital's Radiology Department. An upper G.I. tract barium radiograph led to his diagnosis of oesophageal hiatus hernia. After administration of an infusion and the injection of Yimao Seoul, etc., were without effectiveness and he came for treatment. On examination, the patient's acute sickness was apparent from his face. His pulse was wiry and slippery. His tongue coating was white and his tongue was red. This was stomach qi deficiency of the elderly. The stomach qi deficiency induced the qi mechanism to rebel upward, therefore, he had nausea, vomiting and could not eat. The treatment plan was to tonify qi, harmonise the stomach and descend rebellion.

Prescription

太子参	*tài zǐ shēn*	30g	Radix Pseudostellariae
代赭石	*dài zhě shí*	24g	Haematitum
旋覆花	*xuán fù huā*	15g	Flos Inulae
半夏	*bàn xià*	10g	Rhizoma Pinelliae
百合	*bǎi hé*	20g	Bulbus Lilii
白及	*bái jí*	10g	Rhizoma Bletillae
麦门冬	*mài mén dōng*	15g	Radix Ophiopogonis
沉香	*chén xiāng*	9g	Lignum Aquilariae Resinatum
甘草	*gān cǎo*	6g	Radix Glycyrrhizae

This was decocted with water and divided to be taken twice, one pack per day.

Second Visit

March 26th: After taking four packs of the above formula, his nausea and vomiting were stopped. He could eat, but his acid regurgitation and stomach discomfort were the same as before. Based on the above formula, *sāng piāo xiāo* (Endoconcha Sepiae) 15g, *dà bèi mǔ* (Bulbus Fritillariae Thunbergii) 10g and *wū yào* (Radix Linderae) 12g were added. He was instructed to continue taking it.

Third Visit

April 2nd: His appetite was nearly back to normal. His acid regurgitation and stomach discomfort also basically disappeared. But he still had a stuffy sensation in his stomach, his tongue coating was slightly turbid, he had had no bowel movement for a couple of days and he still had stomach discomfort after eating. The treatment plan was to harmonise the stomach, reduce the food, descend rebellion and free bowel movement.

Prescription

太子参	*tài zǐ shēn*	15g	Radix Pseudostellariae
旋覆花	*xuán fù huā*	15g	Flos Inulae
半夏	*bàn xià*	9g	Rhizoma Pinelliae
当归	*dāng guī*	12g	Radix Angelicae Sinensis
大贝母	*dà bèi mǔ*	10g	Bulbus Fritillariae Thunbergii
焦大黄	*jiāo dà huáng*	9g	Radix et Rhizoma Rhei scorch-fried
砂仁	*shā rén*	8g	Fructus Amomi
焦三仙	*jiāo sān xiān*	30g	Fructus Hordei Germinatus et Crataegi et Massa Fermentata Medicinalis (scorch-fried)
鸡内金	*jī nèi jīn*	12g	Endothelium Corneum Gigeriae Galli
甘草	*gān cǎo*	6g	Radix Glycyrrhizae

Fourth Visit

April 5th: After taking three packs of the above formula, his bowel movement was freed. His tongue coating turned thin and white. He stopped taking his formula and recovered.

Note

Oesophageal regurgitation oesophagitis is a common disease in clinical practice. The author formulated his own *Hé Wèi Jiàng Nì Tāng* (Harmonise the Stomach and Downbear the Counterflow Decoction, 和胃降逆汤) to treat this disease for a couple of cases and all had effective treatment. It is listed below for reference.

党参	*dǎng shēn*	15g	Radix Codonopsis
半夏	*bàn xià*	10~15g	Rhizoma Pinelliae
百合	*bǎi hé*	24g	Bulbus Lilii
白及	*bái jí*	10~15g	Rhizoma Bletillae

大贝母	*dà bèi mǔ*	10g	Bulbus Fritillariae Thunbergii
海螵蛸	*hǎi piāo xiāo*	15g	Endoconcha Sepiae
丹参	*dān shēn*	20g	Radix et Rhizoma Salviae Miltiorrhizae
甘草	*gān cǎo*	6g	Radix Glycyrrhizae

Application

This was decocted with water and divided to be taken twice, once each in the morning and evening, one pack per day.

Modifications

1. For the patients with severe nausea and vomiting, add *dài zhě shí* (Haematitum) and *xuán fù huā* (Flos Inulae), etc.

2. For patients with bleeding, add *sān qī fěn* (Powdered Radix et Rhizoma Notoginseng) and *jiāo dà huáng* (焦大黄 , scorch-fried Radix et Rhizoma Rhei), etc.

3. For patients with severe pain, add *yán hú suǒ* (Rhizoma Corydalis) and *wǔ líng zhī* (Faeces Trogopterori).

4. For constipated patients, add *dà huáng* (Radix et Rhizoma Rhei) and *dāng guī* (Radix Angelicae Sinensis), etc.

Compatibility Analysis

Bàn xià (Rhizoma Pinelliae), *bǎi hé* (Bulbus Lilii), *dǎng shēn* (Radix Codonopsis) and *gān cǎo* (Radix Glycyrrhizae) harmonise the stomach and descend rebellions, tonify qi, nourish yin and tonify the middle *jiao*. If the stomach is harmonised and the qi is descending then the stomach fluid will not rebel upward or flux. If the middle *jiao* qi is plenty, then the stomach and intestines will function normally and can improve the oesophagus' distal end sphincter function to prevent food and stomach fluid flux to treat the root of the problem. *Hǎi piāo xiāo* (Endoconcha Sepiae) and *dà bèi mǔ* (Bulbus Fritillariae Thunbergii) have the function to inhibit stomach acid secretion. They are acid inhibitors in Chinese medicine. They can reduce the stomach acid to stimulate the oesophagus and improve the clinical symptoms. *Dān shēn* (Radix et Rhizoma Salviae Miltiorrhizae), *bái jí* (Rhizoma Bletillae) and *dà bèi mǔ* (Bulbus Fritillariae Thunbergii) can transform blood stasis and generate new blood, and protect the oesophageal mucosal membrane. *Dān shēn* (Radix et Rhizoma Salviae Miltiorrhizae) invigorates blood and transforms blood stasis, and promotes blood circulation. It can help the injured oesophageal mucosal membrane to recover. *Bái jí* (Rhizoma Bletillae) reduces swellings and generates flesh. It can protect the mucosal membrane. *Dà bèi mǔ* (Bulbus Fritillariae Thunbergii) also functions to treat abscesses, generate flesh and protect the mucosal membrane. It also can moisten the intestines to free bowel movement and help the stomach qi in descending. All the above herbs together, can achieve functions to tonify qi, harmonise the stomach, descend rebellion, transform blood stasis, inhibit acid and alleviate pain.

5. *Bàn Xià* (Rhizoma Pinelliae) ——
Bǎi Hé (百合, Bulbus Lilii), *Huáng Qín* (黄芩, Radix Scutellariae)

Function

To harmonise the stomach, benefit the gallbladder, descend rebellion and alleviate pain.

Application

This is used for bile reflux gastritis.

Compatibility Analysis

Bile reflux gastritis is classified under stomach conditions in Chinese medicine. It manifests clinically as a continuously burning pain sensation in the upper abdomen. The pain is is worse after meals and is accompanied by vomiting sour or bitter fluid; there was no relief even after taking acid inhibitor medicines. The disease most often occurs after a stomachectomy, or the pyloric orifice has insufficient function, or due to other causes inducing pyloric orifice functional disorder and inducing bile reflux to the stomach, damaging the gastric mucosal barrier and thus inducing inflammation. As bile is a basic fluid, using acid inhibitors to treat it will not relieve the situation. Chinese medicine postulates two aetiologies for the disease: One is stomach deficiency could not descend. If the stomach is deficient, it cannot control the pyloric sphincter, therefore, its opening and closing function is disturbed and this induces bile reflux. Stomach qi should flow downwards. If the stomach is deficient and it cannot descend then, it will rebel upward and move bile to reflux and rebel upward too. The second one is the gallbladder channel of the foot *shaoyang* stagnated fire rebelling upward. The gallbladder is the wood and the stomach belongs to the earth. If earth is deficient, then wood will overact on its deficiency and attack it. Wood fire attacked the stomach, therefore, the stomach and abdomen had burning pain. Therefore, the treatment plan should harmonise the stomach, descend rebellion, soothe the liver and benefit the gallbladder.

Bàn xià (Rhizoma Pinelliae) combined with *bǎi hé* (Bulbus Lilii) has functions to harmonise the stomach, tonify the middle *jiao* and descend rebellion. *Huáng qín* (Radix Scutellariae) is bitter and cold. It can clear the excess heat in the gallbladder channel of the foot *shaoyang* meridian. It also can freely descend without rebelling upward. They have a mutually supplementing power, and so the symptoms can be relieved.

Case Study

Initial Visit
A male patient, 42 years old, September 24th, 1996.

The patient had stomachache for more than a year. He took all kinds of stomach medicines without significant effect, and it was getting worse. He was afraid he might have a tumour and came for treatment and a check up. The patient complained that his stomach had a burning pain sensation, it was worse after meals, and was accompanied by a dry and bitter mouth, regurgitation of sour fluid, and sometimes vomiting. His liver and gallbladder B-ultrasound test did not show any abnormality. His stomach endoscope reported: The stomach membrane had severe hyperaemia and oedema, it had a flower shape appearance with red protrusions and white concavities, and the pyloric orifice was opening continuously. He was diagnosed with bile reflux gastritis. His pulse was deep. His tongue was dark and tongue coating was white turbid and thick. This was the gallbladder and stomach disharmony, with damp-heat stagnating internally syndrome. The treatment plan was to harmonise the stomach, descend rebellion, benefit the gallbladder and clear heat.

Prescription

黄芩	huáng qín	12g	Radix Scutellariae
柴胡	chái hú	12g	Radix Bupleuri
半夏	bàn xià	10g	Rhizoma Pinelliae
百合	bǎi hé	20g	Bulbus Lilii
乌药	wū yào	12g	Radix Linderae
砂仁	shā rén	8g	Fructus Amomi
代赭石	dài zhě shí	20g	Haematitum
丹参	dān shēn	20g	Radix et Rhizoma Salviae Miltiorrhizae
蒲公英	pú gōng yīng	15g	Herba Taraxaci
白术	bái zhú	12g	Rhizoma Atractylodis Macrocephalae
苏梗	sū gěng	9g	Caulis Perillae
甘草	gān cǎo	6g	Radix Glycyrrhizae

This was decocted with water to be taken one pack per day.

Second Visit

October 7th: After taking seven packs of the above formula, his symptoms of stomachache, vomiting, etc. all improved a lot. But after stopping the formula, he had a bitter mouth and stomachache again. He continued with the original formula again.

Third Visit

October 30th: After taking ten more packs, all of his symptoms disappeared. Recently, due to drinking alcohol, he had stomach fullness and pain, and wanted to vomit. The original formula was used to treat him and he recovered.

Bǎi hé's (百合 , Bulbus Lilii) taste and property are sweet and neutral. It enters three meridians: the heart channel of the hand *shaoyin*, the lung channel of the hand *taiyin* and the spleen channel of the foot *taiyin*. It is an herb to nourish yin, clear heat, moisten the lung and stop cough. In clinical practice, it is more often used for lung yin deficiency induced chronic cough, dry cough or haemoptysis, etc. As it enters the heart channel of the hand *shaoyin* meridian, it has the function to nourish the heart and anchor the spirit. It is also the king herb to treat Lily disease. Currently in clinical practice, *bǎi hé* (Bulbus Lilii) is also used to treat stomach diseases, because it can enter the middle *jiao* and nourish the spleen and the stomach. In the *Divine Husbandman's Classic of the Materia Medica* it states that *bǎi hé* (Bulbus Lilii) can "*tonify the middle jiao and qi*". In the *Materia Medica of the Ming Dynasty* (大明本草 , *Dà Míng Běn Cǎo*) it states that it can "*nourish the five zang organs*". *Bǎi hé* (Bulbus Lilii) has been used since ancient times as an herb to nourish the spleen and stomach.

1. *Bǎi Hé* (Bulbus Lilii) ——
Wū Yào (乌药, Radix Linderae)

Function

To nourish the stomach, calm the middle *jiao* and alleviate pain.

Application

This is used for gastritis, gastric and duodenal ulcers, etc. that belong to stomach yin deficiency with symptoms of stomach distention and fullness, pain, anorexia, a thin and white tongue coating or a scanty tongue coating.

Compatibility Analysis

Bǎi hé (Bulbus Lilii) is sweet, neutral and moist. It can nourish the stomach, generate fluid, and tonify the middle *jiao* and qi. The stomach is the sea of water and grains; it governs acceptance, and likes moistness and hates dryness. *Bǎi hé* (Bulbus Lilii) is sweet and moist which makes it an important herb to nourish the stomach. *Wū yào* (Radix Linderae) is acrid, warm and promotes movement of qi. It can disperse all kinds of qi. It is an herb to regulate qi and alleviate pain. Huang Gong-xiu said that: "*All diseases that belong to qi rebellion that have symptoms of chest and abdomen discomfort, are all pertinent to use it.*" *Wū yào* (Radix Linderae) enters the lung channel of the hand *taiyin*, the spleen channel of the foot *taiyin*, the stomach channel of the foot *yangming* and the kidney channel of the foot *shaoyin* meridians. In the upper *jiao*, it can regulate the qi in the chest and diaphragm. In the middle *jiao*, it can regulate the basal qi of the stomach and the spleen. In the lower *jiao*, it can free the kidney channel of the foot *shaoyin* meridian and eliminate the cold qi in the lower *jiao*. *Bǎi hé* (Bulbus Lilii) combined with *wū yào* (Radix Linderae), can nourish the stomach and calm the middle *jiao*, regulate qi, alleviate pain and especially regulate the qi mechanism in the middle *jiao*. Meanwhile, *bǎi hé* (Bulbus Lilii) uses the acrid and warm properties of *wū yào* (Radix Linderae), to prevent stagnation from its sweet and moist properties. *Wū yào* (Radix Linderae) uses the sweetness and moistness of *bǎi hé* (Bulbus Lilii), to prevent injury to yin from its aromatic penetrating properties. It is useful for any kind of stomach disease.

Case 1

Duodenal ulcer.

Initial Visit

A male patient, 33 years old, August 5th, 1987.

The patient complained of stomach distention, fullness and pain for a couple of months accompanied by anorexia, and discomfort in both hypochondria. He had a barium radiography examination in a hospital and was diagnosed with a duodenal ulcer. His pulse was wiry, thready and slippery. His tongue was red and slightly dark. His tongue coating was white. The treatment plan was to tonify the stomach, invigorate blood, regulate qi and alleviate pain.

Prescription

百合	bǎi hé	24g	Bulbus Lilii
乌药	wū yào	10g	Radix Linderae
檀香	tán xiāng	8g	Lignum Santali Albi
砂仁	shā rén	3g	Fructus Amomi
丹参	dān shēn	20g	Radix et Rhizoma Salviae Miltiorrhizae
海螵蛸	hǎi piāo xiāo	15g	Endoconcha Sepiae
枳壳	zhǐ qiào	10g	Fructus Aurantii
白术	bái zhú	12g	Rhizoma Atractylodis Macrocephalae
枸杞	gǒu qǐ	15g	Fructus Lycii
三棱	sān léng	6g	Rhizoma Sparganii
莪术	é zhú	6g	Rhizoma Curcumae
甘草	gān cǎo	6g	Radix Glycyrrhizae

This was decocted with water and divided to be taken twice, once each in the morning and evening, one pack per day.

Second Visit

August 15th: After taking six packs of the above formula, his stomach distention, fullness and pain all reduced a lot. His appetite increased. Only his stool was slightly dry. Based on the above formula dà bèi mǔ (Bulbus Fritillariae Thunbergii) 10g, was added and he continued to take it.

Third Visit

August 21st: All of his symptoms were reduced. His bowel movement was normal. Again, the original formula was used to treat him.

Fourth Visit

August 27th: All of his symptoms basically disappeared. His appetite was normal. Three packs of the original formula were used to complete the treatment.

Case 2

Stomachache.

Initial Visit

A female patient, 28 years old, November 23rd, 1987.

More than ten days ago, after the patient had had an argument with her neighbour, she immediately felt stomach and abdominal distention and fullness discomfort. After that, pain and anorexia gradually appeared which was more significant after eating. She took dry yeast, etc. without effect and so came for treatment. On palpation her stomach and abdomen were soft and there was no mass. Her stool tended towards dryness. Her tongue coating was thin and white. Her pulse was wiry and thready. This was a syndrome of qi stagnation not soothing, and liver and stomach disharmony. The treatment plan was to soothe the liver, harmonise the stomach, regulate qi and alleviate pain.

Prescription

百合	*bǎi hé*	24g	Bulbus Lilii
乌药	*wū yào*	12g	Radix Linderae
檀香	*tán xiāng*	6g	Lignum Santali Albi
砂仁	*shā rén*	3g	Fructus Amomi
丹参	*dān shēn*	20g	Radix et Rhizoma Salviae Miltiorrhizae
高良姜	*gāo liáng jiāng*	6g	Rhizoma Alpiniae Officinarum
香附	*xiāng fù*	11g	Rhizoma Cyperi
杭白芍	*háng bái sháo*	15g	Radix Paeoniae Lactiflorae
柴胡	*chái hú*	12g	Radix Bupleuri
何首乌	*hé shǒu wū*	12g	Radix Polygoni Multiflori
甘草	*gān cǎo*	6g	Radix Glycyrrhizae

This was decocted with water and divided to be taken twice, once each in the morning and evening, one pack per day.

Second Visit

November 30th: After taking three packs of the above formula, her stomach fullness, distention and pain were reduced a lot. Her bowel movement became normal. It was only after eating that she still felt stomach distention and fullness. She was late for her second visit due to family problems. For the few days before, she had had distention and pain again. Using the original formula as a base, *hé shǒu wū* (Radix Polygoni Multiflori) was eliminated and *bái zhú* (Rhizoma Atractylodis Macrocephalae) 10g, was added. She continued to take it.

Third Visit

December 4th: After taking four packs of the above formula, her stomach distention and pain were eliminated. However at night time she had stomach distention and pain sensations which were significant if pressed on. This was accompanied by blood stasis.

Using the orignal formula, *dāng guī* (Radix Angelicae Sinensis) 12g, *táo rén* (Semen Persicae) 6g, *hóng huā* (Flos Carthami) 6g, *guì zhī* (Ramulus Cinnamomi) 10g, and *háng bái sháo* (Radix Paeoniae Lactiflorae) 30g, were added. She took another four packs and she recovered.

2. *Bǎi Hé* (Bulbus Lilii) ——
Kuǎn Dōng Huā (款冬花, Flos Farfarae)

Function

To nourish yin, moisten the lung, stop cough and calm asthma.

Application

This is used for deficiency and injury inducing cough, dry mouth and quick breathing, with hot flushes or blood in the phlegm. This also can be used for chronic tracheitis.

Compatibility Analysis

Bǎi hé (Bulbus Lilii) combined with *kuǎn dōng huā* (Flos Farfarae) in the *Formulae to Aid the Living* (济生方 , *Jì Shēng Fāng*) is called "*Bǎi Huā Gāo* (百花膏)". The taste and property of *bǎi hé* (Bulbus Lilii) are sweet and neutral. It enters the lung channel of the hand *taiyin* meridian. It functions to nourish the yin, moisten the lung and stop cough. It is a peaceful yin nourishing herb. The taste and property of *kuǎn dōng huā* (Flos Farfarae) are acrid and warm. It enters the lung channel of the hand *taiyin* meridian. It functions to descend qi, stop cough and calm asthma. This is a commonly used herb to warm and moisten the lung channel of the hand *taiyin* meridian. When these two herbs are combined together, *kuǎn dōng huā* (Flos Farfarae) gets *bǎi hé* (Bulbus Lilii) to reduce its warmth and increase its moistening properties. *Bǎi hé* (Bulbus Lilii) gets *kuǎn dōng huā* (Flos Farfarae) to reduce its coldness and increase its power to stop cough and calm asthma. These two herbs neutrally regulate cold and heat which makes them even more pertinent for use for lung channel of the hand *taiyin* meridian deficiency and injury induced chronic cough.

Case Study

Initial Visit

A male patient, 36 years old , August 4th, 1985.

The patient's right upper lung lobe had pulmonary tuberculosis for five years. After treatments he improved. Last month, due to overexertion plus an exterior invasion, he had afternoon hot flushes and cough with a small amount of silk like blood in his phlegm. He was treated with anti tuberculosis medicine for more than a month, but did not see any significant progress and so he came to visit. His chest x-ray showed: Right upper lung lobe had tuberculosis infiltration, the right lung chest membrane was thicker, the rib and the costophrenic angle were blunt (indicating a past episode of pleurisy), his

ESR 25mm/h. His pulse was deficient, thready and forceless, both *cun* were even worse. his tongue was red with a scanty tongue coating. This was the syndrome of lung yin deficiency, with the lung qi having lost its dispersing function. The treatment plan was nourish the yin, moisten the lung, stop cough and clear heat.

Prescription

百部	*bǎi bù*	15g	Radix Stemonae
百合	*bǎi hé*	20g	Bulbus Lilii
款冬花	*kuǎn dōng huā*	15g	Flos Farfarae
紫菀	*zǐ wǎn*	15g	Radix et Rhizoma Asteris
白及	*bái jí*	10g	Rhizoma Bletillae
桔梗	*jié gěng*	9g	Radix Platycodonis
功劳叶	*gōng láo yè*	30g	Folium Ilex
陈皮	*chén pí*	9g	Pericarpium Citri Reticulatae
川贝	*chuān bèi*	10g	Bulbus Fritillariae Cirrhosae
知母	*zhī mǔ*	10g	Rhizoma Anemarrhenae
地榆	*dì yú*	15g	Radix Sanguisorbae
甘草	*gān cǎo*	6g	Radix Glycyrrhizae

This was decocted with water and divided to be taken twice, once each in the morning and evening, one pack per day.

Second Visit

August 11th: His cough was reduced but there was still silk blood in his phlegm. Using the original formula as a base, *shān yú ròu* (Fructus Corni) 15g and *sān qī fěn* (Powdered Radix et Rhizoma Notoginseng) 4g were added. He continued to take it again.

Thrid Visit

October 7th: After taking a total of more than sixty packs of the above formula, all of his symptoms disappeared. His chest x-ray showed: Right upper lung lobe pulmonary tuberculosis was calcified and absorbed, ESR 5mm/h. He was instructed to continuously take Isoniazid to complete the treatment.

Dāng Guī's (当归 , Radix Angelicae Sinensis) taste and property are sweet, bitter, acrid and warm. It enters the liver channel of the foot *jueyin*, the spleen channel of the foot *taiyin* and the heart channel of the hand *shaoyin* meridians. It functions to tonify blood, invigorate blood, regulate menstruation, moisten dryness and moisten the intestines. In the *Divine Husbandman's Classic of the Materia Medica* it lists it as a middle class herb. It is an important herb for blood diseases, especially important for gynaecology. Any women's irregular menstruation, dysfunctional uterine bleeding, blood deficiency inducing amenorrhoea, pregnancy and childbirth, all syndromes use it as a king herb. External sores and any of the meridians or the collaterals not flowing smoothly inducing wind-damp, stuffiness and pain, blood deficiency induced constipation, etc., also are treated effectively with it. Modern animal laboratory research has proved that *dāng guī* (Radix Angelicae Sinensis) can function to regulate the uterus. When the uterus has internal pressure, *dāng guī* (Radix Angelicae Sinensis) has an exciting function on the uterus to make it contract regularly and to increase the contractile strength. When there is no pressure inside the uterus, *dāng guī* (Radix Angelicae Sinensis) also has an inhibiting function on the uterus.

In addition, *dāng guī* (Radix Angelicae Sinensis) is also used for relentless chronic cough.

1. *Dāng Guī* (Radix Angelicae Sinensis) ——
Chuān Xiōng (川芎, Rhizoma Chuanxiong)

Function

To invigorate blood, expel stasis, regulate menstruation and nourish blood.

Application

This can be used for women's irregular menstrual flow, heart area and abdominal fullness and pain as well as all kinds of diseases of or before pregnancy or after childbirth in women. It can be used for any woman's blood level diseases, or deficiency or excess or stasis.

Compatibility Analysis

Dāng guī (Radix Angelicae Sinensis) with *chuān xiōng* (Rhizoma Chuanxiong) is also named as *Fó Shǒu Sǎn* (Hand of Buddha Powder, 佛手散) or *Huó Xuè Sǎn* (Blood-Quickening Powder, 活血散). This is the *Sì Wù Tāng* (Four Agents Decoction) formula with the elimination of *shú dì huáng* (Radix Rehmanniae Praeparata) and *bái sháo* (Radix Paeoniae Alba), two herbs that nourish and retain yin by astringing. Therefore, its power to invigorate blood and regulate menstruation is stronger than *Sì Wù Tāng* (Four Agents Decoction), and so it is called *Huó Xuè Sǎn* (Blood-Quickening Powder). A contemporary doctor Xie Li-heng said that: "*Dāng guī* (Radix Angelicae Sinensis) and *chuān xiōng* (Rhizoma Chuanxiong) are the king herbs for the blood level. Their property is warm and taste is acrid. Warmth can harmonise the blood, sweetness can tonify the blood, acidity can disperse the blood...... As chuān xiōng (Rhizoma Chuanxiong) which is acrid, invigorates blood and promotes movement of qi, it can quickly raise and disperse. If over dosed then it will injure the qi. It cannot be taken individually or for too long. However, if it is used for qi stagnation or blood stasis, it can have great effectiveness, therefore, if assisted by dāng guī (Radix Angelicae Sinensis) it can eliminate blood stasis and engender new blood, as well as harmonise blood. Once* blood is harmonised, then blood can reside within its vessels, and then all kinds of blood diseases will recover......" For clinical practice, Xie Li-heng again said that: "*If pregnancy and the foetus are disturbed, then the foetus is injured and there is haemorrhaging. If it is not due to blood stagnation injuring the foetus, then, it will be due to blood haemorrhaging uncontrollably. Take the two herbs: Dāng guī (Radix Angelicae Sinensis) and chuān xiōng (Rhizoma Chuanxiong), then if the blood is out of control and the foetus is not disturbed, once blood flows smoothly, then the pain will stop. If it is a case that the blood is obstructed and the foetus is not injured, once the blood can flow smoothly, then the foetus will calm down...... If the foetus is disturbed and injured, once the blood flows smoothly, then the foetus will descend. If it is due to a trauma with a large blood loss inducing dizziness, it also can be used to tonify. If the foetus has died in utero causing serious pain in the abdomen, also use it to expel the foetus. The above conditions are all due to blood diseases in patients without qi deficiency. When qi deficiency induces a difficult labour, with blood separating after childbirth, yellowish white lips and face, shortness of breath, irritability, and dizziness on exertion, if misused it will cause immediate death. A double dosage of rén shēn (Radix et Rhizoma Ginseng) is then needed to secure the intangible qi in order to save the tangible blood. For difficulty in opening the iliac bone, add guī jiǎ (Carapax et Plastrum Testudinis) to help the foetus move down to the cervix. If there is cold, add gāo liáng jiāng (Rhizoma Alpiniae Officinarum) and ròu guì (Cortex Cinnamomi). If there is heat, add huáng qín (Radix Scutellariae). If there is sweating, add guì zhī (Ramulus Cinnamomi). If there are spasms add jīng jiè suì (Spica Schizonepetae). Then modify according to symptom elimination. The formula name is not composed of dāng guī (Radix Angelicae Sinensis) and chuān xiōng (Rhizoma Chuanxiong) but instead it is called 'hand of buddha', it means that for all kinds of diseases before pregnancy or after childbirth, the treatment results are like miracles from the Buddha's hands.*" For the compatibility of *dāng guī* (Radix Angelicae Sinensis) and *chuān xiōng* (Rhizoma Chuanxiong) and its clinical applications, Xie has a wealth of experience and ospecial understanding.

Case 1

Abdominal pain during pregnancy.

Initial Visit

A female, 32 years old, October 12th, 1987.

The patient was three months pregnant. After visiting her relative in Cangzhou and the exhaustion from the traveling, she had lower abdominal pain for three days. There was no colporrhagia. Her appetite, urination and bowel movement were normal. Her pulse was wiry and her tongue coating was thin and white. The tiredness disturbed her qi and blood as well as affecting the foetal qi inducing the symptoms. The plan was to nourish blood, invigorate blood, soothe qi and calm the foetus to regulate and treat.

Prescription

当归	*dāng guī*	15g	Radix Angelicae Sinensis
川芎	*chuān xiōng*	6g	Rhizoma Chuanxiong
杭白芍	*háng bái sháo*	20g	Radix Paeoniae Lactiflorae

黄芪	huáng qí	6g	Radix Astragali
白术	bái zhú	6g	Rhizoma Atractylodis Macrocephalae
砂仁	shā rén	3g	Fructus Amomi
桑寄生	sāng jì shēng	15g	Herba Taxilli
甘草	gān cǎo	6g	Radix Glycyrrhizae

This was decocted with water and divided to be taken warm, twice, one pack per day.

Second Visit

October 15th: After taking two packs of the above formula her abdominal pain stopped. The patient was afraid about infecting the foetus and requested to take some packs of formula to nourish and calm the foetus. The formula was changed to a Modified *Tài Shān Pán Shí Sǎn* (Rock of Taishan Foetus-Quietening Powder, 泰山磐石散) for treatment.

Case 2

Postpartum endless lochia.

Initial Visit

A female patient, 29 years old, July 19th, 1995.

The patient delivered two full-term babies more than a month previously. However, the colporrhagia persisted, and was accompanied by mild lower abdominal pain, lustreless face, lassitude, and reduced lactation. Her pulse was deep and choppy. Her tongue was dark and tongue coating was thin and white. This was post childbirth blood stasis obstructing internally, therefore, there was persistent colporrhagia. Prolonged bleeding must injure the qi and blood, therefore, her face was lustreless and she had lassitude. The treatment plan was to first use a formula to invigorate blood, eliminate blood stasis, nourish blood and regulate menstruation.

Prescription

当归	dāng guī	30g	Radix Angelicae Sinensis
川芎	chuān xiōng	12g	Rhizoma Chuanxiong
桃仁	táo rén	10g	Semen Persicae
炮姜	páo jiāng	6g	Rhizoma Zingiberis Praeparatum
枳实	zhǐ shí	15g	Fructus Aurantii Immaturus
川牛膝	chuān niú xī	20g	Radix Cyathulae
熟地	shú dì	20g	Radix Rehmanniae Praeparata
地榆	dì yú	15g	Radix Sanguisorbae
甘草	gān cǎo	6g	Radix Glycyrrhizae
黄酒	huáng jiǔ	50ml	Yellow Rice Wine (as a guide)

This was decocted with water to be taken one pack per day.

Second Visit

July 23rd: After taking three packs of the above formula, her bleeding stopped and her pain was eliminated. After taking four packs, she recovered. However, her lactation was still less than it should be. The formula was changed to nourish blood, tonify qi, free the meridian and increase lactation to treat her.

2. *Dāng Guī* (Radix Angelicae Sinensis) ── *Cè Bǎi Yè* (侧柏叶, Cacumen Platycladi)

Function

To transform blood stasis, constrain the lung, descend rebellion and stop cough.

Application

This is used for chronic cough without recovery.

Compatibility Analysis

Dāng guī (Radix Angelicae Sinensis) is a noble herb among the blood herbs. It can harmonise, invigorate and tonify blood. Any prolonged sickness will enter the collaterals and cause blood stasis, including cough. With prolonged cough, the lung qi loses it dispersing function. The collaterals of the lung shall have blood stasis. Blood stasis in the collaterals is not beneficial to the lung qi, therefore, there was prolonged cough difficult to recover from. *Dāng guī* (Radix Angelicae Sinensis) invigorates blood and frees the collaterals. If the collaterals are freed and the blood stasis is eliminated, then the cough can be stopped. In the *Divine Husbandman's Classic of the Materia Medica* it talks about the functions of *dāng guī* (Radix Angelicae Sinensis), the first sentence says that it "*governs cough and the upward movement of rebellious qi*". As *dāng guī* (Radix Angelicae Sinensis) is an important herb in gynaecology, therefore, the later generations' everyday herbal publications did not talk about *dāng guī* (Radix Angelicae Sinensis) stopping cough and rebellion, so, few people use it this way. When the author treats chronic cough, he adds in *dāng guī* (Radix Angelicae Sinensis), mostly to great effect. *Cè bǎi yè* (Cacumen Platycladi) is a cooling blood and stopping bleeding herb. In clinical practice, it also has few people using it to stop cough, however, it is certainly effective in clinical practice, if used to stop cough or asthma. *Cè bǎi yè* (Cacumen Platycladi) is bitter and slightly cold. It enters the lung channel of the hand *taiyin* meridian. It can constrain the lung and stop cough. In the *Essentials of the Materia Medica* it states that it can "*tonify the lung*". In Korean folk empirical formulae, it is recorded that *cè bǎi yè* (Cacumen Platycladi), a single herb decoction can treat cough and asthma, which shows that it is effective for *cè bǎi yè* (Cacumen Platycladi) to stop cough and asthma. *Dāng guī* (Radix Angelicae Sinensis) and *cè bǎi yè* (Cacumen Platycladi) combined can treat prolonged persistent coughing. It functions to transform blood stasis, descend rebellion, constrain the lung and stop cough. The property of *dāng guī* (Radix Angelicae Sinensis) is warm and its taste is acrid, aromatic and beneficial for moving. *Cè bǎi yè* (Cacumen Platycladi) is slightly cold. It tastes bitter and astringent. It is beneficial for constraining. Of these two herbs, one is warm and the other is cold. They can mutually control harmoniously. One is moving and the other is constraining which just matches the opening, dispersing and descending functions of the lung qi. This compatibility has the characteristics to move blood stasis but does not injure the normal qi, it constrains qi but does not leave blood stasis. These two herbs added to a formula based on syndrome differentiation can produce good results, regardless of whether it is due to cold, heat, deficiency or excess.

Case 1

Night cough.

Initial Visit

A male patient, 6 years old, July 5th, 1995.

The child had flu, fever and cough more than twenty days previously. After the treatment, his fever dropped but his cough had not been stopped. He took various medicines like cough syrup, dioxane C triazine and antibiotics, etc. but they did not work and so he came for treatment. On auscultation, the breathing sounds of both his lungs were rough but no rales were heard. He had a dry cough without phlegm, his night time cough was worse. His pulse was thready and fast. His tongue coating was thin and white. If there is night cough and persistent chronic cough, most of the time it has blood stasis. Prolonged sickness enters the collaterals. The lung collaterals had blood stasis, and so lost their dispersing and descending functions, therefore, his cough could not stop. A little child is an immature yang body. Exterior pathogens attacking the lung are easiest to transform into heat and injure the yin, therefore, he had a dry cough without phlegm, a thready and fast pulse. The treatment plan was to moisten the lung, to stop cough, transform blood stasis and descend rebellion.

Prescription

百部	*bǎi bù*	10g	Radix Stemonae
桔梗	*jié gěng*	8g	Radix Platycodonis
当归	*dāng guī*	12g	Radix Angelicae Sinensis
侧柏叶	*cè bǎi yè*	12g	Cacumen Platycladi
陈皮	*chén pí*	8g	Pericarpium Citri Reticulatae
紫菀	*zǐ wǎn*	9g	Radix et Rhizoma Asteris
杏仁	*xìng rén*	8g	Semen Armeniacae Dulcis
川贝	*chuān bèi*	10g	Bulbus Fritillariae Cirrhosae
荆芥	*jīng jiè*	8g	Herba Schizonepetae
蝉衣	*chán yī*	8g	Periostracum Cicadae
甘草	*gān cǎo*	5g	Radix Glycyrrhizae

This was decocted with water, to 150ml and divided to be taken three times, warm, one pack per day.

Second Visit

July 8th: After taking three packs of the above formula, his cough was eliminated by more than half. He took a further three packs of the original formula.

Third Visit

July 11th: After taking another three packs, his cough stopped. He was instructed to continuously take two more packs to clear it up and he recovered.

Case 2

Initial Visit

A male patient, 27 years old, November 20th, 1995.

The patient had a persistent cough for more than a month after he had a cold. He took medicines like the static infusion of ampicillin and cephalosporin, etc. without effect, and his cough became worse daily. He had chest tightness, little phlegm and dry mouth. His chest x-ray indicated that both lungs' interstitial markings had become thicker. His blood routine test was within the normal range. He had a wiry, thready and fast pulse, a red tongue, and a thin and white tongue coating. The syndrome differentiation was that after the exterior invasion, the residual pathogen was not totally cleared. The residual pathogen stagnated for a long time before being transformed to heat and damaging the lung collaterals, and injuring the lung yin. This manifested in his dry mouth, reduced phlegm, persistent chronic cough, and a thready and fast pulse. The treatment plan should clear heat, moisten the lung and stop cough. Modified *Zhǐ Sòu Sǎn* (Cough-Stopping Powder, 止嗽散) was used to treat him.

Prescription

百部	*bǎi bù*	15g	Radix Stemonae
桔梗	*jié gěng*	10g	Radix Platycodonis
紫菀	*zǐ wǎn*	12g	Radix et Rhizoma Asteris
陈皮	*chén pí*	9g	Pericarpium Citri Reticulatae
当归	*dāng guī*	15g	Radix Angelicae Sinensis
侧柏叶	*cè bǎi yè*	15g	Cacumen Platycladi
杏仁	*xìng rén*	10g	Semen Armeniacae Dulcis
川贝	*chuān bèi*	10g	Bulbus Fritillariae Cirrhosae
黄芩	*huáng qín*	9g	Radix Scutellariae
冬花	*dōng huā*	15g	Flos Farfarae
蝉衣	*chán yī*	8g	Periostracum Cicadae
甘草	*gān cǎo*	6g	Radix Glycyrrhizae

This was decocted with water and divided in two to be taken twice, once each in the morning and evening, one pack per day.

Second Visit

November 25th: After taking five packs of the above formula, his cough was reduced a lot. His symptoms of chest tightness and dry mouth, etc. were all eliminated. As it was effective, the prescription did not change and he continued on the original formula.

Third Visit

November 30th: After taking a further five packs of the above formula, his cough basically disappeared. To improve the treatment effectiveness, he was given a further three packs of the original formula for treatment.

Xuè jié's (血竭 , Sanguis Draconis) taste and property are sweet, salty and neutral. It enters the heart channel of the hand *shaoyin* and the liver channel of the foot *jueyin* meridians. It functions to invigorate blood, stop pain, expel blood stasis and generate new blood. Taken internally it can invigorate blood, expel blood stasis and stop pain. It can be used for any internal injury inducing blood gathering, amenorrhoea, postpartum blood stasis causing pain, falling or fighting fracture or injury, and all kinds of pain of the heart or abdomen. With external use it can treat severe ulcers, carbuncles, abscesses, chronic sores and knife injury induced bleeding. For internal use, it should be ground into powder, dissolved in water and drank.

1. *Xuè Jié* (Sanguis Draconis) ——
Dà Huáng (大黄, Radix et Rhizoma Rhei)

Function

To invigorate blood, eliminate blood stasis and alleviate pain.

Application

This can be used for women's endometriosis syndrome inducing painful menstruation.

Compatibility Analysis

Xuè jié (血竭 , Sanguis Draconis) is an important herb to expel blood stasis and alleviate pain. In the *Grand Materia Medica*, it states that it can "*expel all kinds of pain induced by blood obstruction*". In the *Essentials of the Materia Medica* it also mentions that it can "*harmonise blood, constrain sores, specialise in eliminating pain induced by blood problems, expel blood stasis and generate new blood. It is a miracle herb to harmonise blood.*" *Dà huáng* (Radix et Rhizoma Rhei) is bitter and cold. It is also a commonly used herb to invigorate blood and eliminate blood stasis. In the *Divine Husbandman's Classic of the Materia Medica* it states that it can "*drain blood stasis and blood blockage inducing cold or heat, break up the abdominal abscess and accumulation......*"

These two herbs are both blood invigorating herbs. *Xuè jié* (Sanguis Draconis) invigorates blood and its power to expel blood stasis is strong. *Dà huáng* (Radix et Rhizoma Rhei) invigorates blood and its power to break up blood stasis and stagnation is excellent. *Dà huáng* (Radix et Rhizoma Rhei) is beneficial for moving and freeing blood stasis and stagnation. The property of *xuè jié* (Sanguis Draconis) is dispersing. It can eliminate blood stasis and stagnation. *Dà huáng's* (Radix et Rhizoma Rhei) ability to invigorate blood and expel blood stasis, can help *xuè jié* (Sanguis Draconis) to expel blood stasis and alleviate pain. *Xuè jié's* (Sanguis Draconis) ability to invigorate blood and expel blood stasis also benefits *dà huáng* (Radix et Rhizoma Rhei) in expelling blood stasis and generating new blood. This compatibility is in the category of mutually enhancing. The author always uses this compatibility with *Shào Fù Zhú Yū Tāng* (Lesser Abdomen Stasis-Expelling Decoction) to treat women's endometriosis syndrome and stubborn painful menstruation. He especially uses it for the endometriosis induced painful menstruation that cannot be treated by the regular invigorating blood, transforming blood stasis, regulating qi and alleviating pain formulae. Due to endometrial shedding of fragments on the ovary surface or in the pelvic cavity, it cannot be eliminated without expelling the blood stasis and freeing the stagnation.

Case Study

Initial Visit

A female patient, 37 years old, March 20th, 2000.

Three years ago, the patient due to ovarian cysts had an ovarectomy operation. After the operation, she had lower abdominal pain. When menstruating, her was pain worse. She had been investigated in a hospital. Both a C.T. report and B-ultrasound report suspected that she had a tumour in her abdominal cavity; the previous January she had had surgery in an hospital. Pathology reported it as: Endometriosis syndrome. After the operation, she had persistent lower abdominal pain. Before and after menstruation, it was still very painful and hard to bear so, she came for treatment.

On examination, the patient was physically well developed. The patient had given birth when she was fifteen years old. Her lower abdominal pain was more significant in her right groin. On palpation, there was a string like mass, around 12cm × 4cm. When pressed upon, she felt mild pain. Her pulse was deep and wiry. Her tongue coating was thin and white. This was lower abdominal blood stasis inducing an abdominal abscess, obstructing and stagnating both the qi and blood flow. The meridians and the blood vessels were blocked, therefore, she had the pain. When her menstruation came, it was blocked, blood was not expelling smoothly, and therefore, she had severe pain during menstruation. The treatment plan was to use a formula to invigorate blood, eliminate blood stasis and alleviate pain.

Prescription

丹参	*dān shēn*	30g	Radix et Rhizoma Salviae Miltiorrhizae
当归	*dāng guī*	15g	Radix Angelicae Sinensis
乌药	*wū yào*	15g	Radix Linderae
小茴香	*xiǎo huí xiāng*	9g	Fructus Foeniculi
炮姜	*páo jiāng*	6g	Rhizoma Zingiberis Praeparatum
大黄	*dà huáng*	10g	Radix et Rhizoma Rhei
延胡索	*yán hú suǒ*	15g	Rhizoma Corydalis
血竭	*xuè jié*	2g	Sanguis Draconis (grind into powder, dissolve and drink)
赤芍	*chì sháo*	12g	Radix Paeoniae Rubra
五灵脂	*wǔ líng zhī*	9g	Faeces Trogopterori
甘草	*gān cǎo*	6g	Radix Glycyrrhizae

This was decocted with water and divided to be taken twice, one pack per day.

Second Visit

March 26th: After taking six packs of the above formula, her abdominal pain basically disappeared. As her last menstruation had been more than twenty days previously, her next menstruation was due. The patient was worried that she could have severe pain and so came to visit. She was prescribed Modified *Shào Fù Zhú Yū Tāng* (Lesser Abdomen Stasis-Expelling Decoction) for treatment.

Prescription

小茴香	*xiǎo huí xiāng*	9g	Fructus Foeniculi
炮姜	*páo jiāng*	9g	Rhizoma Zingiberis Praeparatum
延胡索	*yán hú suǒ*	10g	Rhizoma Corydalis
蒲黄	*pú huáng*	9g	Pollen Typhae
五灵脂	*wǔ líng zhī*	9g	Faeces Trogopterori
赤芍	*chì sháo*	12g	Radix Paeoniae Rubra
制没药	*zhì mò yào*	8g	Myrrha (processed)
大黄	*dà huáng*	9g	Radix et Rhizoma Rhei
血竭	*xuè jié*	2g	Sanguis Draconis (grind into powder, dissolve and drink)
乌药	*wū yào*	15g	Radix Linderae
生苡米	*shēng yì mǐ*	30g	Semen Coicis (raw)
甘草	*gān cǎo*	6g	Radix Glycyrrhizae

This was decocted with water and divided to be taken twice, one pack per day.

Third Visit

April 2nd: Her menstruation had come the day before and her right lower abdomen felt mildly painful again, but it was not severe. Her pulse was still deep and wiry. Again, the above formula was used to regulate her. She was instructed to take seven packs of the above formula every month, three days before her menstruation.

Note

The patient in this case took the above formula for four more menstrual cycles and eliminated her pain.

2. *Xuè Jié* (Sanguis Draconis) ——
Bái Huā Shé (白花蛇, Agkistrodon)

Function

To invigorate blood, expel blood stasis, penetrate the bone, expel wind and alleviate pain.

Application

This is used for hyperosteogeny.

Compatibility Analysis

Xuè jié (Sanguis Draconis) expels blood stasis, reduces swelling and alleviates pain. *Bái huā shé* (白花蛇 , Agkistrodon) is sweet, salty and warm. It enters the liver channel of the foot *jueyin* and the kidney channel of the foot *shaoyin* meridians. It can expel wind, penetrate bone and free the collaterals. It is the king herb to treat joint immobility, wind-damp arthralgia and hemiplegia. In the *Materia Medica of the Kaibao Era* it states that *bái huā shé* (Agkistrodon) treats *"bone joint pain and foot weakness that prevent standing for a long time"*. In the *Essentials of the Materia Medica* it states that it can *"penetrate bone, expel wind"*. They explained that this herb is a specialised herb to free the collaterals, free joint mobility and treat joint pain. *Xuè jié* (Sanguis Draconis) is beneficial for expelling blood stasis, invigorating blood and alleviating pain. *Bái huā shé* (Agkistrodon) is beneficial for expelling wind, penetrating the bone, and treating joint immobility. These two herbs combined together have an even stronger power to free joint immobility, eliminate blood stasis and alleviate pain, particularly for hyperosteogeny induced joint pain of the waist, knees, cervical spine, etc. The blood stasis and obstruction for hyperosteogeny cannot be expelled, without *xuè jié* (Sanguis Draconis). The blood stasis and obstruction in the joints cannot be penetrated without *bái huā shé* (Agkistrodon). When the author, in his clinical practice, uses these two herbs, with syndorme differentiation to treat hyperosteogeny, he always has good results. This compatibility was from *Fēng Shī Wēi Líng Fāng* (Rheumatoid Chinese Clematis Formula, 风湿威灵方), in the *Journal of Chinese Medicine* (1985, February). It is a good formula for treating hyperosteogeny.

Case Study

Initial Visit

A male patient, 60 years old, April 12th, 1987.

The patient had low back pain for half a year and it was gradually getting worse. Recently, due to his low back pain, he could not stand erect and he had to bend his back to walk like a bow. He had been treated locally for kidney deficiency induced low back pain but without effect. He was also treated by massage and acupuncture but it was still as painful as before, so, he came for a check up and treatment. His lumbar x-ray frontal view and lateral view reported: The anterior and lateral edges of the first to the fifth vertebrae of his lumbar spine had visible hyperosteogeny above the middle degree. On palpation his lower back was tense and had ankylosis. His lumbar vertebrae were tender on pressing. His pulse was deep and wiry. His tongue was red and his tongue coating was thin and white. This was kidney deficiency and blood vessels stagnation and obstruction syndrome. The kidney governs the bone. As his kidney was deficient then his bone lost nourishment. His blood vessels stagnated and obstructed inducing blockage and then he had pain. The lower back is the abode of the kidneys, therefore, for the elderly, degenerated joint problems most often occur in the lower back. The treatment plan was to tonify the kidney and strengthen the bone, expel blood stasis and alleviate pain.

Prescription

| 山萸肉 | *shān yú ròu* | 15g | Fructus Corni |
| 土鳖虫 | *tǔ biē chóng* | 10g | Eupolyphaga seu Steleophaga |

威灵仙	wēi líng xiān	20g	Radix et Rhizoma Clematidis
当归	dāng guī	12g	Radix Angelicae Sinensis
鹿角霜	lù jiǎo shuāng	15g	Cornu Cervi Degelatinatum
血竭	xuè jié	2g	Sanguis Draconis (grind into powder, dissolve and drink)
乌蛇	wū shé	12g	Zaocys
透骨草	tòu gǔ cǎo	10g	Caulis Impatientis
赤芍	chì sháo	15g	Radix Paeoniae Rubra
川断	chuān duàn	15g	Radix Dipsaci Asperi
杜仲	dù zhòng	15g	Cortex Eucommiae

This was decocted with water and divided to be taken twice, one pack per day.

Second Visit

April 17th: After taking seven packs of the above formula, his low back pain reduced a lot. He could walk upright. As the patient was living on a farm and it was inconvenient for him to decoct herbs, he requested a ready made formula for treatment, and so was prescribed a powder formula.

Prescription

威灵仙	wēi líng xiān	70g	Radix et Rhizoma Clematidis
白花蛇	bái huā shé	2 pcs	Agkistrodon
乌蛇	wū shé	35g	Zaocys
透骨草	tòu gǔ cǎo	35g	Caulis Impatientis
当归	dāng guī	35g	Radix Angelicae Sinensis
土鳖虫	tǔ biē chóng	35g	Eupolyphaga seu Steleophaga
杭白芍	háng bái sháo	35g	Radix Paeoniae Lactiflorae
延胡索	yán hú suǒ	35g	Rhizoma Corydalis
血竭	xuè jié	35g	Sanguis Draconis
川断	chuān duàn	35g	Radix Dipsaci
杜仲	dù zhòng	35g	Cortex Eucommiae

The above herbs were all ground together into fine powder, 5g to be taken each time with either warm water or huáng jiǔ (Yellow Rice Wine), in the morning and at night.

Third Visit

May 20th: After taking one pack of the above formula, his low back pain disappeared and he could move as he wished. He was instructed to follow the same formula and take one more pack to strengthen the result.

Follow up visits showed the patient in this case was still healthy and there was no reoccurrence of his low back pain.

Note

In the *Journal of Chinese Medicine*, in the article *Fēng Shī Wēi Líng Fāng*, the original formula was *bái huā shé* (Agkistrodon) 4 pieces, *wēi líng xiān* (Radix et Rhizoma

Clematidis) 72g, *fáng fēng* (Radix Saposhnikoviae), *dāng guī* (Radix Angelicae Sinensis), *tǔ biē chóng* (Eupolyphaga seu Steleophaga), *xuè jié* (Sanguis Draconis), *tòu gǔ cǎo* (Caulis Impatientis) 32g each, grind together into fine powder, to be taken twice a day, each time take 3g with warm water. This mainly treats wind-damp *bi* syndrome with joint pain, and hyperosteogeny. The author uses it with good effects. If there is cervical vertebral hypertrophy, *gé gēn* (Radix Puerariae Lobatae) can be added. For the lower limbs knee joint hypertrophy, add *chuān niú xī* (Radix Cyathulae). For the lumbar vertebral hypertrophy, add *dù zhòng* (Cortex Eucommiae) and *chuān duàn* (Radix Dipsaci Asperi). If it does not belong to wind-damp, it may or may not use *fáng fēng* (Radix Saposhnikoviae).

Dì lóng (地 龙 ,Pheretima) is first recorded in the *Divine Husbandman's Classic of the Materia Medica*. The original name is white neck earth worm. The name of *dì lóng* (Pheretima) was first seen in the *Illustrated Classic of the Materia Medica*. The taste and property of *dì lóng* (Pheretima) are salty and cold. It enters the stomach channel of the foot *yangming*, the spleen channel of the foot *taiyin* and the kidney channel of the foot *shaoyin* meridians. It functions to clear heat, calm convulsion, invigorate the collaterals and promote urination. It can be used for heat syndromes inducing irritability, cough, rapid asthma, convulsion, dysuria, hemiplegia, etc. It is a commonly used herb in clinical practice. Its compatibilities and applications are broader, such as with *huáng qí* (Radix Astragali) it can tonify qi, invigorate blood, free the collaterals to treat hemiplegia and chronic nephritis. If combined with *gé gēn* (Radix Puerariae Lobatae) it can invigorate blood, free the collaterals, clear heat, stop spasms and convulsions to effectively treat high blood pressure and cervical syndromes.

1. *Dì Lóng* (Pheretima) ——
Táo Rén (桃仁, Semen Persicae)

Function

To transform blood stasis, free the collaterals, clear heat and arrest asthma.

Application

This is used for chronic cough and asthma without recovery.

Compatibility Analysis

Dì lóng (Pheretima) is salty and cold. It can clear heat, stop spasms and convulsions. Modern pharmacological research discovered that it has the function to expand the bronchi. *Táo rén* (Semen Persicae) is bitter and neutral. It can invigorate blood, moisten the lung and stop cough. These two herbs combined together function to transform blood stasis, free the collaterals, clear heat and stop asthma. They are effective for any chronic cough and asthma without recovery. A prolonged cough must injure the lung collaterals; if the lung collaterals are injured then the lung channel of the hand *taiyin* meridian and its blood vessels will be stagnated and obstructed. Prolonged blood stasis transforming to heat must cause injury to the lung collaterals again, therefore, the cough becomes worse and prolonged without recovery and it becomes a vicious circle. *Dì lóng* (Pheretima) with *táo rén* (Semen Persicae) can transform blood stasis, free the collaterals, clear heat, and arrest asthma. They treat both the manifestations and the root of the problem, so it corresponds directly with the pathogenesis therefore, they can stop asthma. They can be added in according to the syndrome differentiation, most of the time good treatment results can be achieved.

Case Study

Initial Visit

A male patient, 7 years old, October 20th, 1997.

The child had cough and asthma for two months without recovery and so his mother brought him to visit. His mother said that two months previously, the patient had had an exterior invasion with fever and cough. After treatment, his fever dropped but he had a persistent cough and it gradually got worse, eventually becoming cough with asthma. He had been treated with oral antibiotics and medicines to stop cough, etc. but they were ineffective. One week previously, he went to a hospital to check it out and for the treatment. He was diagnosed with bronchial asthma. He was treated with cephalosporin, dexamethasone intravenous infusion and an asthma inhaler, etc. His cough and asthma improved but after stopping the medicine, his cough and asthma were still like before. On examination, the child was breathing rapidly with his mouth opened and his shoulders raised, and phlegm sounds came from his throat. There were significant wheezing sounds in both his lungs. He had a red tongue and white tongue coating, and a thready and fast pulse. After the exterior invasion the residual pathogen had not been cleared, and with mistreatment this induced the residual pathogen to be transformed into heat and enter internally, which injured the lung collaterals. The collaterals were injured and blood stasis obstructed the qi pathway of the lung channel of the hand *taiyin* meridian. Therefore, his cough and asthma were prolonged without recovery. The treatment plan was to use a formula to clear heat and disperse the lung, transform blood stasis, and arrest asthma.

Prescription

麻黄	*má huáng*	4g	Herba Ephedrae
白果	*bái guǒ*	6g	Semen Ginkgo
地龙	*dì lóng*	10g	Pheretima
桃仁	*táo rén*	8g	Semen Persicae
桑皮	*sāng pí*	12g	Cortex Mori
冬花	*dōng huā*	9g	Flos Farfarae
半夏	*bàn xià*	6g	Rhizoma Pinelliae
川贝	*chuān bèi*	8g	Bulbus Fritillariae Cirrhosae
杏仁	*xìng rén*	5g	Semen Armeniacae Dulcis

This was decocted with water and divided to be taken three times, one pack per day.

Second Visit

October 24th: After taking four packs of the above formula, his cough and asthma improved. He did not need the asthma inhaler. Based on the original formula, *chán tuì* (Periostracum Cicadae) 6g, was added and he continued to take it again.

Third Visit

October 28th: After taking another four packs of the original formula, his cough and asthma had recovered by more than half. He could go to school, and his appetite increased. He was instructed to follow the same formula continuously.

Fourth Visit

November 15th: After taking another ten packs, his cough and asthma disappeared. He stopped his formula for one week and did not see any reoccurence.

Hé huān pí's (合欢皮 , Cortex Albiziae) taste and property are sweet and neutral. It enters the heart channel of the hand *shaoyin*, the liver channel of the foot *jueyin* and the lung meridian of the hand *taiyin* meridians. Both the qi and taste of this herb are thin. Its property is slow and its power is weak. It can nourish the heart blood, relax the heart qi, break up stagnation and clumps, calm the five *zang* organs and it has functions to calm the heart, harmonise the spirit and tonify the yin. In clinical practice, it is more often used for anxiety induced irritability and insomnia syndromes. In addition, it functions to reduce abscesses, promote menstruation and invigorate blood. It also can treat lung abscesses and the falling or fighting injury induced blood stasis, stagnation, swelling and pain. The author was inspired by Zhu Liang-chun, in the *Journal of Chinese Medicine*, Vol. 8 (1984, August), where Zhu Liang-chun used *hé huān pí* (合欢皮 , Cortex Albiziae) with *tài zǐ shēn* (Radix Pseudostellariae) to treat both qi and yin deficiency induced coronary heart disease and arrhythmia. In clinical practice, he uses them with good effect.

1. *Hé Huān Pí* (Cortex Albiziae) ——
Tài Zǐ Shēn (太子参, Radix Pseudostellariae)

Function

To tonify qi and yin, regulate and harmonise the heart vessels.

Application

This can be used for either or both qi and yin deficiency induced syndromes of palpitations, shortness of breath, chest tightness, lassitude, pain anterior to the heart area and due to deficiency induced unsmooth blood flow, the heart vessels disharmony induced palpitations, fright, anxiety, etc. This can also be used to treat biomedicine's clinically diagnosed coronary heart disease, myocarditis, arrhythmia, etc. that belong to the qi mechanism stagnation and clumping, with both qi and yin deficiency.

Compatibility Analysis

Emotion, the will and the blood vessels are all governed and regulated by two organs, the heart and the liver. The heart governs the blood vessels, and the liver governs storing blood while also ensuring the free flow of qi, therefore, if the heart vessels are in disorder, it can be due to the heart's affliction, or also can be related to the free flow of liver qi. If qi is stagnated then there is blood stasis. If there is blood stasis, then the heart vessel will lose its smooth flow, therefore, when treating heart diseases, one should pay attention to treat both the heart and the liver at the same time.

Hé huān pí (Cortex Albiziae) is sweet and neutral. It specialises in calming the heart and cheering the will, relieving stagnation and calming the spirit. In the *Divine Husbandman's Classic of the Materia Medica* it states that it *"calms the five zang organs, harmonises the heart and the will, allows people to feel happy without worry."* The heart is the Emperor. If heart is calm, then the five *zang* organs will be calm and in harmony. The property of *tài zǐ shēn* (Radix Pseudostellariae) is not warm and not dry, not excessive and not slippery. Its function is between the tonifying of *dǎng shēn* (Radix Codonopsis) and the moistening of *shā shēn* (沙参 , Radix Adenophorae seu Glehniae). It is a good herb to tonify qi and generate fluid. These two herbs combined together function to regulate the liver, relieve stagnation, tonify the qi and harmonise the yin, and to regulate the smooth flow of the heart vessels. They really have a smoothing and tonifying effect with two advantages in that they are gentle and neutral. They have a mutually enhancing effect. Zhu Liang-chun said that: *"Using these two herbs means tonifying qi, harmonising yin, and soothing and harmonising the heart vessels, to let the heart qi to flow unrestrained. If the wood qi is soothed and in harmony, then the heart pain of the chest impediment can be eliminated."*

Case Study

Initial Visit

A female patient, 54 years old, April 2nd, 1987.

The patient presented with chest tightness and feeling short of breath which she blamed on tiredness, and ignored it. Then she had an argument with her husband over their child's wedding. This induced distention, fullness and pain in both her hypochondriac areas. Her left chest tightness also extended to her back and she felt a heaviness, shortness of breath and liked to sigh. The patient was afraid she might have coronary heart disease and came for a check up. The EKG reported: Her cardiac muscle was short of blood. Her pulse was deep, thready and choppy. Her tongue was red and her tongue coating was thin. The syndrome differentiation was: Both qi and yin deficiency in the heart channel of the hand *shaoyin* meridian with qi stagnation; and the heart vessels not flowing smoothly inducing chest impediment syndrome. Both qi and yin deficiency in the heart channel of the hand *shaoyin* meridian, caused her shortness of breath, and her deep and thready pulse. The qi stagnation caused blood in the heart vessels to not move smoothly, therefore, she had chest and hypochondriac pain with a choppy pulse. The treatment plan was to use a formula to tonify qi, harmonise yin, soothe stagnation, and free the blood vessels.

Prescription

合欢皮	*hé huān pí*	15g	Cortex Albiziae
太子参	*tài zǐ shēn*	15g	Radix Pseudostellariae
丹参	*dān shēn*	12g	Radix et Rhizoma Salviae Miltiorrhizae
当归	*dāng guī*	12g	Radix Angelicae Sinensis
川芎	*chuān xiōng*	10g	Rhizoma Chuanxiong
桃仁	*táo rén*	9g	Semen Persicae
川牛膝	*chuān niú xī*	12g	Radix Cyathulae
桔梗	*jié gěng*	8g	Radix Platycodonis
甘草	*gān cǎo*	6g	Radix Glycyrrhizae

This was decocted with water and divided to be taken twice, one pack per day.

Second Visit

April 6th: After taking four packs of the above formula, her appetite increased. Her chest tightness and shortness of breath both improved. She continued with the original formula.

Third Visit

April 13th: After taking another seven packs of the above formula, her symptoms reduced, but sometimes she had hiccough and hypochondriac pain. To the original formula *bàn xià* (Rhizoma Pinelliae) 10g, and *yán hú suǒ* (Rhizoma Corydalis) 15g, were added in.

Fourth Visit

April 20th: Her symptoms had reduced a lot, her chest tightness and pain, and shortness of breath basically disappeared. Again, she used five more packs of the same formula to regulate it and she recovered.

Chì sháo's (赤芍 , Radix Paeoniae Rubra) tastes and property are sour, bitter and slightly cold. It enters the liver channel of the foot *jueyin* meridian. Its functions can clear and sedate the liver fire, expel stasis, invigorate blood and alleviate pain. It is more often used for blood stasis inducing all kinds of diseases, such as amenorrhoea, postpartum blood stasis and accumulation, irregular menstruation and abscesses, sores, swellings and toxins, etc. Our ancestors did not differentiate between *chì sháo* (Radix Paeoniae Rubra) and *bái sháo* (Radix Paeoniae Alba). It was only from the *Illustrated Classic of the Materia Medica* that they began to differentiate between them. (This was discussed previously under the *bái sháo* - Radix Paeoniae Alba compatibilities.) In clinical application *chì sháo* (Radix Paeoniae Rubra), aside from treating blood stasis syndromes, also has good results for blood stasis inducing pain and irritability. It can clear and expel stagnated heat in the blood and alleviate pain.

1. *Chì Sháo* (Radix Paeoniae Rubra) ——
Jīn Yín Huā (金银花, Flos Lonicerae Japonicae)

Function

To clear heat, cool blood, transform blood stasis and reduce swellings.

Application

This is used for the patients with lower limbs phlebitis with redness, swelling and a heaviness sensation. It also can be used for patients with sores, or abscesses with redness, swelling and pain.

Compatibility Analysis

Chì sháo (Radix Paeoniae Rubra) is red in colour and enters the blood level. It is an herb to invigorate blood, unblock the collaterals and expel blood stasis. Its property is cold. It also can clear the stagnated heat in the blood, but its invigorating blood power is stronger than that of clearing heat. *Jīn yín huā* (Flos Lonicerae Japonicae) is sweet and cold. It is specialised in clearing and detoxifying toxins. Its qi is aromatic. It also enters the blood level. It can expel heat clumps in the blood, but its clearing heat power is superior to its expelling clumps power. If these two herbs are combined together, the effectiveness of clearing heat, expelling clumps, cooling blood, and eliminating blood stasis can be mutually enhanced and the power to clear and expel stagnated heat in the blood is even stronger. This compatibility is from many years of experience of the famous senior Chinese medicine doctor, Dr. Jin Can-zhang, from Cangzhou, Hebei Province. The author discovered it at an academic conference and later applied it in clinical practice with good effect.

Case 1

Lower limbs phlebitis.

Initial Visit

A male patient, 62 years old, October 18th, 1987.

The patient's right lower limb was heavy, swollen and distended. His lower leg had dull pain and became worse when walking. The local area was hot, red, and swollen for two weeks. He was diagnosed in a hospital with phlebitis. He had been treated with antibiotics but without significant effect, and so he came to visit. His pulse was wiry and fast. His tongue was red and his tongue coating was thin and white. This was caused by stagnated heat in the blood obstructing the meridians and the blood vessels. The treatment plan was to clear heat, cool blood, invigorate blood and expel blood stasis.

Prescription

金银花	jīn yín huā	30g	Flos Lonicerae Japonicae
赤芍	chì sháo	15g	Radix Paeoniae Rubra
玄参	xuán shēn	24g	Radix Scrophulariae
当归	dāng guī	20g	Radix Angelicae Sinensis
丹参	dān shēn	24g	Radix et Rhizoma Salviae Miltiorrhizae
川牛膝	chuān niú xī	15g	Radix Cyathulae
木瓜	mù guā	15g	Fructus Chaenomelis
桑枝	sāng zhī	30g	Ramulus Mori
葛根	gé gēn	20g	Radix Puerariae Lobatae
甘草	gān cǎo	9g	Radix Glycyrrhizae

This was decocted with water and divided to be taken twice, one pack per day.

Second Visit

October 25th: After taking seven packs of the above formula, her right lower limb redness and swelling had lessened and her pain was reduced a lot. Again, she used five more packs of the original formula for treatment.

Third Visit

October 30th: Her leg's swelling was eliminated and her pain stopped, all of her symptoms basically disappeared. Again she used a further three packs of the same formula to complete the treatment.

Case 2

Initial Visit

A female patient, 61 years old, January 9th, 1996.

The patient's right lower limb was red and swollen. She had heaviness, distention and pain sensations for a week and they were worse while walking. It was better in the morning and got worse in the evening. On palpation of her right lower limb, the fingers sank in. Her appetite was sufficient. Her bowel movement was not smooth. Her pulse was wiry. Her tongue coating was white but turbid. This was damp-heat flowing downward and obstructing the meridians and blood vessels syndrome. The treament plan was to clear heat, resolve dampness, invigorate blood and expel blood stasis.

Prescription

金银花	*jīn yín huā*	30g	Flos Lonicerae Japonicae
赤芍	*chì sháo*	30g	Radix Paeoniae Rubra
川牛膝	*chuān niú xī*	15g	Radix Cyathulae
丹参	*dān shēn*	24g	Radix et Rhizoma Salviae Miltiorrhizae
槟榔片	*bīng láng piàn*	15g	Semen Arecae (sliced)
桑枝	*sāng zhī*	30g	Ramulus Mori
茯苓	*fú líng*	20g	Poria
泽泻	*zé xiè*	15g	Rhizoma Alismatis
当归	*dāng guī*	15g	Radix Angelicae Sinensis
木瓜	*mù guā*	15g	Fructus Chaenomelis
葛根	*gé gēn*	20g	Radix Puerariae Lobatae
生薏仁	*shēng yì rén*	30g	Semen Coicis (raw)
甘草	*gān cǎo*	9g	Radix Glycyrrhizae

This was decocted with water and divided to be taken twice, one pack per day.

Second Visit

January 16th: After taking seven packs of the above formula, her bowel movement occurred a couple of times a day and the distention, swelling and pain on her right lower limb were reduced. Her oedema was also reduced a lot. Based on the above formula, *bīng láng piàn* (sliced Semen Arecae) and *dāng guī* (Radix Angelicae Sinensis) were omitted and *sū gěng* (Caulis Perillae) 9g, and *chē qián zǐ* (Semen Plantaginis) 15g, were added. She continued to take another seven packs.

Later, her son came and said that after his mother took seven packs of the above formula, and she recovered.

Cāng zhú's (苍术 , Rhizoma Atractylodis) taste and property are sweet, acrid and warm. It enters the spleen channel of the foot *taiyin* and the stomach channel of the foot *yangming* meridians. Its functions can strengthen the spleen, dry dampness, expel wind and dispel turbidity. The properties of *cāng zhú* (Rhizoma Atractylodis) are warm and dry. Its aromatic qi is extremely thick, therefore, externally, it can release wind-cold; and internally, it can dispel turbid dampness. It is not only a specialised herb to resolve the spleen and stomach dampness obstruction, but it is also an important herb to treat pathogenic invasion in all seasons. *Cāng zhú* (Rhizoma Atractylodis) and *bái zhú* (Rhizoma Atractylodis Macrocephalae) both can strengthen the spleen and dry dampness, however, *bái zhú* (Rhizoma Atractylodis Macrocephalae) has stronger power to strengthen the spleen and *cāng zhú* (Rhizoma Atractylodis) has stronger power to dry dampness. Zhang Yin-an said that: "*The property of bái zhú (Rhizoma Atractylodis Macrocephalae) is superior to that of cāng zhú (Rhizoma Atractylodis). If one wants to tonify the spleen, then use bái zhú (Rhizoma Atractylodis Macrocephalae). If one wants to activate the spleen, then use cāng zhú (Rhizoma Atractylodis).*" In clinical practice, *cāng zhú* (Rhizoma Atractylodis) aside from drying dampness, it is also effective for strengthening the spleen, expelling wind and dispelling turbidity, for nephritis induced oedema and diabetes.

1. *Cāng Zhú* (Rhizoma Atractylodis) ——
Chuān Niú Xī (川牛膝, Radix Cyathulae)

Function

To strengthen the spleen, dry dampness, soothe the sinews and unblock impediment.

Application

This mainly treats lower limbs atrophy and impediment syndrome, and lassitude of the feet and knees.

Compatibility Analysis

In clinical practice atrophy of the lower limbs and impediment syndrome are most often due to dampness, deficiency or blood stasis. If there is a dampness pathogen obstructing and stagnating, the sinews and the blood vessels do not get nourishment; therefore, there is atrophy, weakness and lassitude. Deficiency patients are due to spleen and kidney deficiency. The spleen governs the muscles of the four limbs, the kidney governs the lower back and knees, when both the spleen and the kidney are deficient, there is atrophy of the lower limbs, lower back and knees, flaccidity and lassitude. If there is blood stasis then the sinews and the blood vessels are not nourished. This also can develop into atrophy. The taste and property of *cāng zhú* (Rhizoma Atractylodis) is sweet, acrid and warm. It enters the spleen channel of the foot *taiyin* and the stomach channel of the foot *yangming* meridians. It has functions to strengthen the spleen and dry dampness. Li Dong-yuan said that *cāng zhú* (Rhizoma Atractylodis) is "*an important herb to eliminate dampness, induce sweating, strengthen the stomach, calm the spleen and treat atrophy*". Our ancestors had a saying that to treat atrophy only treat the *yangming* meridians. *Cāng zhú* (Rhizoma Atractylodis) eliminates dampness and strengthens the spleen which makes it a good herb to treat atrophy. *Chuān niú xī* (Radix Cyathulae) is sweet, sour and neutral.

It enters the liver channel of the foot *jueyin* and the kidney channel of the foot *shaoyin* meridians. It is effective for invigorating blood, promoting menstruation, soothing the sinews and unblocking impediment. In the *Divine Husbandman's Classic of the Materia Medica* it states that it *"governs wind-dampness induced atrophy and impediment"*. The property of *niú xī* (Radix Cyathulae) is beneficial for moving downward. It moves but can tonify. In the *Essentials of the Materia Medica* it states that it *"can tonify the liver and the kidney, strengthen the sinews and the bones, treat lower back, knees and bone pain, foot atrophy and sinews spasm ......"* therefore, *chuān niú xī* (Radix Cyathulae) treats lower limbs atrophy and weakness. It also can guide herbs downward. If combined with *cāng zhú* (Rhizoma Atractylodis) together they can achieve the merit of strengthening the spleen, drying dampness, soothing the sinews and unblocking impediment syndrome and it is best for treating patients with lower limbs atrophy or impediment. If due to damp-heat lodging inducing lower limbs atrophy or arthralgia with decreased function, *huáng bǎi* (Cortex Phellodendri Chinensis) can be added and is then called *Sān Miào Wán* (Mysterious Three Pill, 三妙丸). This is Zhu Dan-xi's famous formula.

Case Study

Initial Visit

A male patient, 45 years old, January 18th, 1991.

In 1966, the patient had a sudden episode where his lower limbs became atrophied, and flaccid, and he could not move them. The exact cause was not clear, but after massage or taking some vitamins, it improved. It usually took a couple of hours to one day and he would recover back to normal. Every year or half year, there was an episode. It often happened due to anger, or resting after overexertion. He had it investigated in some hospitals but they all could find no abnormality. Eventually, he was diagnosed with neurosis. On examination, the patient's spirit and physical development were all normal. This time, his lower limbs had atrophy and flaccidity for three days. When walking, he needed two people to hold him. He could not walk or stand by himself. He had massage treatment and took vitamins but they did not work. His lower limbs had no visible muscle atrophy and the muscle tension was also normal, but there was low libido, impotence, and the scrotum was damp etc. His pulse was slow. His tongue coating was white and slightly turbid. The syndrome differentiation classed it as the downward flow of dampness, with the sinews and the blood vessels having lost their nourishment. Dampness lodging in his lower *jiao* caused his scrotum to be damp. Dampness obstructing his sinews and blood vessels inducing them to lose their nourishment, therefore he had impotence and lower limbs atrophy, flaccidity and lassitude. His slow pulse and turbid tongue coating were all induced by a dampness pathogen. The treatment plan was to strengthen the spleen, resolve dampness, tonify the kidney and unblock the blood vessels.

Prescription

| 川牛膝 | *chuān niú xī* | 40g | Radix Cyathulae |
| 苍术 | *cāng zhú* | 40g | Rhizoma Atractylodis |

鸡内金	jī nèi jīn	30g	Endothelium Corneum Gigeriae Galli
茯苓	fú líng	40g	Poria
生山药	shēng shān yào	50g	Rhizoma Dioscoreae (raw)
桑螵蛸	sāng piāo xiāo	24g	Oötheca Mantidis (fried)
黄柏	huáng bǎi	30g	Cortex Phellodendri Chinensis
丹参	dān shēn	40g	Radix et Rhizoma Salviae Miltiorrhizae
杭白芍	háng bái sháo	40g	Radix Paeoniae Lactiflorae
葛根	gé gēn	40g	Radix Puerariae Lobatae

The above ten herbs, were all ground together into a fine powder. Each time, 10g to be taken with warm water, three times a day. As it was not convenient for him to decoct and drink, therefore, a powdered formula was used to treat him.

Second Visit

February 2nd: After taking the above formula for more than ten days, his lower limbs could move freely as normal. His scrotum moistness was also eliminated, and his sex life also improved. He was instructed to continuously take the above powdered formula for one more batch to strengthen the result.

At a follow up visit a year later, he reported there was no reoccurence.

2. *Cāng Zhú* (Rhizoma Atractylodis) —— *Fáng Fēng* (防风, Radix Saposhnikoviae)

Function

To release the exterior, induce sweating, raise the yang and dispel dampness.

Application

This is used for exterior wind-cold invasion, raw and cold injured interior, fever, aversion to cold, and no sweating; or patients with an exterior invasion with diarrhoea and poor digestion.

Compatibility Analysis

Cāng zhú (Rhizoma Atractylodis) is sweet, acrid and warm. Its qi is aromatic. Its properties are warm and dry. It can expel wind and dispel cold exteriorly as well as dispel dampness and eliminate the turbid interiorly. It is an important herb to treat pathogenic qi in all seasons. *Fáng fēng* (Radix Saposhnikoviae) is acrid, sweet and warm. It is also an herb to release the exterior, induce sweating, eliminate wind and transform dampness. These two herbs combined together have an even stronger power to expel wind, release the exterior, raise the yang and dispel dampness. This compatibility is from Wang Hai-cang's *Shén Zhú Sǎn* (Wondrous Atractylodes Powder, 神术散). Wang Ang said that: "*Fáng fēng* (Radix Saposhnikoviae) is acrid, warm, raising and floating. It can eliminate wind and dry dampness. It is a king herb to treat taiyang disease. Cāng zhú (Rhizoma Atractylodis) is sweet, warm and extremely acrid. It expels cold, induces sweating, expels pathogens and raises the yang.*"

Wang Hai-chang made this formula as a substitute for *Má Huáng Tāng* (Ephedra Decoction, 麻黄汤) and *Guì Zhī Tāng* (Cinnamon Twig Decoction, 桂枝汤). Modern clinics use it for GI tract flu and autumn or winter flu in patients without sweating. If there is sweating, *bái zhú* (Rhizoma Atractylodis Macrocephalae) can be substituted for *cāng zhú* (Rhizoma Atractylodis), and the result is also good.

Case Study

Initial Visit

A male patient, 30 years old, November 20th, 1989.

The patient visited for a fever and headache that was induced by an exterior invasion. The previous night he came back after an evening's drinking, he felt soreness all over his whole body, he had a headache, fever, aversion to cold without sweating, a stuffy nose and cough, with stifling discomfort sensation in his chest, and he wanted to vomit. He took two flu capsules by himself and got slight relief. On the morning of his visit, his fever became worse and all of his symptoms were still not relieved. His body temperature was 39℃. His pulse was floating and wiry. His tongue coating was white, moist and slightly turbid. This was the syndrome of exterior wind-cold invasion with turbid dampness inside. The treatment plan was to release the exterior, induce sweating, transform dampness and expel turbidity.

Prescription

苍术	*cāng zhú*	15g	Rhizoma Atractylodis
防风	*fáng fēng*	12g	Radix Saposhnikoviae
荆芥	*jīng jiè*	10g	Herba Schizonepetae
藿香	*huò xiāng*	10g	Herba Agastachis
川芎	*chuān xiōng*	12g	Rhizoma Chuanxiong
桔梗	*jié gěng*	9g	Radix Platycodonis
薄荷	*bò he*	9g	Herba Menthae
甘草	*gān cǎo*	6g	Radix Glycyrrhizae
板蓝根	*bǎn lán gēn*	15g	Radix Isatidis

This was decocted with water and divided to be taken twice warm, one pack per day.

Second Visit

December 23rd: After taking one pack of the above formula, his sweat came out and his fever dropped. After three packs, his exterior syndrome disappeared, but he still had a stuffy sensation in his stomach and a poor appetite. The treatment plan was changed to harmonise the stomach, strengthen the spleen, eliminate dampness and transform turbidity.

Prescription

| 藿香 | *huò xiāng* | 9g | Herba Agastachis |
| 枳壳 | *zhǐ qiào* | 10g | Fructus Aurantii |

苍术	*cāng zhú*	10g	Rhizoma Atractylodis
焦三仙	*jiāo sān xiān*	30g	Fructus Hordei Germinatus et Crataegi et Massa Fermentata Medicinalis (scorch-fried)
半夏	*bàn xià*	9g	Rhizoma Pinelliae
厚朴	*hòu pò*	9g	Cortex Magnoliae Officinalis
茯苓	*fú líng*	12g	Poria
苏梗	*sū gěng*	9g	Caulis Perillae
砂仁	*shā rén*	6g	Fructus Amomi
甘草	*gān cǎo*	6g	Radix Glycyrrhizae

This was decocted with water, to be taken, one pack per day.
After taking three packs of the above formula, he recovered.

Zé Lán (泽兰 , Herba Lycopi)

Zé lán's (泽兰 , Herba Lycopi) taste and property are bitter, acrid and slightly warm. It enters the liver channel of the foot *jueyin* and the spleen channel of the foot *taiyin* meridians. It is a blood level herb for the *jueyin* meridian. Its property is warm and lightly aromatic. It can soothe liver qi, harmonise the nutritive qi and blood, invigorate blood and expel blood stasis. It is a more gentle blood invigorating herb.

1. Zé Lán (泽兰, Herba Lycopi) ——
Hǔ Zhàng (虎杖, Rhizoma et Radix Polygoni Cuspidati)

Function

To invigorate blood, transform stasis, soothe the liver and detoxify toxins.

Application

This is used for all kinds of hepatitis.

Compatibility Analysis

Zé lán (Herba Lycopi) is a blood invigorating and blood stasis expelling herb. Its property is more gentle. In the Chinese medicinal textbook of the High Level Medical and Pharmaceutical College it states that: "*Zé lán (Herba Lycopi) is acrid, dispersing, warm and unblocking. It is not cold and not dry. Its property is more gentle. It moves, but not strongly. It can soothe the liver qi and unblock the meridians and the blood vessels. It has characteristics to eliminate blood stasis and expel clumps, but does not injure the normal qi; therefore, if there is any blood stasis or stagnation in the blood vessels, the menstruation is not smooth. It is a commonly used herb.*" *Hǔ zhàng* (Rhizoma et Radix Polygoni Cuspidati) is also an herb to invigorate blood and expel blood stasis, but its taste and property are bitter and cold. It also functions to clear heat, eliminate dampness and detoxify toxins. Modern pharmacological research indicates that *hǔ zhàng* (Rhizoma et Radix Polygoni Cuspidati) has an anti-viral function. *Zé lán* (Herba Lycopi) and *hǔ zhàng* (Rhizoma et Radix Polygoni Cuspidati) both enter the liver channel of the foot *jueyin* meridian and both can invigorate blood and transform blood stasis. One is warm and the other is cold. These two herbs combined together can neutrally regulate cold and heat, and also can inhibit the other's bias in order to increase their effectiveness to invigorate blood and transform blood stasis. *Zé lán* (Herba Lycopi) also can soothe the liver qi and *hǔ zhàng* (Rhizoma et Radix Polygoni Cuspidati) also can detoxify toxic pathogens. These two herbs combined together can achieve functions to invigorate blood, transform blood stasis, soothe the liver and detoxify toxins. In clinical practice, using them to treat viral hepatitis is effective. The author thought that to treat viral hepatitis one cannot use the traditional Chinese medicinals' method to clear the damp-heat and detoxify toxins, especially for the hepatitis B patients, as hepatitis B infections are due to a toxic pathogen (hepatitis B virus) invading the human body's blood and body fluid but not due to the six exterior pathogens invading from the exterior to the interior. Hepatitis B virus also only attacks and attaches in the liver. The liver governs storing blood, it stores but does not excrete, and therefore, once the virus has invaded the liver, it is hard to expel it. When the virus enters the liver it must also injure the liver. It will induce disorder in the qi dispersion function and the blood stasis, therefore, to soothe the liver and transform blood stasis is the key method to treat hepatitis B. If the liver qi is freely flowing, blood circulation is free, this combined with detoxifying toxins and pathogen expelling herbs can then clear the virus and recover the liver functions. During the treatment, one should not over use acrid and dry, bitter and cold or attacking herbs to avoid injury to the normal qi. To support the normal qi or take care of the normal qi is also important in treating hepatitis. To select *zé lán* (Herba Lycopi) and *hŭ zhàng* (Rhizoma et Radix Polygoni Cuspidati) to treat hepatitis satisfies the rule to invigorate blood, transform stasis, soothe the liver and detoxify toxins. Meanwhile, both herbs are gentle, not cold and not hot. They transform blood stasis, soothe the liver and detoxify toxins without injury to the normal qi. All can use this compatibility regardless of whether it is hepatitis B or the early stage of the liver cirrhosis or a hepatitis B carrier. The other types of hepatitis are also suitable for using it.

Case 1

Hepatitis B carrier.

Initial Visit

A male patient, 6 years old, February 15th, 1992.

The child recently manifested symptoms of anorexia, fatigue, etc. He came to hospital for tests and treatment. His liver function test was normal except that his Hbs Ag was positive, 1:64. The child's development was good. His parents' laboratory tests both showed no signs of hepatitis B or hepatitis B carrier. The treatment plan was to tonify qi, strengthen the spleen, soothe the liver and detoxify toxins.

Prescription

白花蛇舌草	*bái huā shé shé cǎo*	15g	Herba Hedyotis Diffusae
蒲公英	*pú gōng yīng*	10g	Herba Taraxaci
泽兰	*zé lán*	9g	Herba Lycopi
虎杖	*hŭ zhàng*	9g	Rhizoma et Radix Polygoni Cuspidati
柴胡	*chái hú*	9g	Radix Bupleuri
生黄芪	*shēng huáng qí*	10g	Radix Astragali (raw)
丹参	*dān shēn*	10g	Radix et Rhizoma Salviae Miltiorrhizae
白术	*bái zhú*	9g	Rhizoma Atractylodis Macrocephalae
枳壳	*zhǐ qiào*	9g	Fructus Aurantii
生薏仁	*shēng yì rén*	12g	Semen Coicis (raw)
甘草	*gān cǎo*	6g	Radix Glycyrrhizae
砂仁	*shā rén*	3g	Fructus Amomi

This was decocted with water, 150ml to be taken three times a day, one pack per day.

Second Visit

March 15 th: After totally taking twenty eight packs of the modifications of the above formula, his anorexia and fatigue were both eliminated. His HbsAg was 1:8. He continued to take it.

Third Visit

June 15 th: After continuously taking the above medicine for three months, he had his liver function tested three times, his liver functions were all normal and his Hbs Ag turned negative.

Case 2

Acute hepatitis B.

Initial Visit

A male patient, 45 years old, December 20th, 1997.

The patient recently had symptoms of lassitude, anorexia, stomach and abdominal distention and fullness, etc. It was followed by brownish urine with all symptoms becoming worse, and so he came to hospital for tests and treatment. The liver function test showed: Total bilirubin 180μmol/L, direct bilirubin 100μmol/L, indirect bilirubin 80μmol/L, ALT 1928U/L. Of the seven indications for hepatitis B: HbsAg (+), HbeAg (+) and anti-HBc (+). His B-ultrasound test: The liver physical echo was in medium speed, light spots were rough, portal vein 1.3cm, ascites (−). The gallbladder and the spleen were all normal. On physical examination, the patient's sclera were significantly yellowish. His face was a yellow orange colour. His tongue was red and tongue coating was white and turbid. This was acute hepatitis B with toxic pathogen exuberant syndrome. He was immediately treated with a formula for clearing heat, detoxifying toxins, protecting the liver, reducing enzyme and eliminating yellowness.

Prescription

白花蛇舌草	*bái huā shé shé cǎo*	30g	Herba Hedyotis Diffusae
蒲公英	*pú gōng yīng*	15g	Herba Taraxaci
泽兰	*zé lán*	15g	Herba Lycopi
虎杖	*hǔ zhàng*	15g	Rhizoma et Radix Polygoni Cuspidati
丹参	*dān shēn*	20g	Radix et Rhizoma Salviae Miltiorrhizae
柴胡	*chái hú*	15g	Radix Bupleuri
猪苓	*zhū líng*	15g	Polyporus
土茯苓	*tǔ fú líng*	30g	Rhizoma Smilacis Glabrae
茵陈	*yīn chén*	24g	Herba Artemisiae Scopariae
五味子	*wǔ wèi zǐ*	9g	Fructus Schisandrae Chinensis
垂盆草	*chuí pén cǎo*	15g	Herba Sedi
白术	*bái zhú*	12g	Rhizoma Atractylodis Macrocephalae
枳壳	*zhǐ qiào*	12g	Fructus Aurantii
仙鹤草	*xiān hè cǎo*	24g	Herba Agrimoniae
甘草	*gān cǎo*	6g	Radix Glycyrrhizae

After immersing the above herbs in clear water for half an hour, they were decocted for another half hour and filtered out. Then clear water was added to decoct for another twenty minutes. The two decoctions were mixed together, and divided to be taken twice warm, one pack per day.

Second Visit

January 3rd, 1998: After taking fourteen packs of the above formula, his symptoms of abdominal distention, anorexia, lassitude were all improved. The liver function test: Total bilirubin 85μmol/L, direct bilirubin 60μmol/L, indirect bilirubin 25μmol/L, ALT 237U/L. He continued to take it.

Third Visit

January 17th: His appetite was back to normal. His brownish urine colour was becoming lighter, but he still had abdominal distention and discomfort. His liver function test showed: Total bilirubin 39.9μmol/L, direct bilirubin 16.4μmol/L, indirect bilirubin 23.5μmol/L, ALT 59u/L. To the above formula, *wū yào* (Radix Linderae) 12g, was added. He continued to take it.

Fourth Visit

February 4th: All of his symptoms disappeared. His liver function and the seven laboratory test indicators for hepatitis B were all normal. He was instructed to take one pack every other day for a month and to do a double check.

Fifth Visit

March 12th: His liver function and the seven indicators for hepatitis B all tested normal. He stopped taking his formula.

Note

The author has had reasonable results using Chinese herbs to treat hepatitis B in clinic. Below, for reference, he introduces his treatment formula *Gān Níng Tāng* (Liver-Quietening Decoction), that he has used from experience of over ten years.

Ingredients

白花蛇舌草	*bái huā shé shé cǎo*	15～30g	Herba Hedyotis Diffusae
蒲公英	*pú gōng yīng*	10～15g	Herba Taraxaci
泽兰	*zé lán*	10～15g	Herba Lycopi
虎杖	*hǔ zhàng*	10～15g	Rhizoma et Radix Polygoni Cuspidati
丹参	*dān shēn*	15～20g	Radix et Rhizoma Salviae Miltiorrhizae
柴胡	*chái hú*	10～15g	Radix Bupleuri
枳壳	*zhǐ qiào*	8～12g	Fructus Aurantii
白术	*bái zhú*	10～15g	Rhizoma Atractylodis Macrocephalae
茯苓	*fú líng*	15～20g	Poria
甘草	*gān cǎo*	4～6g	Radix Glycyrrhizae

Method and Administration

The above herbs are immersed in clear water for half an hour, a high heat is used to boil them, then a low heat is used to decoct for 30 minutes. The decoction is then filtered. Clear water is added to the herbs to decoct for another 15 ~ 20 minutes. The decoction is filtered out again and poured in with the first one. These two decoctions should ideally total around 300 ~ 400ml. Divide, to be taken 2 ~ 3 times.

Function

To invigorate blood, transform stasis, soothe the liver and detoxify toxins, tonify the liver, strengthen the spleen.

Application

To be used for all kinds of acute or chronic hepatitis, hepatitis B carriers and the early stages of liver cirrhosis.

Modifications

1. This formula can be used to treat patients who have pure Hbs Ag positive or who have three big positives in the liver five indicators test. If after long term treatment, the test does not turn to negative, herbs can be added like *huáng qí* (Radix Astragali), *tù sī zǐ* (Semen Cuscutae), *líng zhī* (灵芝 , Ganoderma), etc. that can tonify qi, support the yang, strengthen the spleen and tonify the kidney as well as herbs like *fēng fáng* (蜂房 , Nidus Vespae) and *hēi mǎ yǐ* (黑蚂蚁 , Lasius Niger), etc. that can expel blood stasis and unblock the collaterals.

2. For patients who have a liver function disorder, with ALT elevated, add *wǔ wèi zǐ* (Fructus Schisandrae Chinensis), *tǔ fú líng* (Rhizoma Smilacis Glabrae) and *chuí pén cǎo* (Herba Sedi), etc.

3. For patients with significant jaundice, brownish urine, constipation, where both dampness and heat are severe, add *yīn chén* (Herba Artemisiae Scopariae), *dà huáng* (Radix et Rhizoma Rhei) and *huáng bǎi* (Cortex Phellodendri Chinensis), etc. If the patient whose dampness is greater than the heat, add *yīn chén* (Herba Artemisiae Scopariae), *fú líng* (Poria) and *zhū líng* (Polyporus), etc.

4. If the patient has accompanying symptoms of anorexia, stomach and abdominal distention and fullness, add *shā rén* (Fructus Amomi), *jiāo sān xiān* (scorch-fried Fructus Hordei Germinatus et Crataegi et Massa Fermentata Medicinalis) and *jī nèi jīn* (Endothelium Corneum Gigeriae Galli), etc.

5. If the patient has accompanying symptoms of spleen deficiency, lassitude, and shortness of breath add *huáng qí* (Radix Astragali), *xiān hè cǎo* (Herba Agrimoniae), *dà zǎo* (Fructus Jujubae) and *tài zǐ shēn* (Radix Pseudostellariae), etc.

6. For the patient who has liver area pain, distention, and discomfort, add *yù jīn* (Radix Curcumae) and *yán hú suǒ* (Rhizoma Corydalis).

7. For the patient who has liver area pain and discomfort with a thready pulse that belongs to liver yin deficiency, add *hé shǒu wū* (Radix Polygoni Multiflori), *háng bái sháo* (Radix Paeoniae Lactiflorae), *dāng guī* (Radix Angelicae Sinensis) and *gǒu qǐ* (Fructus Lycii), etc.

8. If the patient has accompanying symptoms of kidney deficiency inducing lower back soreness, fatigue and lassitude, add *tù sī zǐ* (Semen Cuscutae), *xiān líng pí* (Herba Epimedii) and *shā yuàn zǐ* (Semen Astragali Complanati), etc.

9. If the patient has accompanying symptoms of insomnia and a lot dreams, add *yè jiāo téng* (Caulis Polygoni Multiflori) and *zǎo rén* (Semen Ziziphi Spinosae), etc.

10. If there is diarrhoea, add *fú líng* (Poria), *chē qián zǐ* (Semen Plantaginis) and *shān yào* (Rhizoma Dioscoreae).

11. For the early stage of liver cirrhosis, add *biē jiǎ* (Carapax Trionycis), *dāng guī* (Radix Angelicae Sinensis), *shān jiǎ* (Squama Manis), *táo rén* (Semen Persicae) and *jī xuè téng* (Caulis Spatholobi), etc.

12. For the patient who has protein inversion or low whey protein, add *huáng qí* (Radix Astragali), *jī xuè téng* (Caulis Spatholobi), *dāng guī* (Radix Angelicae Sinensis), *zǐ hé chē* (紫河车 , Placenta Hominis), *gǒu qǐ* (Fructus Lycii) and *hé shǒu wū* (Radix Polygoni Multiflori), etc. The above herbs have the function to promote protein synthesis. They can be added in according to the syndrome differentiation.

13. For patients in the early stage of liver cirrhosis with ascites, add *huáng qí* (Radix Astragali), *fáng jǐ* (Radix Stephaniae Tetrandrae), *wū yào* (Radix Linderae), *yù lǐ rén* (Semen Pruni), *dà fù pí* (Pericarpium Arecae), *shuǐ hóng huā zǐ* (水红花子 , Fructus Polygoni Orientalis), etc.

14. If the patient has accompanying symptoms of gum bleeding or has a possible haemorrhage, add *ē jiāo* (阿胶 , Colla Corii Asini), *sān qī fěn* (Radix et Rhizoma Notoginseng powder) and *xiān hè cǎo* (Herba Agrimoniae), etc.

In clinical practice, there are a large variety of symptoms. The above is only a couple of simple modification examples. Flexibility is required in order to extrapolate what a specific patient needs, from the above examples. The syndrome differentiation needs to be followed for modifications in order to get satisfactory treatment results.

Formula Explanation

The above discussion about the treatment principles for liver diseases indicate that the key to treating hepatitis is to invigorate blood, transform stasis, soothe the liver and detoxify toxins. From the aspect of medication, you should support and take care of the normal qi, by not thoughtlessly using attacking herbs. Hence, in the formula, *dān shēn* (Radix et Rhizoma Salviae Miltiorrhizae), *zé lán* (Herba Lycopi), and *hǔ zhàng* (Rhizoma et Radix Polygoni Cuspidati) are used as three herbs to invigorate blood, and transform blood stasis. Meanwhile, they can nourish blood, soothe the liver, and detoxify toxins. They are really gentle herbs to invigorate blood and transform blood stasis. Although they can transform blood stasis, they do not injure the normal qi. These three herbs also can prevent liver fibrosis. *Bái huā shé shé cǎo* (Herba Hedyotis Diffusae) and *pú gōng yīng* (Herba Taraxaci) are clearing heat and detoxifying herbs. *Bái huā shé shé cǎo* (Herba Hedyotis Diffusae) is sweet, bland and slightly cold. Although it belongs to clearing heat and detoxifying herbs, its property is gentle. Modern pharmacological research has discovered that it can increase the macrophage's phagocytosing power as well as having anti-cancer and anti-viral functions. *Pú gōng yīng* (Herba Taraxaci) is sweet, bitter and cold. It is also a clearing heat and detoxifying herb. It also functions to protect the liver, benefit the gallbladder and reduce transaminase. The property of these two herbs are gentle and their taste is sweet, therefore, they can clear heat and detoxify toxins without injury to the normal qi. *Chái hú* (Radix Bupleuri) and *zhǐ qiào* (Fructus Aurantii) soothe the liver, resolve stagnation and regulate qi in order to make the liver qi maintain a free flow and help to expel the pathogens out of the body, meanwhile, they guide all herbs to reach the affected spot in the liver. *Bái zhú* (Rhizoma Atractylodis Macrocephalae), *fú líng* (Poria) and *gān cǎo* (Radix Glycyrrhizae) are herbs to strengthen the spleen and tonify the middle *jiao*. This means that with liver disease one should strengthen the spleen first. If the spleen is strengthened, then the normal qi will be plenty to help the liver to recover and clear the pathogens.

Yù jīn's (郁金 , Radix Curcumae) taste and property are acrid, bitter and cold. It enters the heart channel of the hand *shaoyin*, the lung channel of the hand *taiyin* and the liver channel of the foot *jueyin* meridians. It functions to soothe the liver, benefit the gallbladder, cool blood and move blood stasis. Miao Zhong-chun said that: *"Yù jīn (Radix Curcumae) originally belonged to the qi herbs in the blood level. That it can treat all kinds of blood symptoms follows from the saying that blood travelling upward belongs to the internal heat and fire exuberance. This herb can descend qi, if qi is descended, then fire is also descended. Its property also enters the blood, therefore, it can descend fire downward and prevent the blood from moving restlessly."* Therefore, within the moving blood stasis function of *yù jīn* (Radix Curcumae), it also can cool blood and stop bleeding, meanwhile, it stops bleeding without leaving blood stasis. In clinical prcatice, the applications for *yù jīn* (Radix Curcumae) are roughly as follows: Firstly, it can be used for diseases of the liver and the gallbladder system such as chronic hepatitis, cholecystitis that have pain symptoms in the liver area. It has functions to soothe the liver, benefit the gallbladder, expel blood stasis and alleviate pain. It is more often used with *chái hú* (Radix Bupleuri), etc. as a compatibility. Secondly, it is used for blood stasis induced menstrual pain especially for premenstrual pain patients who tend towards qi stagnation, and blood stasis with heat. It should be combined with *dāng guī* (Radix Angelicae Sinensis) and *huáng qín* (Radix Scutellariae), etc. as in Fu Qing-Zhu's *Xuān Yù Tōng Jīng Tāng* (Depression-Resolving and Channel-freeing Decoction, 宣郁通经汤). Thirdly, as it enters the heart channel of the hand *shaoyin*, it can clear the heart and transform blood stasis, therefore, it also can treat manic disorder. It combined with *bái fán* (白矾 , Alumen) becomes *Bái Jīn Wán* (Alum and Curcuma Pill, 白金丸). There is also *yù jīn* (Radix Curcumae) with *shí chāng pú* (Rhizoma Acori Tatarinowii) that treats febrile disease induced coma. It has functions to clear the heart and open the orifices.

1. *Yù Jīn* (Radix Curcumae) ——
Dà Huáng (大黄, Radix et Rhizoma Rhei)

Function

To benefit the gallbladder, reduce jaundice, expel stones and alleviate pain.

Application

This is used for cholecystitis and gallstone syndromes.

Compatibility Analysis

Yù jīn (Radix Curcumae) is acrid, bitter and cold. It enters the liver channel of the foot *jueyin* and the gallbladder channel of the foot *shaoyang* meridians. It functions to invigorate blood, transform blood stasis, soothe the liver, benefit the gallbladder, expel stagnation and alleviate pain. It is a qi within blood herb. Modern pharmacological research found that *yù jīn* (Radix Curcumae) contains curcumin that can promote bile secretion and excretion. *Dà huáng* (Radix et Rhizoma Rhei) is bitter and cold. It is a heat clearing and bowel movement freeing herb. It also has the function to increase the bile flow amount, therefore, these two herbs combined together function to benefit the gallbladder and expel stones. In addition, these two herbs both can invigorate blood and transform blood stasis. This also can help reduce cholecystitis symptoms and its recovery. Meanwhile, *dà huáng* (Radix et Rhizoma Rhei) contains rhein, emodin and aloe emodin. It has very strong anti-bacterial and anti-inflammatory functions, therefore, it is also effective for treating cholecystitis.

Case Study

Initial Visit

A female patient, 36 years old, June 3rd, 1986.

The patient had right upper abdomen pain for three months. She was investigated with a B-ultrasound test in a hospital. There was a 0.8cm stone in her gallbladder. It could move when she changed her posture. She had it treated with stone-expelling medicine for more than twenty days but without effect, so she came to our hospital for clinical treatment. The patient complained: Aside from her upper right abdominal pain, there was stomach and abdominal distention, which got worse after eating, frequent hiccough, constipation, dizziness in her head, a bitter mouth, etc. On examination, she had wiry pulse, a red tongue with a white slightly turbid tongue coating. This was the gallbladder and the stomach heat stagnation with qi blocked in the *fu* organs syndrome. The treatment plan was to sedate the *fu* organs, clear heat, benefit the gallbladder and expel the stone.

Prescription

金钱草	*jīn qián cǎo*	40g	Herba Lysimachiae
郁金	*yù jīn*	20g	Radix Curcumae
大黄	*dà huáng*	10g	Radix et Rhizoma Rhei
柴胡	*chái hú*	15g	Radix Bupleuri
枳壳	*zhǐ qiào*	15g	Fructus Aurantii
半夏	*bàn xià*	9g	Rhizoma Pinelliae
木香	*mù xiāng*	9g	Radix Aucklandiae
虎杖	*hǔ zhàng*	15g	Rhizoma et Radix Polygoni Cuspidati
威灵仙	*wēi líng xiān*	15g	Radix et Rhizoma Clematidis
鸡内金	*jī nèi jīn*	20g	Endothelium Corneum Gigeriae Galli
甘草	*gān cǎo*	6g	Radix Glycyrrhizae

This was decocted with water and divided to be taken twice, one pack per day.

Second Visit

April 7th: After taking four packs of the above formula, she then had four to five bowel movements per day with mild abdominal pain. Her symptoms of abdominal distention and dizziness in her head, etc. were all reduced a lot. On examination, her pulse was wiry but deficient and slightly forceless. Based on the above formula, *dà huáng* (Radix et Rhizoma Rhei) was reduced to 6g and *shēng huáng qí* (raw Radix Astragali) 30g, was added to tonify her qi in order to increase the contraction ability of her gallbladder.

Third Visit

June 23rd: After totally taking more than twenty packs of the modifications of the above formula, all of her symptoms disappeared and no stone was visible in her B-ultrasound test.

Bǔ gǔ zhī (补骨脂 , Fructus Psoraleae) is acrid and warm. It enters the spleen channel of the foot *taiyin*, the kidney channel of the foot *shaoyin* and the pericardium channel of the hand *jueyin* meridians. It can warm and tonify the spleen and the kidney. It also has reducing urination and stopping enuresis functions.

1. *Bǔ Gǔ Zhī* (Fructus Psoraleae) ——
Yì Zhì Rén (益智仁, Fructus Alpiniae Oxyphyllae)

Function

To warm and tonify the spleen and the kidney, to secure and bind to stop enuresis.

Application

This is used for the spleen or lung qi deficiency, with the kidney primary qi not secure thus inducing enuresis and urinary incontinence.

Compatibility Analysis

Zhen Quan said that *bǔ gǔ zhī* (Fructus Psoraleae) can *"stop urination"*. In the *Essentials of the Materia Medica* it states that it can *"strengthen the primary yang qi and reduce urination."* *Yì zhì rén* (Fructus Alpiniae Oxyphyllae) is acrid and warm. It enters the spleen channel of the foot *taiyin*, the stomach channel of the foot *yangming* and the kidney channel of the foot *shaoyin* meridians. It is also an herb to warm the middle *jiao*, strengthen the spleen, secure the kidney and warm the yang qi. In the *Essentials of the Materia Medica* it states that it *"can bind essence and secure qi......reduce urination"*. These two herbs combined together, their power to warm the middle *jiao*, strengthen the spleen, warm the yang qi and secure the kidney is even stronger. It is in the category of mutual accentuation. If the spleen is strengthened, then the lung qi will be plenty. This means that the earth is cultivated in order to generate the metal. If the spleen and the lung qi are plenty, then the water pathway is controlled and there is no possibility of urinary incontinence or enuresis. If the kidney qi is warm and secured, then the qi transformation of the gallbladder will be normal. If the water and fluid are controlled then there is no possibility of enuresis.

Case Study

Initial Visit

A male patient, 5 years old, April 7th, 1987.

The child had enuresis since he was young. He was treated again and again without effect and so came to visit. His mother complained that for more than four years he had had enuresis. If he was exerted, it became worse. His appetite was normal. On examination, the child's development was as normal. He had no anaemia or poor development symptoms. His tongue was slightly pale, and his tongue coating was thin and white. His pulse was wiry and thready. This was kidney deficiency causing the urinary bladder qi to not be secure. The treatment plan was to warm the kidney, secure and bind the the urinary bladder qi. Zhang Jing-yue's *Gǒng Tí Wán* (Dyke-Strenthening Pill, 巩堤丸) was modified to treat him.

Prescription

补骨脂	*bǔ gǔ zhī*	9g	Fructus Psoraleae
益智仁	*yì zhì rén*	10g	Fructus Alpiniae Oxyphyllae
山药	*shān yào*	15g	Rhizoma Dioscoreae
菟丝子	*tù sī zǐ*	10g	Semen Cuscutae
茯苓	*fú líng*	9g	Poria
熟地	*shú dì*	10g	Radix Rehmanniae Praeparata
乌药	*wū yào*	9g	Radix Linderae
山萸肉	*shān yú ròu*	9g	Fructus Corni
桑螵蛸	*sāng piāo xiāo*	10g	Oötheca Mantidis (fried)
甘草	*gān cǎo*	6g	Radix Glycyrrhizae

This was decocted with water and divided to be taken twice, one pack per day.

Meanwhile, *wǔ bèi zǐ* (五倍子 , Galla Chinensis) and *bái zhǐ* (Radix Angelicae Dahuricae) were used in equal amounts and ground into a powder for external use. Each time, 10g were taken and warm water was used to mix as a paste to apply on the child's umbilicus. Every night before sleep, it was applied on his umbilicus, using gauze to cover it and adhesive tape or bandage to fix it in place. When he woke up the next morning, it was removed.

Second Visit

April 15th: Throughout the period of using the medicine, the child only wet the bed once. As it was effective the prescription was not changed and he continued to take the original formula again.

After taking more than thirty packs of the above formula, the child's bed wetting was eliminated.

Note

To apply *wǔ bèi zǐ* (Galla Chinensis) on the umbilicus, it also can treat night sweating in small children. Please refer to *huáng qín* (Radix Scutellariae) with *sāng yè* (Folium Mori) for the details.

The taste and property of *wǔ bèi zǐ* (Galla Chinensis) are sour and neutral. It enters the lung channel of the hand *taiyin*, the kidney channel of the foot *shaoyin* and the large intestine channel of the hand *yangming* meridians. Due to its sour taste it can constrain and its astringency can retain, therefore, it has functions to retain the lung with astringency, retain the intestines and stop sweating. This case used *wǔ bèi zǐ* (Galla Chinensis) with *bái zhǐ* (Radix Angelicae Dahuricae), with the acrid, warm, aromatic, free and strong penetrating power of *bái zhǐ* (Radix Angelicae Dahuricae) assisting *wǔ bèi zǐ* (Galla Chinensis) to get a better treatment effect. If the child is too young and cannot take the decoction, only using this formula for treating enuresis still gives effective results.

Yīn chén (茵陈 , Herba Artemisiae Scopariae) is bitter and slightly cold. It enters the spleen channel of the foot *taiyin*, the stomach channel of the foot *yangming*, the liver channel of the foot *jueyin* and the gallbladder channel of the foot *shaoyang* meridians. Its functions can clear damp-heat, and clear jaundice. It is a main herb to treat jaundice, regardless of whether it is yin jaundice or yang jaundice, both can use it for treatment. Modern pharmacological research indicates that *yīn chén* (Herba Artemisiae Scopariae) also has clearing heat, benefiting the gallbladder, anti-viral, anti-bacterial and anti-fungal functions. In recent years, there are reports that it can effectively treat mouth ulcers. The author tried it with reasonably good results. If combined with *qīng dài* (Indigo Naturalis), the result is even better.

1. Yīn Chén (Herba Artemisiae Scopariae) ——
Qīng Dài (青黛, Indigo Naturalis)

Function

To clear heat, expel dampness, detoxify toxins and alleviate pain.

Application

This is used for mouth ulcers.

Compatibility Analysis

Mouth ulcers are most often due to the stagnated heat in the heart channel of the hand *shaoyin* and the spleen channel of the foot *taiyin* meridians. The spleen opens its orifice at the mouth. The tongue is the sprout of the heart. The damp-heat of the spleen channel of the foot *taiyin* meridian or the stagnated fire of the heart channel of the hand *shaoyin* meridian both can induce mouth and tongue ulcers. *Yīn chén* (Herba Artemisiae Scopariae) is bitter and cold. It enters the spleen channel of the foot *taiyin* meridian, and clears the damp-heat in the spleen channel. It also can promote urination and guide the heart fire downward. *Qīng dài* (Indigo Naturalis) is salty and cold. It has clearing heat and detoxifying toxins functions. It is a king herb to reduce swelling, and alleviate pain from mouth ulcers. These two herbs combined together, *yīn chén* (Herba Artemisiae Scopariae) clears damp-heat, and descends the heart fire to treat the root problem and *qīng dài* (Indigo Naturalis) reduces swelling and alleviates pain to treat the manifestations. To treat both the root problem and manifestations a good effect can be had very quickly. When clinically applied, water can be used to decoct them and the decoction can be used as a mouth wash; or based on the syndrome differentiation, add in the other herbs to take orally.

Case Study

Initial Visit

A female patient, 33 years old, June 20th, 1995.

The patient had a history of mouth ulcers. There were a couple of episodes each year. Recently, there were a couple of round ulcers on her lower lip, inside the cheek and on the tongue. They were in the size of a soy bean and had a burning pain sensation. It became worse when drinking hot water or eating. She had treated them with anti-inflammatory medicines but they did not work and so she came to visit. Her pulse was wiry and fast. Her tongue was red and her tongue coating was thin and white. This was caused by the heart and spleen fire stagnation streaming upward to the mouth and tongue. The treatment plan was to clear heat, detoxify toxins, clear the heart and sedate the spleen.

Prescription

茵陈	*yīn chén*	40g	Herba Artemisiae Scopariae
青黛	*qīng dài*	3g	Indigo Naturalis
蒲公英	*pú gōng yīng*	15g	Herba Taraxaci
栀子	*zhī zǐ*	9g	Fructus Gardeniae
藿香	*huò xiāng*	9g	Herba Agastachis
薄荷	*bò he*	9g	Herba Menthae
甘草	*gān cǎo*	6g	Radix Glycyrrhizae

This was decocted with water, and divided in two. One part was used orally and the other was used as a mouth wash, for a couple of times. One pack per day.

Second Visit

June 24th: After talking four packs of the above formula, her mouth ulcers basically disappeared and there was not much pain. The patient was instructed to take two packs of the above formula again and she got better.

Sāng yè's (桑叶 , Folium Mori) tastes and property are sweet, bitter and cold. It enters the lung channel of the hand *taiyin* and the liver channel of the foot *jueyin* meridians. *Sāng yè* (Folium Mori) was collected after falling frost. The dark green leaves that have not withered were picked. After drying under the sun then it can be used as a medicine. As it holds the lung qi in the autumn, it can disperse the light and the clear. Its power to expel wind-heat is even stronger. Its taste and property are sweet and cold. It also functions to clear the liver and brighten the eyes. It is a commonly used herb to expel wind and clear heat in clinical practice.

1. *Sāng Yè* (Folium Mori) ——
Jú Huā (菊花, Flos Chrysanthemi)

Function

To soothe wind, clear heat, clear the liver and brighten the eyes.

Application

This is used for exterior wind-heat invasion, dizziness in the head and eyes, fever, cough and the liver fire flaring upward inducing redness, swelling, pain and dizziness, etc.

Compatibility Analysis

Sāng yè (Folium Mori) is sweet and cold. It enters the lung channel of the hand *taiyin* and the liver channel of the foot *jueyin* meridians. Its qi is light, and clear. It can disperse. Its taste and property are sweet and cold and it can clear heat. It enters the lung channel of the hand *taiyin* meridian to clear the lung heat and can stop cough. It enters the liver channel of the foot *jueyin* meridian to draw the liver heat and can clear the head and the eyes. *Jú huā* (Flos Chrysanthemi) is sweet, bitter and slightly cold. It enters three meridians of the liver channel of the foot *jueyin*, the lung channel of the hand *taiyin* and the kidney channel of the foot *shaoyin*. Its property is clear, cool and it is beneficial for expelling wind-heat in the head and the eyes. Li Shi-zhen said that it: *"can tonify the two organs of the metal and the water. The water is tonified in order to control the metal and to tonify the metal will calm the wood. If the wood is calmed, the wind will be extinguished. If fire is descended, then the heat will be eliminated."* *Sāng yè* (Folium Mori) and *jú huā* (Flos Chrysanthemi) are both sweet and cold herbs. Their properties are clearly dispersing. However, *sāng yè* (Folium Mori) tends towards entering the lung channel of the hand *taiyin* meridian and *jú huā* (Flos Chrysanthemi) tends towards entering the liver channel of the foot *jueyin* meridian. One is more soothing and the other is more clearing. These two herbs combined together can enhance their effectiveness to soothe wind, clear heat, calm the liver and brighten the eyes, therefore, interior wind-heat invasion or the initial stage of the wind febrile induced cough more often use this compatibility as in *Sāng Jú Yǐn* (Mulberry Leaf and Chrysanthemum Beverage, 桑菊饮). In addition, using this compatibility for liver fire flaring upward also has more effect.

Case Study

Initial Visit

A male patient, May 14th, 1986.

Recently, the patient felt feverish on his face, heaviness and dizziness of his head, painful eyes and there was a small ulcer on the left side of his nose. His pulse was wiry, thready and fast. His tongue was red and his tongue coating was thin and white. This was the upper *jiao* heat stagnation syndrome. The treatment plan was to use a soothing and expelling wind-heat method.

Prescription

桑叶	*sāng yè*	24g	Folium Mori
菊花	*jú huā*	20g	Flos Chrysanthemi
金银花	*jīn yín huā*	20g	Flos Lonicerae Japonicae
川芎	*chuān xiōng*	10g	Rhizoma Chuanxiong
升麻	*shēng má*	10g	Rhizoma Cimicifugae
荷叶	*hé yè*	10g	Folium Nelumbinis
栀子	*zhī zǐ*	10g	Fructus Gardeniae
甘草	*gān cǎo*	9g	Radix Glycyrrhizae
连翘	*lián qiáo*	12g	Fructus Forsythiae

This was decocted with water and divided to be taken twice, one pack per day.

Second Visit

May 31st: The patient said that after taking four packs of the above formula, he had recovered.

Chái hú's (柴胡 , Radix Bupleuri) tastes and property are bitter, bland and slightly cold. It enters the liver channel of the foot *jueyin*, the gallbladder channel of the foot *shaoyang*, the pericardium channel of the hand *jueyin* and the *san jiao* channel of the hand *shaoyang* meridians. It is an important herb to harmonise, reduce heat, soothe the liver and break up stagnation. In clinical practice, it is most often used in three ways: Firstly, it is used to harmonise, reduce heat, and alternate the heat and cold of the *shaoyang* syndrome. Without this herb, they cannot be treated, for example *Xiǎo Chái Hú Tāng* (Minor Blupurum Decoction) treats the *shaoyang* syndrome using *chái hú* (Radix Bupleuri) as the king herb. Secondly, it is used to soothe the liver and break up the stagnation. Any syndrome where the liver has lost its free flowing function, without *chái hú* (Radix Bupleuri) it cannot be soothed, for example *Xiāo Yáo Sǎn* (Free Wanderer Powder) and *Sì Nì Sǎn* (Counterflow Cold Powder), etc. both use *chái hú* (Radix Bupleuri) as the king herb. Thirdly, it is used to raise yang and raise the sinking qi. For any qi sinking of the middle *jiao* or where the clear yang cannot be raised up, *chái hú* (Radix Bupleuri) also can raise up the clear yang, for example Li Dong-yuan's *Bǔ Zhōng Yì Qì Tāng* (Centre-Supplementing Qi-Boosting Decoction, 补中益气汤) and Zhang Xi-chun's *Shēng Xiàn Tāng* (Fall-Upbearing Decoction, 升陷汤), both use *chái hú* (Radix Bupleuri) to raise up the yang qi but with a small dosage. If using a large dosage, then it has the power to soothe and disperse but loses it raising function.

1. *Chái Hú* (Radix Bupleuri) ——
Qīng Hāo (青蒿, Herba Artemisiae Annuae)

Function

To clear heat, cool blood, soothe the liver, eliminate steaming.

Application

This is used for chronic fever without relief of unknown aetiology.

Compatibility Analysis

Chái hú (Radix Bupleuri) is bitter, bland and slightly cold. It has functions to harmonise, break up stagnation and reduce fever. If the heat pathogen is in the half exterior and half interior, without *chái hú* (Radix Bupleuri) it cannot be dispersed and resolved. *Chái hú* (Radix Bupleuri) also can soothe the liver and break up stagnation and promote soothing and expelling of internal heat stagnation. *Qīng hāo* (Herba Artemisiae Annuae) is bitter and cold. It can clear heat and cool blood, eliminate hidden heat and bone steaming. In the *Divine Husbandman's Classic of the Materia Medica* it states that it *"treats heat staying in the bone joints"*. Modern clinical research found that *qīng hāo* (Herba Artemisiae Annuae) contains Artemisinin. It has a very good heat relieving function. It is effective for treating the chronic fever of unknown origin. When these two herbs are combined together, the power to clear heat and cool blood is even stronger and can soothe the liver and break up the stagnation to let the internal stagnated heat be dispelled and the hidden heat pathogen be cleared. Therefore, the compatibility can be used for treating unrelieved chronic fever of unknown aetiology.

Initial Visit

A male patient, 33 years old, June 22nd, 2000.

The patient had fever for more than forty days without relief and so came to visit. One and half months ago, due to an exterior invasion, he had fever, headache and whole body soreness. After treatment (medicine unknown), his symptoms of headache, etc. were relieved, except for the fever. It was sometimes high and sometimes better. His body temperature was usually around 37℃ and sometimes reached as high as 39℃. He had laboratory tests in an hospital and they did not find any abnormality with his heart and lungs. A routine blood test was also in the normal range. He got infusions of antibiotics and ribavirin, etc. but his body heat still had not reduced. His current symptoms included a dry mouth, thirst, body heat, body temperature 37.5℃, a red tongue, a white tongue coating, and a wiry and fast pulse. The syndrome differentiation was that due to mistreatment of an exterior invasion, the residual pathogens were not cleared and both qi and yin were injured. The residual pathogens had not been cleared; therefore, his fever was not reducing. The treatment should not injure qi and yin. If qi is deficient without securing, then there is sweating. The sweating injured the yin; therefore, he had dry mouth and thirst. The residual pathogens had not been cleared and went internally due to the deficiency and hid between half exterior and half interior. The pathogens could not be dispelled or vented out to get relief; therefore, he had chronic fever without relief. The treatment plan was to clear heat, cool blood, tonify qi and nourish yin, soothe the liver and vent out the pathogens.

Prescription

柴胡	*chái hú*	15g	Radix Bupleuri
青蒿	*qīng hāo*	20g	Herba Artemisiae Annuae
生地	*shēng dì*	15g	Radix Rehmanniae
丹皮	*dān pí*	9g	Cortex Moutan
太子参	*tài zǐ shēn*	15g	Radix Pseudostellariae
生石膏	*shēng shí gāo*	24g	Gypsum Fibrosum (raw)
知母	*zhī mǔ*	10g	Rhizoma Anemarrhenae
羚羊粉	*líng yáng fěn*	3g	Cornu Saigae Tataricae Powder (dissolved)
生薏仁	*shēng yì rén*	15g	Semen Coicis (raw)
甘草	*gān cǎo*	6g	Radix Glycyrrhizae

This was decocted with water and divided to be taken twice, one pack per day.

Second Visit

June 28th: After taking six packs of the above formula, his sweating was stopped and his fever was reduced. His dry mouth and thirst were also better, but he still had lassitude and insomnia symptoms. The above formula was modified for continued treatment.

Prescription

柴胡	*chái hú*	10g	Radix Bupleuri
青蒿	*qīng hāo*	15g	Herba Artemisiae Annuae
太子参	*tài zǐ shēn*	15g	Radix Pseudostellariae
黄芪	*huáng qí*	20g	Radix Astragali
夜交藤	*yè jiāo téng*	30g	Caulis Polygoni Multiflori
枣仁	*zǎo rén*	15g	Semen Ziziphi Spinosae
砂仁	*shā rén*	8g	Fructus Amomi
仙鹤草	*xiān hè cǎo*	30g	Herba Agrimoniae
甘草	*gān cǎo*	6g	Radix Glycyrrhizae
生石膏	*shēng shí gāo*	20g	Gypsum Fibrosum (raw)

This was decocted with water, to be taken as above.

Third Visit

July 3rd: After taking another five packs, all of his symptoms basically recovered. He was instructed to rest well and stop taking the formula for observation. Later, at a follow up visit, he said there was no reoccurence.

Gǔ Suì Bǔ (骨碎补 , Rhizoma Drynariae)

Gǔ Suì Bǔ's (骨碎补 , Rhizoma Drynariae) taste and property are bitter and warm. It enters the liver channel of the foot *jueyin* and the kidney channel of the foot *shaoyin* meridians. It functions to tonify the kidney, strengthen the bone, invigorate blood and alleviate pain. It is an important herb for the Traumatology Department. Meanwhile, it also can warm the kidney to stop diarrhoea as well as treat symptoms of kidney deficiency induced toothache and tinnitus, etc. It is a constantly used herb.

1. *Gǔ Suì Bǔ* (Rhizoma Drynariae) ——
Dì Gǔ Pí (地骨皮, Cortex Lycii)

Function

To tonify the kidney qi, regulate the yin and yang, to alleviate toothache.

Application

This is used for the kidney deficiency induced toothache and gum bleeding, teeth loss and toothache that is worse at night.

Compatibility Analysis

The kidney governs the bone. Teeth are the surplus of the bone, teeth and bone are from the same source. Teeth also need the kidney essence to nourish them, therefore, in the *Wondrous Lantern for Peering into the Origin and Development of Miscellaneous Disease - Mouth, Teeth and Tougue Disease* (*Zá Bìng Yuán Liú Xī Zhú - Kǒu Chǐ Chún Shé Bìng Yuán Liú,* 杂病源流犀烛·口齿唇舌病源流) it states that: "*Teeth are the manifestation of the kidney and the root of the bone.*" The growth or falling out of the tooth is closely related to the kidney essence being plenty or deficient. If the kidney qi is plenty, then teeth will be firm and secure. If the kidney qi is deficient, then teeth will be loose, falling out or painful. *Gǔ suì bǔ* (骨 碎 补 , Rhizoma Drynariae) is bitter and warm. It tonifies the kidney, and strengthens the bone. It also can alleviate pain. It is an important herb to treat toothache induced by kidney deficiency. *Dì gǔ pí* (Cortex Lycii) is sweet and cold. It enters the kidney channel of the foot *shaoyin* meridian. It can nourish yin and enrich the kidney. These two herbs combined together can regulate and tonify the yin and yang of the kidney channel of the foot *shaoyin* meridian, and tonify the kidney essence. *Gǔ suì bǔ* (Rhizoma Drynariae) warms the kidney yang to help generate and transform the essence. *Dì gǔ pí* (Cortex Lycii) nourishes the kidney yin and restricts the deficiency yang to let it return to its place and recover the inter-transforming of the yin and yang situation, therefore, it is effective for kidney deficiency, where the qi of the kidney essence could not flow upward to nourish the tooth inducing toothache or loose teeth.

Case Study

Initial Visit

A female patient, 61 years old, April 7th, 1993.

The patient had had a toothache in her right lower jaw for more than half a month. It became worse at night time, and in the daytime it was better. When eating, she could not use her right side teeth to chew food. She was treated with medicines for toothaches etc., but they did not work and so she came to our clinic for treatment. On examination, the patient's right lower tooth was loose. Her gum was slightly swollen. There was no cavity. Her pulse was deep and slightly wiry. Her tongue was red and her tongue coating was thin and white. This was kidney deficiency induced toothache. The kidney governs the bone, and teeth are the surplus of the bone. If the kidney is deficient, it cannot go upward to nourish the teeth; therefore, she had toothache and a loose tooth. The treatment plan was to use a formula to tonify the kidney, strengthen the bone, clear the deficiency heat and alleviate pain.

Prescription

骨碎补	*gǔ suì bǔ*	12g	Rhizoma Drynariae
地骨皮	*dì gǔ pí*	20g	Cortex Lycii
知母	*zhī mǔ*	9g	Rhizoma Anemarrhenae
生地	*shēng dì*	12g	Radix Rehmanniae
补骨脂	*bǔ gǔ zhī*	9g	Fructus Psoraleae
山茱萸	*shān zhū yú*	10g	Fructus Corni
山药	*shān yào*	20g	Rhizoma Dioscoreae
甘草	*gān cǎo*	6g	Radix Glycyrrhizae

This was decocted with water and divided to be taken twice, one pack per day.

Second Visit

April 12th: After taking five packs of the above formula, her toothache was stopped. However, when eating, she still felt a slight pain and discomfort. After taking four more packs of the original formula, the patient recovered.

Guì zhī's (桂枝 , Ramulus Cinnamomi) tastes and property are acrid, sweet and warm. It enters the lung channel of the hand *taiyin*, the heart channel of the hand *shaoyin* and the urinary bladder of the foot *taiyang* meridians. It functions to release the muscle layer and the exterior, warm the meridians and unblock the blood vessels. If it moves upward, it can release the exterior to expel wind-cold. If it travels to the four limbs, it can warm and unblock the meridians and the blood vessels. Its clinical compatibilities are broad. If combined with *bái sháo* (Radix Paeoniae Alba) it can treat exterior deficiency and pathogenic excess syndromes as well as harmonise nutritive and defensive qi, and release the exterior, but without the disadvantage of too much sweating. (Please refer to the previous compatibility of *bái sháo* - Radix Paeoniae Alba). If combined with *xìng rén* (Semen Armeniacae Dulcis) and *hòu pò* (Cortex Magnoliae Officinalis) it can descend qi and stop coughing. If combined with *má huáng* (Herba Ephedrae) and *fù zǐ* (Radix Aconiti Lateralis Praeparata) together they can warm the menstruation and alleviate pain. If combined with *dāng guī* (Radix Angelicae Sinensis) and *sháo yào* (Radix Paeoniae Lactiflorae) they can invigorate blood and unblock the menstruation. If combined with *fú líng* (Poria) and *bái zhú* (Rhizoma Atractylodis Macrocephalae) they can free the yang qi, promote urination and transform the phlegm and the thin mucus.

1. *Guì Zhī* (Ramulus Cinnamomi) —— *Jīn Yín Téng* (金银藤, Lonicerae Ramulus)

Function

To unblock the meridians and invigorate the collaterals, to clear heat, reduce swelling and alleviate pain.

Application

This is used for wind-damp-heat arthralgia and gout that manifests as redness, swelling, heat and pain of the joints.

Compatibility Analysis

Guì zhī (Ramulus Cinnamomi) is red in colour and it enters the blood level. Its property is warm and beneficial for unblocking the meridians and invigorating the collaterals. Its branches are good at reaching the end of the four limbs. In the *Divine Husbandman's Classic of the Materia Medica* it states that it *"promotes joints mobility"*. *Jīn yín huā* (Flos Lonicerae Japonicae) is sweet and cold. It has clearing heat and detoxifying functions. Its vine is also beneficial for unblocking the collaterals and can clear heat stagnation in the meridians and the collaterals to alleviate pain. It is a commonly used herb to treat wind-damp-heat arthralgia There is a folk empirical formula that soaks the *jīn yín téng* (金银藤 , Caulis Lonicerae Japonica) in wine to treat wind-damp-heat arthralgia. It is the sweetness and coldness of the *jīn yín téng* (Caulis Lonicerae Japonica) that relies on the warm and unblocking property of the wine to enhance its effectiveness to unblock the collaterals. The author uses *guì zhī* (Ramulus Cinnamomi) to substitute the wine. This is using the warm and unblocking property of *guì zhī* (Ramulus Cinnamomi) to help *jīn yín téng* (Caulis Lonicerae Japonica) to unblock the meridians and invigorate the collaterals, as *jīn yín téng* (Caulis Lonicerae Japonica) has extra power to clear heat but insufficient power to unblock the meridians. In addition, *guì zhī* (Ramulus Cinnamomi) also can release the exterior and expel wind. If combined with *jīn yín téng* (Caulis Lonicerae Japonica) they can clear heat and detoxify toxins, then they can expel the wind-damp-heat pathogens. In clinical practice, the dosage of *guì zhī* (Ramulus Cinnamomi) should be smaller than *jīn yín téng* (Caulis Lonicerae Japonica). The ratio is 1:3 to 1:5. The correct dosage depends on the severity of the heat pathogen.

Case 1

Initial Visit

A male patient, 35 years old, August 4th, 1987.

The patient had a history of joint pain for more than ten years. It was sometimes better and sometimes worse, depending on changes in the weather. Recently, for more than ten days, joint pain throughout his whole body became worse and was more significant on the joints of his knees and elbows. This was accompanied by redness and swollen nodules, whole body lassitude, stomachache, acid regurgitation and anorexia. He had it investigated at a Tianjin hospital and was diagnosed with wind-damp-heat arthralgia. Due to the stomachache he was afraid to take pain-killers; therefore, he came to ask for Chinese medicine treatment. His pulse was wiry, slippery and fast. His tongue was dark and his tongue coating was white. His ESR was 20mm/h. The treatment plan was to clear heat, unblock the collaterals, expel wind and eliminate dampness.

Prescription

金银藤	*jīn yín téng*	30g	Caulis Lonicerae Japonica
桂枝	*guì zhī*	10g	Ramulus Cinnamomi
杭白芍	*háng bái sháo*	12g	Radix Paeoniae Lactiflorae
知母	*zhī mǔ*	15g	Rhizoma Anemarrhenae
防己	*fáng jǐ*	15g	Radix Stephaniae Tetrandrae
生薏仁	*shēng yì rén*	30g	Semen Coicis (raw)
苍术	*cāng zhú*	10g	Rhizoma Atractylodis
砂仁	*shā rén*	9g	Fructus Amomi
乌药	*wū yào*	9g	Radix Linderae
檀香	*tán xiāng*	6g	Lignum Santali Albi
甘草	*gān cǎo*	6g	Radix Glycyrrhizae

This was decocted with water and divided to be taken twice, one pack per day.

Second Visit

August 8th: After taking four packs of the above formula, his joint pain, swelling and redness were slightly reduced, but there was no change in his stomachache. To the above formula, *yán hú suǒ* (Rhizoma Corydalis) 15g, and *jiāo sān xiān* (scorch-fried Fructus Hordei Germinatus et Crataegi et Massa Fermentata Medicinalis) 30g, were added in. He continued to take it.

Third Visit

August 15th: After taking another seven packs, his joint pain became mild. His redness and swelling were eliminated. His stomachache had reduced a lot, and his appetite had increased. He continued on with the same formula.

Case 2

Initial Visit

A female patient, 34 years old, July 31st, 1995.

The patient had joint pain for a couple of months. There was more severe pain in the finger joints accompanied by redness, swelling, distention and pain sensations. She could hardly flex and extend them and they were hot to the touch. Her rheumatoid factor was positive and her ESR was 30mm/h. Her pulse was wiry and fast. Her tongue was red and her tongue coating was thin and white. This was the wind-damp-heat arthralgia. The treatment plan was to clear heat, unblock the collaterals, tonify the kidney and eliminate the dampness.

Prescription

金银藤	*jīn yín téng*	30g	Caulis Lonicerae Japonica
桂枝	*guì zhī*	10g	Ramulus Cinnamomi
杭白芍	*háng bái sháo*	15g	Radix Paeoniae Lactiflorae
丹参	*dān shēn*	30g	Radix et Rhizoma Salviae Miltiorrhizae
当归	*dāng guī*	15g	Radix Angelicae Sinensis
川牛膝	*chuān niú xī*	15g	Radix Cyathulae
延胡索	*yán hú suǒ*	15g	Rhizoma Corydalis
苍术	*cāng zhú*	10g	Rhizoma Atractylodis
桑枝	*sāng zhī*	30g	Ramulus Mori
土鳖虫	*tǔ biē chóng*	10g	Eupolyphaga seu Steleophaga
杜仲	*dù zhòng*	15g	Cortex Eucommiae
川断	*chuān duàn*	15g	Radix Dipsaci
砂仁	*shā rén*	6g	Fructus Amomi
甘草	*gān cǎo*	6g	Radix Glycyrrhizae

This was decocted with water and divided to be taken twice, one pack per day.

Second Visit

August 9th: After taking nine packs of the above formula her joint pain was significantly reduced. The redness, swelling and hot sensations on her finger joints were also reduced. She continued taking the original formula.

Third Visit

August 31st: Again, after taking more than twenty packs of the above formula, her redness, swelling and pain basically disappeared. Her ESR was 12mm/h and the rheumatoid factor was slightly positive.

2. *Gui Zhī* (Ramulus Cinnamomi) ——
Fú Líng (茯苓, Poria)

Function

To warm the yang qi and promote urination.

Application

This is used for kidney qi deficiency, where water cannot transform inducing dysuria. It is also used for rebellious kidney qi, where water overflow becomes phlegm and flushes rebelliously upward to above the diaphragm. Or where the phlegm and fluid stay internally, the fluid cannot be dispersed upward and so causes dry mouth, thirst without wanting to drink, if water is drank then it is vomited out immediately inducing the phlegm and thin mucus syndrome, etc.

Compatibility Analysis

Guì zhī (Ramulus Cinnamomi) is acrid, sweet and warm. It enters the lung channel of the hand *taiyin*, the heart channel of the hand *shaoyin* and the urinary bladder channel of the foot *taiyang* meridians. It is an herb to warm the yang and unblock the meridians. *Fú líng* (Poria) is sweet and neutral. It enters the heart channel of the hand *shaoyin*, the lung channel of the hand *taiyin*, the spleen channel of the foot *taiyin* and the kidney channel of the foot *shaoyin* meridians. It is an herb to leach out dampness and promote water metabolism. These two herbs combined together function to warm the yang, transform the qi and promote water metabolism. Water, thin mucus and body fluid metabolism is mainly the function of the three meridians of the lung channel of the hand *taiyin*, the spleen channel of the foot *taiyin* and the kidney channel of the foot *shaoyin* for regulating and balancing them. In the *Plain Questions - A Separate Treatise on the Channel Vessels (Sù Wèn Jīng Mài Bié Lùn*, 素问·经脉别论) it states that: "*After the fluid enters the stomach, the refined part is transmitted upward to the spleen. The spleen qi, in turn, spreads it upward to the lungs that regulate the water pathways and descend it down to the urinary bladder. The refined water is spread all over and moves in all meridians.*" Kidney yang has the main function of acting on the absorption, distribution and excretion of the water and fluid. The kidney yang not only can distribute the water and fluid to the whole body but also can transport the extra water, fluid and body waste down to the urinary bladder, as well as control the urine excretion from the urinary bladder, therefore, there is the saying: "*The kidney is the water organ.*" In addition, one of the spleen functions is to help the stomach to transform and transport its fluid. If the spleen is deficient and it cannot transform and transport, its qi mechanism does not function well, then it can make water and dampness stop and stagnate, therefore, in clinical practice, the water and fluid regulation function from the *zang* and *fu* organs is called: "*the qi mechanism*". When the kidney yang is deficient and the spleen is deficient, without transforming and transporting, then a qi mechanism disorder and water metabolism disorder may manifest. *Guì zhī* (Ramulus Cinnamomi) is acrid, sweet and warm. It can warm the *mingmen* (life-gate) fire like adding the firewood under the cauldron. On one hand, it can help the qi mechanism of the urinary bladder, and on the other hand, it helps the spleen qi to steam upward. It makes the water and fluid vapourise and be transformed and transported. *Fú líng* (Poria) is sweet and bland. It can strengthen the spleen, leach out the dampness and promote water metabolism. If it gets the warmth of *guì zhī* (Ramulus Cinnamomi) it can enhance its qi mechanism and strengthen its transforming and transporting functions and regulate the water pathways. *Fú líng* (Poria) also can help excretion after *guì zhī* (Ramulus Cinnamomi) has warmed and transformed the water and dampness. These two herbs combined together are a good compatibility to warm the yang, transform qi and promote water metabolism. The Zhang Zhong-jing's *Wǔ Líng Sǎn* (Poria Five Powder, 五苓散), *Líng Guì Zhú Gān Tāng* (Poria, Cinnamon Twig, Atractylodes Macrocephala, and Liqurice Decoction, 苓桂术甘汤) and the *Guì Líng Wán* (Cinnamon Twig and Poria Pill, 桂苓丸) in the *Comprehensive Medicine According to Master Zhang* (张氏医通 , *Zhāng Shì Yī Tōng*) all use this compatibility.

Case Study

Initial Visit

A male patient, 65 years old, April 20th, 1992.

The patient had lower abdominal distention and fullness as well as stranguria for a couple of days. He had been investigated in the local city hospital but no abnormality was found. For the past half day, his lower abdominal distention and fullness were hard to bear. He had urinary catheterisation in an hospital's Emergency Department, but there was no sign of urine retention. The patient was still in as much pain as before. Through his friend's referral, he came to our hospital's clinic for treatment. On examination, the patient was dispirited, with an expression of pain on his face, a dry mouth and thirst, but he did not want to drink, after he drank water he wanted to vomit it out and he could not eat. His tongue coating was white, turbid and slippery. His tongue was dark. His pulse was wiry and forceful on the left side and the right side was a little slow. This was a syndrome of yang deficiency and the qi mechanism not functioning well, with water and thin mucus stagnating inside, fluid could not be dispersed all over. The *mingmen* (life-gate) had yang deficiency, with no power to perform the qi mechanism; therefore, he had less urine and lower abdominal distention and fullness. Then water and thin mucus stagnated inside, fluid could not be dispersed all over the body and be supplied upward, therefore, he had a dry mouth but did not want to drink and he wanted to vomit after drinking. The pulse was wiry on the left side with a slippery tongue these are indications of the water and fluid stagnating internally. The treatment plan was to warm the yang, transform qi and promote water metabolism.

Prescription

茯苓	*fú líng*	60g	Poria
桂枝	*guì zhī*	10g	Ramulus Cinnamomi
白术	*bái zhú*	15g	Rhizoma Atractylodis Macrocephalae
半夏	*bàn xià*	10g	Rhizoma Pinelliae
泽泻	*zé xiè*	12g	Rhizoma Alismatis
甘草	*gān cǎo*	6g	Radix Glycyrrhizae

This was decocted with water and divided to be taken warm, three times, one pack per day.

Second Visit

April 2nd: After taking two packs of the above formula, his urination was smooth. His lower abdominal distention and fullness were also reduced, and he could eat again. His pulse turned to wiry and thready. Using the above formula, *háng bái sháo* (Radix Paeoniae Lactiflorae) 24g, and *jié gěng* (Radix Platycodonis) 6g, were added, and *bái zhú* (Rhizoma Atractylodis Macrocephalae) was eliminated. As his pulse was thready, there was concern that the formula could injure his yin, therefore, *háng bái sháo* (Radix Paeoniae Lactiflorae) was added to nourish yin and promote urination and eliminate the acridity and dryness of *bàn xià* (Rhizoma Pinelliae). *Jié gěng* (Radix Platycodonis) was added to disperse his lung qi. The lung is the upper source of water, if the lung qi is dispersed and it can flow smoothly, then the water pathway will flow freely and be regulated.

Third Visit

April 27th: All of his symptoms were basically eliminated. His urination and appetite were normal. He was instructed to take two more packs. After half a year's follow up no reoccurence was seen.

Hé Yè (荷叶 , Folium Nelumbinis)

Hé yè's (荷叶 , Folium Nelumbinis) tastes and property are bitter, neutral and slightly cold. It enters the liver channel of the foot *jueyin*, the stomach channel of the foot *yangming* and the spleen channel of the foot *taiyin* meridians. If fresh, it is beneficial for clearing the summer pathogens in the summer. If dried, it can generate and raise the clear yang from the middle *jiao* and has functions to raise the clear and descend the turbid, clear the liver and brighten the eyes. Wang Jie-Gu's *Zhǐ Zhú Wán* (Immature Bitter Orange and Atractylodes Macrocephala Pill, 枳术丸) uses *bái zhú* (Rhizoma Atractylodis Macrocephalae), *zhǐ shí* (Fructus Aurantii Immaturus) and old rice wrapped in a lotus leaf to simmer as pills. It is used to strengthen the spleen and eliminate stagnation. *Hé yè* (Folium Nelumbinis) in this formula has the function to raise up the clear yang of the spleen and stomach, in order to help *zhǐ shí* (Fructus Aurantii Immaturus) and *bái zhú* (Rhizoma Atractylodis Macrocephalae), to strengthen the transforming and transporting function and to eliminate accumulations. As the qi of this herb is clear and aromatic, it can raise the clear and descend the turbid. It is also a commonly used herb to treat any conditions where the clear yang could not rise up or where the liver yang rebelled upward inducing dizziness, or headache.

1. *Hé Yè* (Folium Nelumbinis) ——
Shēng Má (升麻, Rhizoma Cimicifugae)

Function

To raise up the clear yang, clear the head and brighten the eyes.

Application

This can be used for any condition of the upper *jiao* heat stagnating without dispersing, turbid qi obstructing the upper orifices inducing dizziness and a heavy head sensation, for noise inside the brain, ulcers on the face or head, according to the specific syndrome differentiation, adding it in can be effective. It is also effective for stubborn mouth ulcers.

Compatibility Analysis

Hé yè (Folium Nelumbinis) tastes bitter and its properties are neutral and slightly cold. Its qi is clear and aromatic. It enters the liver channel of the foot *jueyin*, the spleen channel of the foot *taiyin* and the stomach channel of the foot *yangming* meridians. It is a good herb to raise the clear, descend the turbid and generate the clear yang. In the *Qīng Zhèn Tāng* (Clearing Invigoration Decoction, 清震汤) in the *Life - Saving Collection* (*Bǎo Mìng Jí*, 保命集) it treats headache with tinnitus. It uses *hé yè* (Folium Nelumbinis) as the king herb. *Shēng má* (Rhizoma Cimicifugae) is bitter, sweet and slightly cold. In the *Divine Husbandman's Classic of the Materia Medica* it states that it can *"detoxify all kinds of toxins"*. Meanwhile, it also has a venting function to treat *"malaria and pathogenic qi"*. After the *Jin* and *Yuan* Dynasties, it was found that *shēng má* (Rhizoma Cimicifugae) also has the function to raise clear yang. It was recognised as a raising and lifting herb. *Hé yè* (Folium Nelumbinis) and *shēng má* (Rhizoma Cimicifugae) both have raising functions. These two herbs combined together, have an even greater power to raise the clear yang. If the clear yang is raised up, then the turbid yin will descend, therefore, it is effective to use this compatibility.

Case Study

Initial Visit

A female patient, 34 years old, April 14th, 1987.

This patient complained of dizziness for more than a month. When dizzy, she felt she was almost falling down but it would pass after a few moments. After that, she felt her head was sinking and heavy; she also had noises inside her head like hearing thunder from far away. After a medical check-up, no abnormality was found and she came to our Chinese medical clinic for treatment. On examination, her pulse was deep, wiry and thready. Her tongue was white and slightly turbid. This was the brain noise syndrome. A pathogen of turbid dampness had covered the upper clear orifices, the clear yang was stagnated and obstructed, which manifested in dizziness with noises in the head. The treatment plan was to raise the clear and descend the turbid.

Prescription

荷叶	*hé yè*	15g	Folium Nelumbinis
升麻	*shēng má*	4g	Rhizoma Cimicifugae
苍术	*cāng zhú*	9g	Rhizoma Atractylodis
茯苓	*fú líng*	15g	Poria
半夏	*bàn xià*	10g	Rhizoma Pinelliae
陈皮	*chén pí*	9g	Pericarpium Citri Reticulatae
竹茹	*zhú rú*	9g	Caulis Bambusae in Taenia
枳实	*zhǐ shí*	10g	Fructus Aurantii Immaturus
甘草	*gān cǎo*	6g	Radix Glycyrrhizae

This was decocted with water and divided to be taken warm, twice, one pack per day.

Second Visit

April 20th: After taking five packs of the above formula, her dizziness was reduced but she still had head heaviness and the noise inside her head. Again, she continued with the same formula.

Third Visit

April 25th: After taking another five packs, her dizziness was eliminated. However, every morning at 11:00AM, she still had head sinking and noise sensations but the severity was reduced. Using the above formula, *shí chāng pú* (Rhizoma Acori Tatarinowii) 10g, *dān shēn* (Radix et Rhizoma Salviae Miltiorrhizae) 30g, and *gé gēn* (Radix Puerariae Lobatae) 30g, were added. She continued taking the formula.

Later, at a follow up visit, she said that she had recovered.

Note

The dosage of *shēng má* (Rhizoma Cimicifugae) when combined with *hé yè* (Folium

Nelumbinis), depends on the condition. If used to raise the yang then a small dosage is used, between 4g to 6g. If used for ulcers on the head or face or where there is heat toxin and fire stagnation, *shēng má* (Rhizoma Cimicifugae) can be used heavily and the dosage should be above 10g.

Yín yáng huò (淫羊藿 , Herba Epimedii) is also called *xiān líng pí* (仙灵脾 , Herba Epimedii). Its tastes and property are sweet, acrid and warm. It enters the liver channel of the foot *jueyin* and the kidney channel of the foot *shaoyin* meridians. It functions to tonify the kidney, strengthen the yang, and strengthen the sinews and the bones. Its property is warm and can tonify the essence, *mingmen* (life-gate), and strengthen the true yang. It is an important herb to treat male yang deficiency induced infertility and female yin deficiency induced infertility. The *Divine Husbandman's Classic of the Materia Medica* lists it as an upper class herb and states that it *"governs yin atrophy, exhaustion or injury, or pain in the penis, it can tonify strength and strengthen the will power."* In the *Grand Materia Medica* it states that it *"can tonify the essence and is an herb for the hand and foot yangming meridians, the san jiao meridian and the mingmen (life-gate). It is appropriate for the true yang deficiency patients."*

1. *Yín Yáng Huò* (Herba Epimedii) ——
Zhī Mǔ (知母, Rhizoma Anemarrhenae)

Function

To strengthen the yang, tonify the yin, to regulate and tonify the yin and yang.

Application

This is used for menopause due to both yin and yang deficiency, for the penetrating vessel and the conception vessel disharmony inducing irritability, hot flushes, sweating, insomnia, fatigue, memory loss, headache, dizziness and high blood pressure, etc.

Compatibility Analysis

From a Chinese medicine perspective, menopause is due to the gradual weakening of kidney qi, the disharmony between the penetrating vessel and the conception vessel, the essence and blood deficiency, and the dew of heaven being almost exhausted with physical changes. At this time, the human body's regulating function of yin and yang is reduced, kidney yang deficiency and kidney yin insufficiency induce both yin and yang deficiency. If the yin is deficient then the deficient yang floats, it cannot be stored and so manifests in symptoms of irritability, being easily angered, hot flushes and sweating, headache, dizziness, etc. If the yang is deficient, then the meridians and the blood vessels lose their warmth and nourishment which induces fatigue, lassitude, and dispiritedness. When both the yin and yang are deficient then the menstrual blood is not regulated. The treatment plan was to tonify both yin and yang, and regulate the yin and yang. *Yín yáng huò* (淫 羊 藿 , Herba Epimedii) is acrid and warm. It can tonify the essence and strengthen the kidney yang. It is an important herb to tonify the *mingmen* (life-gate). The property of *zhī mǔ* (Rhizoma Anemarrhenae) is cold. It enters the kidney channel of the foot *shaoyin* meridian. It functions to nourish the kidney of the water and the yin. In the *Divine Husbandman's Classic of the Materia Medica* it states that it can *"tonify deficiency and qi"*. In the *Grand Materia Medica* it states that *"downward, it can moisten the dryness of the kidney and nourish the yin; upward, it can clear the lung of the metal and sedate the fire.*" It is a good herb to nourish the kidney yin. These two herbs combined together can tonify both yin and yang as well as regulate and tonify both the yin and yang of the kidney channel of the foot *shaoyin* meridian. Zhang Jing-yue said that: "*If one is good at tonifying yang, he should pursue the yang from the yin, then the yang gets help from the yin and there will be unlimited generation and transformation. If one is good at tonifying the yin, he should pursue the yin from the yang, then yin gets yang to rise as a fountain, never to be used up.*" *Yín yáng huò* (Herba Epimedii) warms the yang. If it gets the help from *zhī mǔ* (Rhizoma Anemarrhenae), it can pursue the yang from the yin and the qi of essence will be plenty. *Zhī mǔ* (Rhizoma Anemarrhenae) nourishes yin. If it gets the help from *yín yáng huò* (Herba Epimedii), it can pursue the yin from the yang and the kidney yin will be even more full. Therefore, these two herbs combined together can mutually tonify the yin and the yang and mutually enhance each other. It is the best compatibility to tonify both yin and yang deficiency of the body and regulate the yin and yang balance of an organism. Moreover, the cold of *zhī mǔ* (Rhizoma Anemarrhenae) can inhibit and control the dryness and heat from *yín yáng huò* (Herba Epimedii). The warmth of *yín yáng huò* (Herba Epimedii) also can reduce the bitterness and coldness of *zhī mǔ* (Rhizoma Anemarrhenae). They are stable and balanced without bias.

Case Study

Initial Visit

A female patient, 49 years old, September 7th, 1996.

The patient had had irritability, palpitations, insomnia and dizziness for more than three months accompanied by heat and sweating in the whole body, irritability, and amenorrhoea for half a year. She had been investigated in a hospital and diagnosed with menopausal syndrome. She was treated with *Gēng Nián Kāng* (Healthy Menopause Capsule, 更年康) and oryzanol etc. for more than a month without effect and so came to the Chinese medicine hospital for clinical treatment. On examination, the patient was physically well developed. She had a wiry and thready pulse, a red tongue, with a thin tongue coating, her blood pressure was 150/80 mmHg, and there was no abnormality on her EKG. On inquiry, her appetite, urination and bowel movement were normal. However, sometimes, she felt symptoms of fatigue, lassitude, easily getting irritable or angry, etc. This was the female menopause stage, both yin and yang deficiency, and the penetrating vessel not in harmony syndrome. A modified *Èr Xiān Tāng* (Two Immortal Decoction, 二仙汤) was prescribed to treat her.

Prescription

仙灵脾	*xiān líng pí*	15g	Herba Epimedii
知母	*zhī mǔ*	12g	Rhizoma Anemarrhenae
当归	*dāng guī*	12g	Radix Angelicae Sinensis
枣仁	*zǎo rén*	15g	Semen Ziziphi Spinosae
夜交藤	*yè jiāo téng*	30g	Caulis Polygoni Multiflori
丹参	*dān shēn*	24g	Radix et Rhizoma Salviae Miltiorrhizae

百合	*bǎi hé*	24g	Bulbus Lilii
仙茅	*xiān máo*	10g	Rhizoma Curculiginis
川牛膝	*chuān niú xī*	15g	Radix Cyathulae
山萸肉	*shān yú ròu*	10g	Fructus Corni
生地	*shēng dì*	15g	Radix Rehmanniae
甘草	*gān cǎo*	6g	Radix Glycyrrhizae

This was decocted with water and divided to be taken twice, one pack per day.

Second Visit

September 14th: After taking seven packs of the above formula, her symptoms of irritability, sweating, dizziness and insomnia, etc. were improved and she felt cheerful. She was instructed to continuously take the original formula.

Third Visit

September 21st: After taking another seven packs, all of her symptoms had basically disappeared. She was instructed to take an additional seven packs and she recovered.

Huáng qí's (黄芪 , Radix Astragali) taste and property are sweet and warm. It enters the hand and foot *taiyin* meridians. It is an herb to tonify qi and invigorate the yang. It can strengthen the defensive qi, secure the exterior, warm the qi and raise sinking qi. For the yang deficiency patient who has exterior deficiency with spontaneous sweating, it can strengthen the exterior and constrain the sweating. For patients with insufficiency of qi and blood, both body and spirit fatigue, it can tonify the deficiency and benefit the damage. For the stomach and the spleen deficient and weak patients with poor appetite and diarrhoea, it can cultivate the earth to stop diarrhoea. For patients were the yang qi does not transform and transport, manifesting in oedema and dysuria, it can activate the yang qi and promote water metabolism. For abscesses or carbuncles whose surfaces are falling inward with scanty pus and blood, it can draw internally and drain the pus. For post stroke patients who are emaciated, with limb paralysis, it can warm the meridians and harmonise the blood. For patients with the middle *jiao* qi sinking, uterine bleeding and prolapse of the rectum, it can reinstate them by securing the qi. If the right compaitibility for *huáng qí* (Radix Astragali) is found, its applications are broad, such as when combined with *rén shēn* (Radix et Rhizoma Ginseng) is the *Shēn Qí Gāo* (Ginseng and Astragalus Plaster, 参芪膏); it is a formula to strongly tonify the qi. If combined with *fù zǐ* (Radix Aconiti Lateralis Praeparata) it is the *Qí Fù Gāo* (Astragalus and Aconite Plaster, 芪附膏); it is a strong formula that warmly tonifies. If combined with *bái zhú* (Rhizoma Atractylodis Macrocephalae) it becomes *Qí Zhú Gāo* (Astragalus and Atractylodes Macrocephala Plaster, 芪术膏) that can cultivate and tonify the root of the post heaven. If combined with *dāng guī* (Radix Angelicae Sinensis) it becomes *Bǔ Xuè Tāng* (Blood-Supplementing Decoction, 补血汤) that can tonify both qi and blood. These are famous compatibilities in clinical practice.

1. *Huáng Qí* (Radix Astragali) ——
Jīn Yín Huā (金银花, Flos Lonicerae Japonicae)

Function

To tonify qi, detoxify toxins, retain ulcers and generate flesh.

Application

This is used for prolonged unhealed abscesses and carbuncles or when the surface of an abscess or carbuncle is falling inward, or the initial stage of an ulcer that is mildly red, swollen and belongs to qi deficiency.

Compatibility Analysis

In the *Divine Husbandman's Classic of the Materia Medica* it states that *huáng qí (Radix Astragali)* "governs prolonged ulcers and broken sores, drains pus and alleviates pain, helps severe wind induced sores, five kinds of haemorrhoids, scrofula and testicle atrophy......" We can see that *huáng qí* (Radix Astragali) is a main herb to treat ulcers in ancient times. The taste and property of *jīn yín huā* (金银花 , Flos Lonicerae Japonicae) is sweet and cold. It is beneficial for detoxifying the toxins in the blood. The medical expert Huang Yuan-yu in the *Qing* Dynasty said that it can *"clearly disperse febrile wind, eliminate swollen toxins and treat all kinds of ulcers."* These two herbs are beneficial for treating ulcers; however, *huáng qí* (Radix Astragali) treats deficiency and *jīn yín huā* (Flos Lonicerae Japonicae) treats excess. In clinical practice, for any of the ulceration syndromes, there is more deficiency within the excess in the initial stage, and in the later stage there is more excess within the deficiency. *Huáng qí* (Radix Astragali) helps the normal qi to draw internally and drain pus to prevent the toxic pathogens from going internally. *Jīn yín huā* (Flos Lonicerae Japonicae) clears heat pathogens, reduces swelling, and expels clumps in order to detoxify the toxic pathogens in the blood. These two herbs combined together can effectively tonify qi, detoxify toxins, retain ulcers and expel the clumps. They can treat the early, middle and later stages ulcers, abscesses and carbuncles. For the dosage ratio of these two herbs, if the toxic heat pathogen is severe or in the initial stage of the ulcer, use a large dosage of *jīn yín huā* (Flos Lonicerae Japonicae) and supplement with a little *huáng qí* (Radix Astragali). If it is at a later stage of the ulcer, the normal qi is deficient and the toxic pathogen has invaded inward, then use a large dosage of *huáng qí* (Radix Astragali) and supplement with a little *jīn yín huā* (Flos Lonicerae Japonicae).

Case 1

Initial Visit

A male patient, 19 years old, April 9th, 1985.

Two years ago, when the patient was working on a farm, his left lower leg was cut with a sickle just above the ankle, by accident. He did not treat or take care of it. Later, he got an infection and the wound did not heal for a long time. He used *Bá Dú Gāo* (Toxin-Drawing Plaster, 拔毒膏) to treat it but it did not work and the wound enlarged daily. He went to the local county hospital and was treated with penicillin for more than a month but it still did not improve. Therefore, he looked around for treatments and tried using oral and external medicines to treat it, but it was sometimes better and sometimes worse. The ulcer expanded to be the size of a goose egg. The colour of the ulcer surface was dark and lustreless, dark purple and falling inward. There was no pus and no pain. The patient was physically well developed. His pulse was deficient, big and fast. His tongue coating was thin and white. This was prolonged ulcer with mistreatments inducing both qi and blood deficiency and weakness as well as the ulcer toxin invading internally syndrome. The treatment plan was to tonify qi, nourish the nutritive qi, generate flesh and detoxify toxins.

Prescription

生黄芪	*shēng huáng qí*	50g	Radix Astragali (raw)
金银花	*jīn yín huā*	50g	Flos Lonicerae Japonicae
当归	*dāng guī*	30g	Radix Angelicae Sinensis
元参	*yuán shēn*	30g	Radix Scrophulariae
甘草	*gān cǎo*	30g	Radix Glycyrrhizae

This was decocted with water and divided to be taken twice, one pack per day.

In addition, the patient was instructed to use *zhì rǔ xiāng* (processed Olibanum) and *zhì mò yào* (processed Myrrha) 50g each, to be ground into a fine powder and mixed with sesame oil to be applied on the wound as well, and to be changed once or twice a day.

Second Visit

. May 9th: After the patient applied more than thirty packs of the above formula, the surface of his ulcer turned red. There was more pus and granulation of the muscle swelled up to higher than the surface of the ulcer. Again, a modification of the original formula was used to clear it up.

Prescription

生黄芪	*shēng huáng qí*	24g	Radix Astragali (raw)
金银花	*jīn yín huā*	15g	Flos Lonicerae Japonicae
当归	*dāng guī*	15g	Radix Angelicae Sinensis
川牛膝	*chuān niú xī*	9g	Radix Cyathulae
甘草	*gān cǎo*	10g	Radix Glycyrrhizae
黄酒	*huáng jiǔ*	50g	Yellow Rice Wine (as a guide)

This was decocted with water and divided to be taken twice.

Third Visit

May 20th: After taking more than ten packs again of the above formula the surface of his ulcer was basically retained and recovered. He was instructed to take five more packs of the same formula again to complete the treatment.

Note

This case used *zhì rǔ xiāng* (processed Olibanum) and *zhì mò yào* (processed Myrrha) externally. This is the *Shēng Jī Sǎn* (Flesh-Engendering Powder, 生肌散) in the *Medical Revelations* (医学心悟 , *Yī Xué Xīn Wù*). It is used to treat severe ulcerations, applied externally on the ulcers to let the rotten muscle transform and the new muscle generate by itself and can take out the toxins and close the wound. Li Shi-zhen thought that: "*Rǔ xiāng (Olibanum) invigorates blood and mò yào (Myrrha) dispels blood. Both of them can alleviate pain, reduce swelling and generate flesh, therefore, these two herbs are always used together.*" This formula not only can be used externally but also can be taken orally to treat injury induced blood stasis and pain.

Case 2

Initial Visit

A male patient, 10 years old, April 22nd, 1986.

The patient had a red, swollen nodule under his right chin, about the size of a dove's egg. On palpation it felt hot. If pressed on, he felt pain. He had had it for more than ten days. He was injected with penicillin for a week without effect and so came to visit. The patient did not have body heat, aversion to cold or other discomfort. His pulse was wiry, thready and fast. His tongue was red and tongue coating was thin and white. This was the initial stage of the stagnated pathogenic toxin syndrome. The treatment plan was to use a formula to clear heat, detoxify toxins, reduce swelling and expel clumping.

Prescription

金银花	jīn yín huā	30g	Flos Lonicerae Japonicae
生黄芪	shēng huáng qí	20g	Radix Astragali (raw)
当归	dāng guī	20g	Radix Angelicae Sinensis
丹参	dān shēn	24g	Radix et Rhizoma Salviae Miltiorrhizae
元参	yuán shēn	24g	Radix Scrophulariae
制没药	zhì mò yào	6g	Myrrha (processed)
制乳香	zhì rǔ xiāng	6g	Olibanum (processed)
连翘	lián qiáo	10g	Fructus Forsythiae
升麻	shēng má	10g	Rhizoma Cimicifugae
陈皮	chén pí	9g	Pericarpium Citri Reticulatae
甘草	gān cǎo	10g	Radix Glycyrrhizae

This was decocted with water and divided to be taken three times, one pack per day.

Second Visit

May 20th: After taking four packs of the above formula, the patient's swelling was eliminated and his pain was stopped. He was instructed to take another two packs after which he recovered.

Case 3

Initial Visit

A female patient, 32 years old, June 22nd, 1982.

Nine months after giving birth and during the breast feeding period, she suddenly had left breast distention and pain, and on the upper right, there was a red, swollen clump around the size of a nut. If pressed on, she felt pain. Her appetite was normal. Her pulse was deficient and wiry. Her tongue coating was thin and yellow. This was the initial stage of a breast abscess, with qi and blood stagnation and impediment syndrome. After childbirth, her qi and blood were deficient, therefore, her pulse was deficient and wiry. The treatment plan was to invigorate blood, transform stasis, tonify qi and detoxify toxins.

Prescription

生黄芪	shēng huáng qí	24g	Radix Astragali (raw)
金银花	jīn yín huā	20g	Flos Lonicerae Japonicae
赤芍	chì sháo	12g	Radix Paeoniae Rubra
当归	dāng guī	12g	Radix Angelicae Sinensis
王不留行	wáng bù liú xíng	10g	Semen Vaccariae
橘叶	jú yè	9g	Folium Citir
瓜蒌	guā lóu	15g	Fructus Trichosanthis
柴胡	chái hú	15g	Radix Bupleuri
元参	yuán shēn	15g	Radix Scrophulariae
山甲珠	shān jiǎ zhū	10g	Squama Manis

甘草	*gān cǎo*	10g	Radix Glycyrrhizae
制没药	*zhì mò yào*	6g	Myrrha (processed)
制乳香	*zhì rǔ xiāng*	6g	Olibanum (processed)

This was decocted with water and divided to be taken twice, once each in the morning and evening, one pack per day.

Second Visit

June 25th: After taking three packs of the above formula, her pain was stopped. The redness and swelling were also visibly reduced. The clump became soft, and when pressed on, there was mild pain. Based on the above formula, *chì sháo* (Radix Paeoniae Rubra) was eliminated and *dān shēn* (Radix et Rhizoma Salviae Miltiorrhizae) 20g, was added and she continued to take it.

Third Visit

June 30th: After taking another four packs, all of her symptoms basically disappeared. The patient's formula was stopped and she was instructed to use a hot towel to cover it every night and she recovered.

2. *Huáng Qí* (Radix Astragali) ——
Zhǐ Shí (枳实, Fructus Aurantii Immaturus)

Function

To raise up the sinking, tonify qi, release the middle *jiao* and harmonise the stomach.

Application

This can be used for stomach prolapse, rectum prolapse and metroptosis.

Compatibility Analysis

The property of *huáng qí* (Radix Astragali) is warm. It functions to tonify qi and raise the sinking. *Zhǐ shí* (Fructus Aurantii Immaturus) is an important herb to break up qi and eliminate the glomus. Modern pharmacological research discovered that *zhǐ shí* (Fructus Aurantii Immaturus) can contract smooth muscles. It has a better treatment effect for stomach prolapse and metroptosis, etc. From a Chinese medicine perspective the stomach prolapse and metroptosis both belong to qi deficiency and sinking syndromes. Although *zhǐ shí* (Fructus Aurantii Immaturus) can contract the smooth muscles, if it is used for long time it could break up qi and injure the normal qi, which is harmful for the condition. Therefore, when applying *zhǐ shí* (Fructus Aurantii Immaturus) to treat syndromes of stomach prolapse, etc. it should be combined with qi tonifying and the normal qi supporting herbs. It is combined with *huáng qí* (Radix Astragali) to help support the normal qi and counteract the action of *zhǐ shí* (Fructus Aurantii Immaturus) in breaking up qi, while, *huáng qí* (Radix Astragali) also can raise up the yang qi and lift up the sinking, to accentuate the power of *zhǐ shí* (Fructus Aurantii Immaturus) in contracting smooth muscles, they are mutually beneficial. This compatibility also has functions to raise blood pressure.

Initial Visit

A female patient, February 12th, 1987.

The patient had a history of stomach prolapse for more than five years. Her body was emaciated. She always had stomach distention and fullness, with frequent abdominal pain, and nausea accompanied by lassitude, anorexia, dizziness and palpitations, etc. Her barium x-ray examination showed: Her stomach was in a U-shape, low tension, the lesser gastric curvature angle was about three fingers lower than the anterior sacro-iliac spine（ASIS）. Her pulse was deficient, weak and forceless. Her tongue coating was thin and white. This was the middle *jiao* stomach and spleen deficiency with weakness, and the raising and descending of the qi mechanism was in disorder. When the spleen is deficient, the clear yang could not rise up, and so she had lassitude, dizziness and palpitations. When the stomach has lost its harmony and descending function, it manifests in stomach distention and fullness, nausea and anorexia. The treatment plan was to regulate the middle *jiao*, strengthen the spleen and harmonise the stomach.

Prescription

生黄芪	*shēng huáng qí*	24g	Radix Astragali (raw)
枳实	*zhǐ shí*	20g	Fructus Aurantii Immaturus
百合	*bǎi hé*	20g	Bulbus Lilii
乌药	*wū yào*	10g	Radix Linderae
砂仁	*shā rén*	6g	Fructus Amomi
白术	*bái zhú*	12g	Rhizoma Atractylodis Macrocephalae
檀香	*tán xiāng*	9g	Lignum Santali Albi
半夏	*bàn xià*	9g	Rhizoma Pinelliae
茯苓	*fú líng*	15g	Poria
甘草	*gān cǎo*	5g	Radix Glycyrrhizae

This was decocted with water, one pack per day.

The patient was instructed to eat little and often, take less exercise after eating, and to lie down for ten minutes after meals.

Second Visit

February 25th: After taking five packs of the above formula, her symptoms of stomach distention and fullness, nausea and dizziness, etc. basically disappeared, and her appetite had increased. The patient followed the original formula for another five packs, all of her symptoms became less significant and her pulse was also more forceful than before. For strengthening the effect of treatment, the decoction was changed into a powder formula and she took it often.

Prescription

生黄芪	*shēng huáng qí*	100g	Radix Astragali (raw)
枳实	*zhǐ shí*	50g	Fructus Aurantii Immaturus
百合	*bǎi hé*	100g	Bulbus Lilii
乌药	*wū yào*	50g	Radix Linderae
茯苓	*fú líng*	80g	Poria
鸡内金	*jī nèi jīn*	50g	Endothelium Corneum Gigeriae Galli
升麻	*shēng má*	30g	Rhizoma Cimicifugae
甘草	*gān cǎo*	30g	Radix Glycyrrhizae

The above herbs were ground together into a fine powder, each time she took 6 ~ 9g after meals, three times a day. At a follow up visit half year later, she said there had been no reoccurence of her stomach prolapse.

3. *Huáng Qí* (Radix Astragali) ——— *Fáng Fēng* (防风, Radix Saposhnikoviae)

Function

To expel wind and secure the exterior.

Application

This is used for exterior deficiency and spontaneous sweating, easily contracting an exterior invasion or for patients with repeated exterior invasion with deficiency and a weak physical constitution.

Compatibility Analysis

The taste of *huáng qí* (Radix Astragali) is sweet and its qi is warm. It enters the lung channel of the hand *taiyin* and the spleen channel of the foot *taiyin* meridians. It functions to tonify qi and secure the exterior, tonify the spleen and supplement the yang qi. *Fáng fēng* (Radix Saposhnikoviae) is acrid, sweet and warm. It enters the lung channel of the hand *taiyin*, the spleen channel of the foot *taiyin*, the stomach channel of the foot *yangming* and the liver channel of the foot *jueyin* meridians. It functions to release the exterior, expel wind and induce sweating. Although *fáng fēng* (Radix Saposhnikoviae) is acrid, warm and a wind expelling herb to release the exterior, its power is slow and gentle. Li Gao described it as *"the moistening herb among the wind expelling herbs"*. The compatibility of *huáng qí* (Radix Astragali) and *fáng fēng* (Radix Saposhnikoviae) is the compatibility of two opposites. *"Though they are mutually antagonistic, they are mutually enhancing."* *Huáng qí* (Radix Astragali) helps *fáng fēng* (Radix Saposhnikoviae) to skillfully secure the exterior but without trapping the pathogens. *Fáng fēng* (Radix Saposhnikoviae) helps *huáng qí* (Radix Astragali) to expel pathogens without injuring the normal qi. One expels and the other tonifies simultaneously. It can be said that they are mutually enhancing. They have functions to secure the exterior and stop sweating, tonify qi and release the exterior. It is more suitable for patients with exterior deficiency or poor constitution who have frequent reoccurences of flu. With the exterior deficiency patient, their nutritive and defensive qi are in disharmony, if the defensive qi is deficient then it cannot securely defend against outside pathogens and is therefore, easily invaded. If the nutritive qi is deficient then the body fluid has lost its internal holding, therefore, there is always spontaneous sweating. When prolonged, the true qi is consumed internally and the normal qi cannot defend against the pathogens, therefore, there are recurrent external invasions. If it is frequent without recovery and if only the exterior is secured then it will close the door and trap the pathogens inside. If aimed simply at expelling the pathogens and releasing the exterior, it must consume and injure the normal qi and make the nutritive and defensive qi even weaker. *Huáng qí* (Radix Astragali) combined with *fáng fēng* (Radix Saposhnikoviae) just matches this aetiology. The famous formula *Yù Píng Fēng Sǎn* (Jade Wind-Barrier Powder) uses this compatibility.

Case 1

Poor constitution with exterior invasion.

Initial Visit

A female patient, 24 years old, April 26th, 1986.

The patient was used to having a poor constitution. Each time, when there was infectious flu, she would get it first and had lassitude, shortness of breath and always with spontaneous sweating, anorexia, a loss of taste for foods, stomach distention, fullness with mild pain, diarrhoea twice a day, as well as frequent nausea and vomiting. Her face was lustreless and her voice was low and quiet. Her pulse was thready, weak and forceless. Her tongue was pale and her tongue coating was thin and white. This was the spleen and the lung qi deficiency syndrome.

Prescription

生黄芪	shēng huáng qí	24g	Radix Astragali (raw)
防风	fáng fēng	9g	Radix Saposhnikoviae
白术	bái zhú	12g	Rhizoma Atractylodis Macrocephalae
杭白芍	háng bái sháo	12g	Radix Paeoniae Lactiflorae
陈皮	chén pí	4g	Pericarpium Citri Reticulatae
藿香	huò xiāng	4g	Herba Agastachis
半夏	bàn xià	9g	Rhizoma Pinelliae
苏梗	sū gěng	9g	Caulis Perillae
枳壳	zhǐ qiào	9g	Fructus Aurantii
甘草	gān cǎo	6g	Radix Glycyrrhizae

This was decocted with water, one pack to be taken per day.

Second Visit

April 29th: After taking three packs of the above formula, her appetite increased. Her abdominal pain and diarrhoea had also improved. The above formula was modified and she continued to take it.

Prescription

生黄芪	*shēng huáng qí*	24g	Radix Astragali (raw)
防风	*fáng fēng*	9g	Radix Saposhnikoviae
白术	*bái zhú*	12g	Rhizoma Atractylodis Macrocephalae
陈皮	*chén pí*	9g	Pericarpium Citri Reticulatae
杭白芍	*háng bái sháo*	15g	Radix Paeoniae Lactiflorae
桂枝	*guì zhī*	10g	Ramulus Cinnamomi
柴胡	*chái hú*	10g	Radix Bupleuri
半夏	*bàn xià*	9g	Rhizoma Pinelliae
香附	*xiāng fù*	10g	Rhizoma Cyperi
甘草	*gān cǎo*	6g	Radix Glycyrrhizae

This was decocted with water, one pack per day, and divided to be taken twice.

Third Visit

May 10th: After taking six packs of the above formula, all her symptoms basically disappeared except that there was still always deficiency sweating. The decoction was changed to a powder formula and the patient was instructed to take it often to clear it up.

Prescription

生黄芪	*shēng huáng qí*	100g	Radix Astragali (raw)
防风	*fáng fēng*	30g	Radix Saposhnikoviae
桂枝	*guì zhī*	30g	Ramulus Cinnamomi
杭白芍	*háng bái sháo*	50g	Radix Paeoniae Lactiflorae
白术	*bái zhú*	50g	Rhizoma Atractylodis Macrocephalae
甘草	*gān cǎo*	6g	Radix Glycyrrhizae

The above herbs were ground together into a fine powder, each time 5g were taken, twice a day.

Note

After the patient took two batches of the above formula, her deficiency sweating stopped and her recurrent flu also basically recovered.

This formula is the combination of *Yù Píng Fēng Sǎn* (Jade Wind-Barrier Powder) and *Guì Zhī Tāng* (Cinnamon Twig Decoction). It uses the *Yù Píng Fēng Sǎn* (Jade Wind-Barrier Powder) to secure the exterior and expel wind and uses *Guì Zhī Tāng* (Cinnamon Twig Decoction) to harmonise the nutritive qi and the defensive qi. If the nutritive qi and the defensive qi are harmonised, wind is expelled and the exterior is secured, then the spontaneous sweating will be stopped and the exterior deficiency will recover. The author always uses modifications of this formula to treat low immunity and patients who easily catch colds, most of the time with good results.

Case 2

Yang deficiency with exterior invasion.

Initial Visit

A male patient, 56 years old, November 23rd, 1985.

The patient had a history of stomach prolapse. Three days previously he caught a cold. He took medicine but he did not recover and so came to visit. At the time of the visit, the patient still had a stuffy nose, aversion to cold, and his body was slightly hot. He had sweating but the heat did not reduce and was accompanied by whole body lassitude, palpitations, chest tightness and anorexia. His pulse was very thready and floating. His tongue was pale and tongue coating was thin and white. This was yang deficiency with exterior invasion syndrome. The very thready, floating and forceless pulse indicated yang qi deficiency. With yang qi deficiency and his prolapsed stomach, this meant his middle *jiao* qi was sinking; therefore, he had chest tightness, anorexia and lassitude. The aversion to cold, stuffy nose and body heat were the exterior invasion without releasing syndrome. The treatment plan was to tonify qi, support the yang and release the exterior.

Prescription

生黄芪	*shēng huáng qí*	24g	Radix Astragali (raw)
防风	*fáng fēng*	10g	Radix Saposhnikoviae
柴胡	*chái hú*	9g	Radix Bupleuri
升麻	*shēng má*	6g	Rhizoma Cimicifugae
桂枝	*guì zhī*	8g	Ramulus Cinnamomi
苏梗	*sū gěng*	9g	Caulis Perillae
枳壳	*zhǐ qiào*	20g	Fructus Aurantii
党参	*dǎng shēn*	15g	Radix Codonopsis
砂仁	*shā rén*	4g	Fructus Amomi
藿香	*huò xiāng*	9g	Herba Agastachis
甘草	*gān cǎo*	6g	Radix Glycyrrhizae

This was decocted with water and divided to be taken twice, one pack per day.

Second Visit

November 23rd: After taking three packs of the above formula, his symptoms of aversion to cold, stuffy nose, and chest tightness, etc. were eliminated. His appetite increased but he still had symptoms of lassitude, sweating on exertion and palpitations, etc. Based on the original formula, *bái zhú* (Rhizoma Atractylodis Macrocephalae) 12g, was added and he continued taking the formula.

Third Visit

November 28th: After taking another three packs, his exterior invasion symptoms were all eliminated. His sweating was also stopped. The prescription was also changed to tonify qi and strengthen the spleen to treat his stomach prolapse.

4. *Huáng Qí* (Radix Astragali) ——
Jī Xuè Téng (鸡血藤, Caulis Spatholobi)

Function

To tonify qi and nourish blood.

Application

This is used for hepatitis or liver cirrhosis inducing a lowered albumin level or protein ratio inversion.

Compatibility Analysis

Huáng qí (Radix Astragali) is sweet and warm. It is a good herb to tonify the middle *jiao* and qi. It functions to increase the organism's power of immunity and protect the liver, and prevent glycogen from reducing. It is a commonly used herb to treat liver diseases. *Jī xuè téng* (Caulis Spatholobi) is bitter, sweet and warm. It enters the liver channel of the foot *jueyin* and the kidney channel of the foot *shaoyin* meridians. It functions to tonify, nourish and invigorate blood. It tonifies but does not stagnate. A recent report stated that, in clinical practice, this herb is used to treat aplastic anaemia and treat cancer patients during radiation therapy which has induced a reduction in white blood cells and treatment was effective for all. In clinical practice, the author uses *huáng qí* (Radix Astragali) with *jī xuè téng* (Caulis Spatholobi) to treat hepatitis or liver cirrhosis patients who have low human serum albumin, and gets a reasonable treatment effect. These two herbs combined together have functions to tonify qi and nourish blood with the characteristic of tonifying without stagnation. This could increase an organism's immunity and promote cell regeneration.

Case 1

Initial Visit

A male patient, 50 years old, June 22nd, 1995.

The patient had a history of chronic hepatitis. In recent months, he had symptoms of abdominal distention, anorexia, lassitude and fatigue, etc. again. He had been hospitalised at Beijing and Tianjin hospitals for treatments and was diagnosed with liver cirrhosis. Due to limited family resources, hospitalising away from his hometown was too expensive, so he was transferred to our clinic for treatments. A liver function test showed: TTT 9.7u, GPT 67u/L, total protein 74g/L, albumin 35g/L, globulin 39g/L, HbsAg (+). His liver B-ultrasound examination showed: The liver physical echo increased, the surface was granular, blood vessels inside the liver were not clear, portal vein width expanded to 1.4cm, ascites (–). His face was lustreless. His pulse was wiry and slightly deep. His tongue was dark and his tongue coating was white but slightly turbid. This was blood stasis in the liver channel of the foot *jueyin* meridian. The treatment plan was to soothe the liver, transform blood stasis, invigorate blood and strengthen the spleen.

Prescription

丹参	dān shēn	20g	Radix et Rhizoma Salviae Miltiorrhizae
生黄芪	shēng huáng qí	20g	Radix Astragali (raw)
鸡血藤	jī xuè téng	30g	Caulis Spatholobi
泽兰	zé lán	15g	Herba Lycopi
虎杖	hǔ zhàng	15g	Rhizoma et Radix Polygoni Cuspidati
柴胡	chái hú	15g	Radix Bupleuri
枳壳	zhǐ qiào	12g	Fructus Aurantii
白术	bái zhú	12g	Rhizoma Atractylodis Macrocephalae
当归	dāng guī	15g	Radix Angelicae Sinensis
仙鹤草	xiān hè cǎo	24g	Herba Agrimoniae
蒲公英	pú gōng yīng	15g	Herba Taraxaci
砂仁	shā rén	8g	Fructus Amomi
白花蛇舌草	bái huā shé shé cǎo	30g	Herba Hedyotis Diffusae
甘草	gān cǎo	6g	Radix Glycyrrhizae

This was decocted with water and taken, one pack per day.

Second Visit

July 21st: After taking thirty packs of modifications of the above formula, his liver function test was normal: Total protein 74g/L, albumin 39g/L, globulin 35g/L. The patient felt well. He was instructed to continuously take the modifications of the original formula every other day for half a year and then stop the formula.

He was followed up for two years and no reoccurence was seen.

Case 2

Initial Visit

A male patient, 44 years old, May 11th, 1991.

The patient had a history of hepatitis for more than ten years. Recently due to overexertion, again, he had lassitude of his lower limbs, stomach and abdominal distention and fullness, yellowish urine, and pain in his liver area, and so he came to our hospital for examinations. His liver function test showed: Total bilirubin 24μmol/L, direct bilirubin 12μmol/L, indirect bilirubin 12μmol/L, TTT 13.5U, total protein 68g/L, albumin 34g/L, GPT 65U/L, HbsAg (+). His liver B-ultrasound examination showed: The liver physical distribution was stronger than usual with a horizontal spot shaped echo, the blood vessels inside the liver were not clear, the portal vein was 1.5cm, ascites (–). He was diagnosed with the early stage of liver cirrhosis. His pulse was deep and wiry. His tongue was dark and his tongue coating was white and turbid. This was damp-heat in the liver and the gallbladder, the spleen deficiency could not transform and transport thus inducing the dampness pathogen which obstructed and stagnated, manifesting in blood stasis that could not move. The treatment plan was to strengthen the spleen, expel dampness, tonify qi and invigorate blood.

Prescription

白花蛇舌草	*bái huā shé shé cǎo*	30g	Herba Hedyotis Diffusae
蒲公英	*pú gōng yīng*	15g	Herba Taraxaci
丹参	*dān shěn*	15g	Radix et Rhizoma Salviae Miltiorrhizae
柴胡	*chái hú*	15g	Radix Bupleuri
生黄芪	*shēng huáng qí*	15g	Radix Astragali (raw)
鸡血藤	*jī xuè téng*	30g	Caulis Spatholobi
泽兰	*zé lán*	15g	Herba Lycopi
虎杖	*hǔ zhàng*	15g	Rhizoma et Radix Polygoni Cuspidati
茵陈	*yīn chén*	15g	Herba Artemisiae Scopariae
川军	*chuān jūn*	6g	Radix et Rhizoma Rhei
茯苓	*fú líng*	15g	Poria
泽泻	*zé xiè*	12g	Rhizoma Alismatis
鳖甲	*biē jiǎ*	10g	Carapax Trionycis
甘草	*gān cǎo*	6g	Radix Glycyrrhizae

This was decocted with water and divided to be taken twice, one pack per day.

Second Visit

June 17th: After taking a total of thirty packs of the above formula, his clinical symptoms were all reduced. His liver function test of bilirubins, GPT and TTT were all normal, total protein 69g/L, albumin 38g/L, globulin 31g/L, HBsAg 1:32. The above formula was modified and *shēng huáng qí* (raw Radix Astragali) was increased to 24g. He continued taking it.

Third Visit

July 20th: The above formula was modified again. After the patient took more than thirty packs, all of his symptoms disappeared. His liver functions were normal: HBsAg 1:8, total protein 72g/L, albumin 41g/L, globulin 31g/L.

The patient was instructed to take the same formula every other day in order to strengthen the treatment effectiveness.

Three months later, his liver function examination was normal and so he stopped his formula.

He was followed up until now, and there has been no reoccurence.

5. *Huáng Qí* (Radix Astragali) ——
Dāng Guī (当归, Radix Angelicae Sinensis)

Function

To tonify qi and generate blood.

Application

It can be used for severe blood loss or female uterine bleeding, postpartum bleeding induced anaemia syndrome that manifests in facial colour fading to yellow, fatigue and lassitude or with low grade fever, a deficient and forceless pulse, etc. It also can be used for when after the ulcer breaks down, and has too much pus and blood inducing a deficiency syndrome; or for too much blood loss after an operation inducing an anaemia syndrome. It is also effective for thrombocytopaenia.

Compatibility Analysis

Huáng qí (Radix Astragali) combined with *dāng guī* (Radix Angelicae Sinensis) is also called *Dāng Guī Bǔ Xuè Tāng* (Chinese Angelica Blood-Supplementing Decoction, 当归补血汤). It is originally recorded in the *Secrets from the Orchid Chamber* (兰室秘藏 , *Lán Shì Mì Cáng*). The dosage of *huáng qí* (Radix Astragali) is five time the dosage of *dāng guī* (Radix Angelicae Sinensis). After severe blood loss, it is not only the blood deficiency, but also the yang qi that has lost its attachment, and so is scattered. Under this situation, the tangible blood cannot be generated fast and the intangible qi should be secured immediately. Therefore, it heavily uses *huáng qí* (Radix Astragali) as the king herb to greatly tonify the primary qi with a little *dāng guī* (Radix Angelicae Sinensis) to nourish blood and harmonise the nutritive qi. This comes from the theory that if the yang is generated then the yin will grow. In clinical practice, if the anaemic patient belongs to chronic blood deficiency, the dosage can be changed to 2:1. This uses the theory that yin and yang are inter-transforming.

Case Study

Initial Visit

A female patient, 25 years old, April 29th, 1986.

After giving birth a year previously, the patient had a deficient and weak physical constitution, lassitude, and was easily tired, especially when the weather changed; she felt heaviness, soreness and pain in her back. Although she was treated elsewhere, she did not improve and so came to our clinic for treatments. Her face was fading yellow in colour and lustreless. Her pulse was deep, thready and forceless. Her tongue was pale and the coating was thin and white. Her haemoglobin was 90g/L. Her menstruation came on time but was slightly heavy with a light colour, a thin flow and needed ten days to clear. This was qi and blood deficiency after childbirth syndrome. Her blood was deficient so her facial colour was lustreless and her tongue was pale, and she had a thready pulse. Her qi was deficient so she had lassitude and was easily tired with a deep and forceless pulse. Her qi was deficient so her yang qi could not disperse, therefore, when the weather changed, she had heaviness, soreness and pain sensations in her back. The treatment plan was to use a formula to tonify qi and blood, and raise the yang.

Prescription

生黄芪	*shēng huáng qí*	24g	Radix Astragali (raw)
当归	*dāng guī*	10g	Radix Angelicae Sinensis
桔梗	*jié gěng*	6g	Radix Platycodonis
柴胡	*chái hú*	9g	Radix Bupleuri

升麻	*shēng má*	4g	Rhizoma Cimicifugae
葛根	*gé gēn*	20g	Radix Puerariae Lobatae
党参	*dǎng shēn*	20g	Radix Codonopsis
仙鹤草	*xiān hè cǎo*	24g	Herba Agrimoniae
甘草	*gān cǎo*	6g	Radix Glycyrrhizae

This was decocted with water and divided to be taken twice, once each in the morning and evening, one pack per day.

Second Visit

May 6th: After taking seven packs of the above formula, her lassitude and back heaviness sensation were reduced. Based on the original formula, *ē jiāo* (Colla Corii Asini) 10g, and *huáng jīng* (黄精 , Rhizoma Polygonati) 15g, were added. She continued to take it.

Third Visit

May 27th: The above formula was modified and the patient took more than twenty packs, her clinical symptoms basically disappeared. The colour of her menstruation flow also became red. Her haemoglobin laboratory examination reported 110g/L. She was instructed to follow the same formula for seven more packs again.

6. *Huáng Qí* (Radix Astragali) ——
Sāng Yè (桑叶, Folium Mori)

Function

To tonify qi and secure the exterior, to clearly disperse and stop sweating.

Application

This is used for both qi and yin deficiency induced spontaneous sweating and night sweating.

Compatibility Analysis

Huáng qí (Radix Astragali) is sweet and warm. It is a qi tonifying and yang reinforcing herb. It functions to strengthen the defensive qi, secure the exterior, warm the qi and retain sweat. Using it for the yang deficiency with spontaneous sweating patients has good results. *Sāng yè* (Folium Mori) is sweet and cold. It is influenced the most by the autumn qi. It functions to clear, disperse, expel wind and clear heat. It also can tonify yin and retain sweats. In the *Thoroughly Revised Materia Medica* it states that it can "*stop night sweating*". In the *Divine Husbandman's Classic of the Materia Medica* it states that it can "*eliminate cold and heat, and induce sweating*". These two herbs combined together, can both clear and disperse, and tonify the securing of *wèi* qi, regulate both cold and warmth to make them tonifying but without stagnating, soothing but without leakage. It can be used for both qi and yin deficiency induced spontaneous sweating and night sweating. This compatibility and the *huáng qí* (Radix Astragali) with *fáng fēng* (Radix

Saposhnikoviae) are similar, but *fáng fēng* (Radix Saposhnikoviae) expels wind and is slightly warm. It is used with good effect when the yang deficiency and the defensive qi is not solid, with exterior pathogens inducing spontaneous sweating. However, for this compatibility, *sāng yè* (Folium Mori) also can expel the wind pathogen. Aside from using it for the defensive qi not being solid with exterior pathogens inducing spontaneous sweating, it also can tonify yin and treat night sweating. It regulates both cold and heat. Its applications are even broader and there are no side effects.

Case Study

Initial Visit

A male patient, 11 years old, December 20th, 1998.

The child had spontaneous sweating and night sweating for three years. His development was normal. Every night when he slept, there was sweating. During day time activities and during meals, his sweat dripped out like water accompanied by whole body lassitude, and being easily tired. His appetite, urination and bowel movement were normal. His pulse was deficient and thready. His tongue was thin and white. This was both qi and yin deficiency inducing sweating syndrome. His yang was deficient so he had spontaneous sweating. His yin was deficient so he had night sweating. The treatment plan was to tonify qi, secure the exterior, tonify yin and retain sweats.

Prescription

生黄芪	*shēng huáng qí*	20g	Radix Astragali (raw)
桑叶	*sāng yè*	12g	Folium Mori
白术	*bái zhú*	9g	Rhizoma Atractylodis Macrocephalae
防风	*fáng fēng*	8g	Radix Saposhnikoviae
当归	*dāng guī*	9g	Radix Angelicae Sinensis
生地	*shēng dì*	9g	Radix Rehmanniae
杭白芍	*háng bái sháo*	10g	Radix Paeoniae Lactiflorae
山茱萸	*shān zhū yú*	9g	Fructus Corni
甘草	*gān cǎo*	6g	Radix Glycyrrhizae

This was decocted with water and divided to be taken twice, one pack per day.

In addition: Evey night, just before sleep, he was instructed to use *wǔ bèi zǐ* (Galla Chinensis) 10g and grind it into a powder to mix with warm water to form a paste to apply on his umbilicus, then use a cotton cloth to fix it there, and remove it the next morning.

Second Visit

December 27th: After taking seven packs of the above formula, his night sweating stopped and his daytime spontaneous sweating had also recovered by more than half. As the prescription was effective it was not changed, and he continued with the original formula again. After taking ten more packs again, he recovered.

7. *Huáng Qí* (Radix Astragali) ──
Jì Yú (鲫鱼, Carassius Auratus)

Function

To tonify qi and the middle *jiao*, unblock the meridians and promote lactation.

Application

This is used for postpartum scanty lactation or no lactation.

Compatibility Analysis

For the postpartum scanty lactation or no lactation, the first aetiology is a stagnated or obstructed mammary gland where the mammary ducts flow is not smooth, and milk cannot come down. The second aetiology is qi and blood deficiency after childbirth, where the qi and blood cannot generate and transform into milk and induces soft breasts with scanty lactation or no lactation. Therefore, treatment should be aimed at tonifying qi and the middle *jiao* to cultivate and tonify the post-heaven that is the source of generation and transformation. If the source of generation and transformation is sufficient, then lactation will be generated. In addition, the lactation flow should be unblocked, to make the mammary ducts flow freely so milk can come down.

Huáng qí (Radix Astragali) is sweet and warm. It is a miracle herb to tonify qi and the middle *jiao*, to cultivate and tonify the post-heaven. In the *Divine Husbandman's Classic of the Materia Medica* it states that it can "*tonify deficiency*". In the *Essentials of the Materia Medica* it states that it can "*tonify the middle jiao and the primary qi, warm the san jiao, strengthen the spleen and the stomach.*" *Jì yú* (鲫 鱼 , Carassius Auratus) is sweet and warm. It enters the spleen channel of the foot *taiyin* and the stomach channel of the foot *yangming* meridians. The *Miscellaneous Records of Famous Physicians* lists it as a first class herb. Li Shi-zhen said that it can "*tonify the stomach*". Mong Xi said that it "*governs the weak stomach that cannot eat, regulates the middle jiao and tonifies the five zang organs*". *Jì yú* (Carassius Auratus) is a substance that has blood, muscles and emotion. It has the strongest property to tonify qi and blood and it is the best one to promote lactation and increase lactation. Post childbirth, most women in the north of China use *jì yú* (Carassius Auratus) to boil soup in order to increase their lactation. In addition, the scales of *jì yú* (Carassius Auratus) and the scales of *chuān shān jiǎ* (Squama Manis) have the same property. It also has the function to unblock the meridians and promote lactation. These two herbs combined together can tonify the middle *jiao* qi, tonify the spleen and the stomach, unblock the meridians and promote lactation. If only *huáng qí* (Radix Astragali) and *jì yú* (Carassius Auratus) are used to boil the soup, it has the function to increase lactation. If they are added in according to the syndrome differentiation, the effect is even more significant.

Case Study

Initial Visit
A female, 29 years old, July 23rd, 1995.

A month after giving birth, the patient's lactation was scanty and both breasts were soft. As she had twins, she could not satisfy both babies' needs. She had a lustreless facial colour with lassitude, anorexia, a thin and white tongue coating, a deep and forceless pulse. This was postpartum qi and blood deficiency, the spleen and the stomach were weak and deficient, and there was insufficient generation and transformation. The treatment plan was to tonify qi, strengthen the spleen, unblock the meridians and promote lactation. The author used one of his personal formulae, *Huáng Qí Jì Yú Tāng* (Astragalus and Carassius Auratus Decoction, 黄芪鲫鱼汤) to treat her.

Prescription

生黄芪	*shēng huáng qí*	30g	Radix Astragali (raw)
当归	*dāng guī*	12g	Radix Angelicae Sinensis
山药	*shān yào*	20g	Rhizoma Dioscoreae
茯苓	*fú líng*	15g	Poria
白术	*bái zhú*	15g	Rhizoma Atractylodis Macrocephalae
丹参	*dān shēn*	15g	Radix et Rhizoma Salviae Miltiorrhizae
砂仁	*shā rén*	6g	Fructus Amomi
王不留行	*wáng bù liú xíng*	15g	Semen Vaccariae (fried)
路路通	*lù lù tōng*	15g	Fructus Liquidambaris
甘草	*gān cǎo*	6g	Radix Glycyrrhizae

Huó jì yú (活鲫鱼 , Carassius Auratus), (around 200g) one piece, with scales, with or without the internal organs are all suitable.

First, put the *jì yú* (Carassius Auratus) into a pot with 1 litre of clear water to boil for 20 minutes, then remove the *jì yú* (Carassius Auratus) and decoct the above herbs for 30 minutes in the remaining soup of *jì yú* (Carassius Auratus). Then, pour out the decoction and add water to decoct again for 20 minutes. Then mix the two decoctions, and divide to be taken twice, one pack per day.

Second Visit

July 28th: After taking five packs of the above formula, her breasts had sensations of fullness and distention. Her lactation was plenty and could satisfy one baby's needs. She was instructed to follow the original formula continuously for seven more packs and she recovered.

Note

The author used his own personal formula, *Huáng Qí Jì Yú Tāng* (Astragalus and Golden Carp Decoction) to treat a couple of postpartum scanty lactation or no lactation patients, all with effective results.

Ingredients

生黄芪	*shēng huáng qí*	30g	Radix Astragali (raw)
当归	*dāng guī*	12~15g	Radix Angelicae Sinensis
白术	*bái zhú*	12~15g	Rhizoma Atractylodis Macrocephalae

党参	*dǎng shēn*	15g	Radix Codonopsis
砂仁	*shā rén*	6g	Fructus Amomi
山药	*shān yào*	20g	Rhizoma Dioscoreae
茯苓	*fú líng*	15g	Poria
王不留行	*wáng bù liú xíng*	15g	Semen Vaccariae (fried)
甘草	*gān cǎo*	6g	Radix Glycyrrhizae
活鲫鱼	*huó jì yú*	1piece	Carassius Auratus

Modifications

1. If the patient's breasts have distention and pain, the mammary ducts are blocked and lactation cannot come down, reduce the dosage of *dǎng shēn* (Radix Codonopsis), *bái zhú* (Rhizoma Atractylodis Macrocephalae) and *fú líng* (Poria), and add *shān jiǎ zhū* (Squama Mantis), *lù lù tōng* (Fructus Liquidambaris) and *jú yè* (Folium Citri).

2. If accompanied by anorexia, and poor appetite, add herbs like *huò xiāng* (Herba Agastachis) and *jī nèi jīn* (Endothelium Corneum Gigeriae Galli), etc. that are aromatic and can increase appetite, strengthen the stomach and reduce food accumulation.

Formula Explanation

The spleen and the stomach are the root of the post-heaven and the source for qi and blood generation and transformation. The generation and transformation of the lactation totally relies on the supply of the post-heaven, therefore, the key to treating this disease is to cultivate and tonify the post-heaven in order to fulfill the source of generation and transformation. *Huáng qí* (Radix Astragali), *bái zhú* (Rhizoma Atractylodis Macrocephalae), *shān yào* (Rhizoma Dioscoreae), *dǎng shēn* (Radix Codonopsis), *fú líng* (Poria) and *gān cǎo* (Radix Glycyrrhizae) form the *Sì Jūn Zǐ Tāng* (Four Gentlemen Decoction, 四君子汤) with *huáng qí* (Radix Astragali) and *shān yào* (Rhizoma Dioscoreae) added. It has the function to tonify the middle *jiao* and qi. *Huáng qí* (Radix Astragali) with *dāng guī* (Radix Angelicae Sinensis) is *Dāng Guī Bǔ Xuè Tāng* (Chinese Angelica Blood-Supplementing Decoction). Postpartum, both qi and blood are deficient, therefore, there is no power to generate or transform the lactation. Using this to increase qi and blood, means that the lactation will become plenty by itself. *Jì yú* (Carassius Auratus) is also a deficiency tonifying herb. Meanwhile, its scales also have functions to unblock the collaterals and promote lactation. If combined with *wáng bù liú xíng* (Semen Vaccariae), its power to unblock the collaterals and promote lactation is even stronger to make the mammary ducts flow fluently, without the concern that lactation is being stagnated or obstructed. There is a little amount of *shā rén* (Fructus Amomi) in the formula to regulate qi, ease the middle *jiao*, increase appetite and arouse the spleen to prevent the other herbs from inducing stagnation or obstruction. For the whole formula, all the herbs together can function to tonify qi and the middle *jiao*, unblock the meridians and promote lactation. In clinical practice, for any patients with qi and blood deficiency with scanty lactation, this formula is the most suitable.

8. *Shēng Huáng Qí* (raw Radix Astragali) ——
Tù Sī Zǐ (菟丝子, Semen Cuscutae),
Fú Líng (茯苓, Poria)

Function

To tonify the kidney, strengthen the spleen, promote water metabolism and reduce swelling.

Application

This is used for both the spleen and kidney deficiency, fluid transformation and transportation disorder inducing oedema. In modern clinical practice, it if effective for the patient who has idiopathic oedema.

Compatibility Analysis

Idiopathic oedema is a syndrome that its main expression is a water and salt metabolism disorder. Most patients have accompanying obesity. The cause is still unknown. It more often happens in 20 ~ 50 years old, menstruating women. It is thought that the pathology change of this disease is an increase in the capillaries permeability with damage to the capillary base membrane and the blood vessels expansion and contraction function is in disorder, the oestrogen and the progesterone ratio are also in disorder, with the oestrogen relatively increased; and the nerve and mental factors also playing a certain role to the episode. It manifests in periodic swelling, abdominal distention, getting worse pre-menstrually or oedema, worse for standing or in the late evening. The evening body weight can increase 1 ~ 1.5 kg, and is always accompanied by symptoms of lassitude, thirst, scanty urine, breast distention and pain, headache, mood swings, nerves and narcolepsy, etc. The liver and kidney function tests are normal. This disease is classified under oedema in Chinese medicine. This is induced by both the spleen and kidney deficiency, water and fluid transforming and transporting disorder. The kidney is the root of pre-heaven, it is the key for water and fluid metabolism. If kidney qi is deficient, it cannot warm and transform fluid, then the fluid will overflow all over to become oedema. In the *Plain Questions - Treatise on Disharmony* (素问·逆调论 , *Sù Wèn - Nì Tiáo Lùn*) it states that the *"kidney is a water organ. It governs the body fluid."* In the *Plain Questions - Treatise on Points for Water and Heat* (素问·水热穴论 , *Sù Wèn - Shuǐ Rè Xuè Lùn*), it also states that the *"kidney is the gate of the stomach. If the door is not closed properly, water is gathered together and follows it up and down; they overflow to the skin to become muscle swelling. Muscle swelling is the water gathering and sickness."* The spleen is the root of the post-heaven. It has transforming and transporting water and dampness functions. If the spleen is deficient, it cannot transform and transport, then water stagnated inside the body can occur and manifest as oedema. Therefore, in the *Plain Questions - Great Treatise on the Essentials of Supreme Truth* (素问·至真要大论 , *Sù Wèn - Zhì Zhēng Dà Yào Lùn*) it states that: *"All kinds of dampness, swelling and fullness, all belong to the spleen."* If both the spleen and the kidney are deficient, the water and fluid metabolism must be disordered and so oedema occurs. This disease belongs to an organism's internal regulation disorder. It is not induced by the exterior pathogen invasion, therefore, treatment should be aimed at regulating and tonifying the spleen and the kidney as the main approach.

Shēng huáng qí (raw Radix Astragali) is sweet and warm. It is an herb to tonify qi and reinforce the yang. It can warm and activate the spleen yang to promote water movement. It is a main herb to cultivate and tonify the post-heaven. Modern clinical research discovered that *huáng qí* (Radix Astragali) has significant function in promoting urination. After taking it, the urination can increase 64% (Note: *Huáng qí* - Radix Astragali promotes urination. Its effective dosage range is small. If dosage is smaller, there is no urine promotion function. If dosage is bigger, then conversely, the urination will reduce. The author suggests that, usually, it is best between 20 ~ 40g). The taste and property of *tù sī zǐ* (Semen Cuscutae) are sweet and warm. It enters the liver channel of the foot *jueyin* and the kidney of the foot *shaoyin* meridians. It has the function to tonify the liver and the kidney. In the *Divine Husbandman's Classic of the Materia Medica* it states that: "*It governs connecting the broken off and injured, tonifies deficiency, tonifies strength and make the healthy people fat.*" Later generations more often use this as an important herb to tonify the kidney, strengthen the yang and secure the essence. The property of this herb is more peaceful. Though it warms the kidney, and strengthens the yang, it does not injure the yin. *Fú líng* (Poria) is sweet and neutral. It enters the spleen channel of the foot *taiyin*, the stomach channel of the foot *yangming*, the lung channel of the hand *taiyin*, the kidney channel of the foot *shaoyin* and the heart channel of the hand *shaoyin* meridians. It is the main herb to tonify the spleen, promote water metabolism and leach out the dampness. If combined with *huáng qí* (Radix Astragali) its power to strengthen the spleen and promote water metabolism is even greater. If combined with *tù sī zǐ* (Semen Cuscutae) then it can help the kidney to promote water metabolism. These three herbs combined together can achieve functions to warm and tonify the spleen and the kidney, promote water metabolism and reduce swelling.

Case Study

Initial Visit

A female patient, 45 years old, August 20th, 1991.

The patient had oedema for two years. It was sometimes better and sometimes worse. After exertion, the oedema in her lower limbs became worse. After treatments were ineffective, she came to visit. On examination, the patient's body was slightly obese; her lower limbs had pitting oedema. A routine urine test was normal. She complained that she had oedema in her lower extremities for two years and the oedema was more significant in the afternoon. She also had symptoms of lassitude, a heaviness sensation in her lower limbs, irritability, hot flushes and sweating, etc. If she took some diuretic medicine, then her oedema could be reduced, but if she stopped the medicine, her oedema would be the same as before. She stated her menstruation flow was scanty and menstruation was always late. Her pulse was deep and slow. Her tongue was pale and her tongue coating was thin and moist. This was the spleen and kidney deficiency induced oedema. The treatment plan was to use a formula to warm the kidney, strengthen the spleen, and promote water metabolism.

Prescription

生黄芪	*shēng huáng qí*	30g	Radix Astragali (raw)
菟丝子	*tù sī zǐ*	30g	Semen Cuscutae
茯苓	*fú líng*	20g	Poria
猪苓	*zhū líng*	10g	Polyporus
桂枝	*guì zhī*	10g	Ramulus Cinnamomi
白术	*bái zhú*	15g	Rhizoma Atractylodis Macrocephalae
泽泻	*zé xiè*	10g	Rhizoma Alismatis
仙灵脾	*xiān líng pí*	15g	Herba Epimedii
知母	*zhī mǔ*	10g	Rhizoma Anemarrhenae

This was decocted with water and divided to be taken twice warm, one pack per day.

Second Visit

August 25th: After taking five packs of the above formula, the oedema in her lower extremities was visibly reduced; her lassitude, hot flushes and sweating were also reduced. The prescription was not changed as it was considered effective. She continued to take the original formula again.

Third Visit

July 2nd: After taking another seven packs, all of her symptoms basically disappeared. She was then prescribed *Gēng Nián Kāng* (Healthy Menopause Capsule) and oryzanol to complete the treatment.

9. *Shēng Huáng Qí* (raw Radix Astragali) ——
Tù Sī Zǐ (菟丝子, Semen Cuscutae),
Chē Qián Zǐ (车前子, Semen Plantaginis)

Function

To tonify the spleen and the kidney, to tonify qi and free stranguria.

Application

This is used for patients with taxation stranguria or stranguria that is accompanied by lassitude, shortness of breath, low back and knee soreness. It is also effective for the modern clinical urethral syndrome.

Compatibility Analysis

Female urethral syndrome is a disease, the main symptoms of which are frequent urination, urgent urination, painful urination and dysuria, and a sinking sensation and distention in the lower abdomen. However, routine urine tests and urine cultures are all negative. Its aetiology is still unclear. Most doctors think that it could be related to an allergy to female external contraceptives or to anxiety neurosis, etc. At present, there is no particularly effective treatment. Through syndrome differentiation, it belongs to the scope of the *lao lin* or *qi lin*, as this disease happens more in menstruating females, and the episode occurs or is aggravated around the menstruation period or after intercourse.

It is more often due to the pre-heaven deficiency or overexertion of the post-heaven causing injury. After giving birth, the fatigue injures both the spleen and the kidney, water and dampness transforming and transporting is affected and causes stagnation in the lower jiao inducing this disease. *Shēng huáng qí* (raw Radix Astragali) is sweet and warm. It can strengthen the spleen and tonify qi, warm the yang and promote water metabolism. *Tù sī zǐ* (Semen Cuscutae) is sweet and warm. It can warm the kidney, strengthen the yang and secure the essence. These two herbs warm and tonify the spleen and the kidney to help the water and fluid transform, transport and keep metabolism in normal order to resolve the damp pathogen in the lower jiao to treat the root problem. *Chē qián zǐ* (Semen Plantaginis) is sweet and cold. It has the function to promote water metabolism and free the urination. It can let the stagnated damp pathogen be excreted to treat the manifestations. Meanwhile, *chē qián zǐ* (Semen Plantaginis) still functions to tonify deficiency and has the characteristic of *"promoting urination but without loss of qi"*. These three herbs treat both the root problem and the manifestations, both tonifiying and unblocking simultaneously as well as tonifying the spleen and the kidney, promoting urination and freeing the *lin* syndromes.

Case Study

Initial Visit

A female patient, 37 years old, April 27th, 1994.

The patient recently had frequent urination, urgent urination and impeded urination sensations. She was treated for a urinary tract infection for two days at a local clinic but without effect and so came to visit. Aside from the symptoms of frequent urination, urgent urination, etc, the patient also had lower back soreness and a feeling of lassitude. Her routine urine test was not abnormal. The patient was physically well developed. Her pulse was deep and wiry. Her tongue coating was thin and white. This was the urethral syndrome and belongs to Chinese medicine's qi *lin*. The treatment plan was to tonify qi and the kidney, to promote urination and free the *lin* syndrome.

Prescription

生黄芪	*shēng huáng qí*	24g	Radix Astragali (raw)
菟丝子	*tù sī zǐ*	15g	Semen Cuscutae
车前子	*chē qián zǐ*	15g	Semen Plantaginis (wrapped up)
石韦	*shí wéi*	10g	Folium Pyrrosiae
益母草	*yì mǔ cǎo*	15g	Herba Leonuri
白术	*bái zhú*	12g	Rhizoma Atractylodis Macrocephalae
茯苓	*fú líng*	15g	Poria
白茅根	*bái máo gēn*	15g	Rhizoma Imperatae
知母	*zhī mǔ*	10g	Rhizoma Anemarrhenae

This was decocted with water and divided to be taken twice, one pack per day.

Second Visit

May 4th: Her symptoms of frequent urination and urgent urination, etc. were already eliminated, but she still had lower back soreness and the lassitude sensation. Based on the above formula, *dù zhòng* (Cortex Eucommiae) 5g, and *chuān duàn* (Radix Dipsaci) 15g, were added and she continued taking it.

Third Visit

May 9th: After taking another five packs, all of her symptoms recovered. She stopped taking her formula and was observed, but no reoccurrence was seen.

Huáng lián's (黄连 , Rhizoma Coptidis) taste and property are bitter and cold. It enters the heart channel of the hand *shaoyang*, the liver channel of the foot *jueyin*, the gallbladder channel of the foot *shaoyang*, the stomach channel of the foot *yangming* and the large intestine of the hand *yangming* meridians. It is an herb to clear heat, dry dampness, sedate the fire and detoxify toxins. It is an important herb for treating the eyes and dysentery. Its clinical compatabilities and applications are broad. For example when combined with *mù xiāng* (Radix Aucklandiae) it becomes *Xiāng Lián Wán* (Aucklandia and Coptidis Pill, 香连丸) to treat any kind of white or red dysentery. When combined with *gān jiāng* (Rhizoma Zingiberis) it becomes *Jiāng Lián Sǎn* (Dried Ginger and Coptidis Powder, 姜连散) to treat all kinds of cold or heat induced dysenteries. When combined with *shēng jiāng* (Rhizoma Zingiberis Recens) it becomes *Xiāng Huáng Sǎn* (Aromatic Coptidis Powder, 香黄散) to treat spleen deficiency induced diarrhoea or watery diarrhoea. If combined with *wú zhū yú* (Fructus Evodiae) and *bái sháo* (Radix Paeoniae Alba) it becomes *Wù Jǐ Wán* (Fifth and Sixth Heavenly Stem Pill, 戊己丸) to treat dysentery with abdominal pain. When combined with *fáng fēng* (Radix Saposhnikoviae) and *huáng lián* (Rhizoma Coptidis) it becomes *Jù Jīn Wán* (Gold Gathering Pill, 聚金丸) to treat heat accumulation and bleeding. When combined with goat liver it becomes *Shèng Jīn Wán* (Superior to Gold Pill, 胜金丸) to treat all kinds of eye diseases. When combined with *wú zhū yú* (Fructus Evodiae) it becomes *Zuǒ Jīn Wán* (Left-Running Metal Pill, 左金丸) to harmonise the stomach and stop vomiting, and combined with *ròu guì* (Cortex Cinnamomi) it becomes *Jiāo Tài Wán* (Peaceful Interaction Pill, 交泰丸) that can communicate the heart with the kidney to treat palpitations and insomnia. *Huáng lián* (Rhizoma Coptidis) as an herb is very bitter. Its property is very cold. It can sedate the excess fire, and clear damp-heat in the intestines. Based on modern pharmacological research, it has the strongest action on shigella. It contains berberine which can increase the white blood cells' ability to ingest staphylococcus aureus (Rosenbach Staphyloccocus aureus) and it is more superior to any other antibiotic, without any side effects. In recent years, clinical reports stated that it effectively treats diabetics.

1. *Huáng Lián* (Rhizoma Coptidis) ——
Ròu Guì (肉桂, Cortex Cinnamomi)

Function

To improve communication between the heart and kidney.

Application

This is used for lost communication between the heart and the kidney inducing palpitations and insomnia. It also can increase the treatment effectiveness for other causes of insomnia, if added in with other herbs chosen according to the syndrome differentiation.

Compatibility Analysis

This compatibility is named *Jiāo Tài Wán* (Peaceful Interaction Pill). This formula is from the *Comprehensive Medicine According to Master Han* (韩氏医通 , *Hán Shì Yī Tōng*). It mainly treats the heart and kidney miscommunication inducing palpitations and insomnia. The ratio for their dosages is 6:1, as in 6 times the dosage of *huáng lián* (Rhizoma Coptidis) to one part of *ròu guì* (Cortex Cinnamomi). If used to treat the exuberant heart yang, with the fire bothering the mind and the spirit inducing insomnia, then heavily use *huáng lián* (Rhizoma Coptidis) to clear the heart and sedate the fire with a little *ròu guì* (Cortex

Cinnamomi) to guide the fire to return it to its original place in order to communicate the yin with the yang, so the insomnia can be cured. In clinical practice, it is used for most insomnia as well as for treating the heart and the kidney miscommunication induced insomnia, each with satisfactory treatment results.

In clinical practice, insomnia has many aetiologies and the types are also different. However, in summary, the aetiology is due to the heart. The heart governs the spirit, if the spirit is not calm, then there is no sleep. The spirit not being calm is induced by yin and yang being unable to communicate with each other. In the *Collected Treatises of Zhang Jing-yue - Insomnia* (景岳全书 · 不寐篇 , *Jǐng Yuè Quán Shū - Bù Mèi Piān*) it states that "*patients with excessive thinking and pensiveness injury, shock, fear, worry or no stressing factors but still cannot sleep, more belong to the essence and the blood of the true yin deficiency, yin and yang miscommunication and the spirit cannot quietly settle in its place.*" In the *Systematised Patterns with Specific Treatments - Insomnia* (类证治裁 · 不寐 , *Lèi Zhèng Zhì Cái - Bù Mèi*) it also clearly points out that: "*For insomnia, the illness is due to the yang not communicating with the yin harmoniously.*" If the yang cannot communicate with the yin harmoniously, then the spirit floats and cannot sleep. If there is yin without yang coming inside then the spirit has nowhere to be held, therefore insomnia occurs. In brief, yin and yang not communicating harmoniously covers all the aetiologies for insomnia. The heart and the kidney are the organs of the water and the fire. The water and the fire are the yin and yang. For harmonious communication between the heart and the kidney, the yin and yang must communicate harmoniously. If yin and yang can communicate harmoniously, then one can sleep. In clinical practice, the method to treat the root problem is based on using the aetiology to select the appropriate formula to treat insomnia, however, for some insomnia, especially for the chronic stubborn insomnia, it usually has a slower effect. If add in *huáng lián* (Rhizoma Coptidis) and *ròu guì* (Cortex Cinnamomi), this can make the yin and yang communicate harmoniously quickly. Han Fei-xia said that: "*Huáng lián* (Rhizoma Coptidis) *and ròu guì* (Cortex Cinnamomi) *can help make the heart and the kidney communicate harmoniously at once.*" Based on the author's clinical experience, and according to the syndrome differentiation, he adds in the *Jiāo Tài Wán* (Peaceful Interaction Pill). The purpose of this is not to use *huáng lián* (Rhizoma Coptidis) to clear the heart and sedate the fire, but to balance the heart yang and yin, and uses *ròu guì* (Cortex Cinnamomi) to guide the heart yang to return to its place, in other words to guide the qi of the yin and yang to communicate. Therefore, the dosage ratio for these two herbs should not be big; it is most effective between 2:1 and 3:1.

Case 1

Initial Visit

A female patient, 69 years old, March 15th, 1987.

The patient was working in an education job, long term. She had excessive thinking and pensiveness. It induced her insomnia with symptoms of difficulty in falling asleep or in sleeping again after waking up from sleep. Every night, she could only sleep for two to three hours. She also needed medicines such as diazepam, etc to help her sleep. She had accompanying symptoms of memory loss, palpitations, fatigue, lassitude and anorexia, etc. Her pulse was deep, weak and forceless. Her tongue was pale and her tongue coating was thin and white. This was both the spleen and kidney deficiency inducing insomnia. Excessive thinking and pensiveness will injure the heart and the spleen, if the heart is injured, then the nutritive blood is consumed. If the spleen is injured, then the transforming resource is insufficient and the essence of the water and grains cannot nourish the heart and the spirit, therefore, it induces the insomnia. The treatment plan was to use a formula to tonify qi and strengthen the spleen, to nourish the blood and anchor the spirit. Modified *Guī Pí Tāng* (Spleen-Returning Decoction) was used to treat her.

Prescription

生黄芪	*shēng huáng qí*	20g	Radix Astragali (raw)
党参	*dǎng shēn*	15g	Radix Codonopsis
白术	*bái zhú*	12g	Rhizoma Atractylodis Macrocephalae
陈皮	*chén pí*	10g	Pericarpium Citri Reticulatae
远志	*yuǎn zhì*	10g	Radix Polygalae
木香	*mù xiāng*	4g	Radix Aucklandiae
桂圆肉	*guì yuán ròu*	12g	Arillus Longan
酸枣仁	*suān zǎo rén*	10g	Semen Ziziphi Spinosae
夜交藤	*yè jiāo téng*	30g	Caulis Polygoni Multiflori
甘草	*gān cǎo*	6g	Radix Glycyrrhizae
生姜	*shēng jiāng*	3pcs	Rhizoma Zingiberis Recens
大枣	*dà zǎo*	6pcs	Fructus Jujubae (as a guide)

This was decocted with water and divided to be taken twice, warm, one pack per day.

Second Visit

March 18th: After taking three packs of the formula, her symptoms of fatigue, lassitude, palpitations and shortness of breath improved, but her insomnia was still the same as before. She had had insomnia for many years and it was stubborn. *Guī Pí Tāng* (Spleen-Returning Decoction) was the chosen formula to treat the root problem and it was difficult to get a quick result, therefore, *huáng lián* (Rhizoma Coptidis) 6g, and *ròu guì* (Cortex Cinnamomi) 3g, were added into the original formula and she continued to take it.

Third Visit

March 21st: Her insomnia was slightly improved. She could sleep without taking diazepam, however, her sleep was not assured and she woke up a lot. The prescription was unchanged as it was considered effective, and she continued to take it again.

Fourth Visit

April 15th: After taking another twelve packs of the above formula, her sleep was already stable. She also could take a one hour nap calmly. Her appetite increased and her clinical symptoms basically disappeared. She stopped her formula and changed to *Guī Pí Wán* (Spleen-Returning Pill) to complete the treatment.

Case 2

Initial Visit

A male patient, 29 years old, April 22nd, 1986.

The patient had a medical history of insomnia. It was sometimes better and sometimes worse. Over the previous fortnight a family member was sick, and his job was not going as well as he wished. This, coupled with excessive worry and fatigue, aggravated his insomnia. He felt irritable and heat inside his heart which was worse before sleep. He could never sleep through the night. He was dispirited and had the accompanying symptom of a bitter mouth. His pulse was wiry, thready and fast. His tongue was red and his tongue coating was thin and white. This syndrome was heart yin deficiency, with the heart fire exuberant, and the spirit not settled in its place. His excessive worry and fatigue then consumed and injured his heart yin. His heart yin was deficient so his pulse was wiry and thready, and his heart felt irritable. His heart yin was deficient so his heart yang was exuberant, therefore, he had a fast pulse. He had irritability and heat in his heart, therefore, he had a red tongue and a bitter mouth. His heart fire was exuberant, so his spirit could not settle in its place, therefore, overnight, he could not sleep. The treatment plan was to clear the heart and calm the spirit.

Prescription

莲子心	lián zǐ xīn	12g	Plumula Nelumbinis
栀子	zhī zǐ	10g	Fructus Gardeniae
淡豆豉	dàn dòu chǐ	9g	Semen Sojae Praeparatum
酸枣仁	suān zǎo rén	10g	Semen Ziziphi Spinosae
茯苓	fú líng	20g	Poria
陈皮	chén pí	10g	Pericarpium Citri Reticulatae
生地黄	shēng dì huáng	15g	Radix Rehmanniae
川连	chuān lián	6g	Rhizoma Coptidis
肉桂	ròu guì	3g	Cortex Cinnamomi
竹茹	zhú rú	9g	Caulis Bambusae in Taenia
甘草	gān cǎo	6g	Radix Glycyrrhizae

This was decocted with water and divided, to be taken twice.

Second Visit

April 25th: After taking three packs of the above formula, his irritablity symptoms were eliminated and he could sleep but easily woke up. The same formula was used to treat him again.

Third Visit

May 10th: After taking a total of ten packs of the above modification, he could sleep calmly. His irritability and bitter mouth were also both eliminated.

Case 3

Initial Visit

A male patient, 47 years old, June 12th, 1986.

The patient had had insomnia for more than twenty years, since his college days. Although he tried many different ways of treatment, it still had not recovered. Recently, his insomnia was aggravated. Every day, he only could sleep for around three hours. If he encountered stress from his work, then he was hardly able to sleep for the whole night. There were accompanying symptoms of hypomnesia, forgetfulness, fatigue, lassitude, poor concentration, etc. His pulse was deep and thready. His tongue was slightly red. His tongue coating was thin and white. This was overexertion of the brain, consumming his essence and blood, with the spirit having lost its nourishment. The treatment plan was to nourish yin and blood, calm the spirit and anchor the will.

Prescription

石菖蒲	*shí chāng pú*	12g	Rhizoma Acori Tatarinowii
远志	*yuǎn zhì*	10g	Radix Polygalae
龟甲	*guī jiǎ*	10g	Carapax et Plastrum Testudinis
酸枣仁	*suān zǎo rén*	12g	Semen Ziziphi Spinosae
生龙骨	*shēng lóng gǔ*	15g	Os Draconis (raw)
龙眼肉	*lóng yǎn ròu*	10g	Arillus Lon gan
川连	*chuān lián*	6g	Rhizoma Coptidis
肉桂	*ròu guì*	3g	Cortex Cinnamomi
夜交藤	*yè jiāo téng*	30g	Caulis Polygoni Multiflori
甘草	*gān cǎo*	6g	Radix Glycyrrhizae
小麦	*xiǎo mài*	1pinch	Fructus Tritici (as a guide)

This was decocted with water and divided to be taken twice, three-fifths before sleep, and the rest the next morning while the stomach was still empty.

Second Visit

June 18th: After taking six packs of the above formula, his symptoms of fatigue, lassitude, etc. were improving. His sleep also improved. He could sleep for around five hours every night. He continued taking the same formula.

Third Visit

July 14th: After taking another six packs, he could not continue taking the formula due to going on a business trip, but he still could sleep for six hours. This was a lot better than before. As he had the insomnia for a long time, in order to strengthen the treatment effectiveness, the decoction was changed to capsules and he was instructed to take them often to complete the treatment.

龟甲	guī jiǎ	100g	Carapax et Plastrum Testudinis
生龙骨	shēng lóng gǔ	150g	Os Draconis (raw)
石菖蒲	shí chāng pú	100g	Rhizoma Acori Tatarinowii
远志	yuǎn zhì	100g	Radix Polygalae
川连	chuān lián	60g	Rhizoma Coptidis
肉桂	ròu guì	30g	Cortex Cinnamomi
酸枣仁	suān zǎo rén	50g	Semen Ziziphi Spinosae
琥珀	hǔ pò	100g	Succinum
代赭石	dài zhě shí	100g	Haematitum

The above herbs were ground together into a fine powder, sieved through a 100 mesh fine filter, then baked to dry and disinfected, and put into 0.3g capsules. Each time 8 capsules were taken with warm water once in the morning, and once at night.

Note

The above formula is *Kǒng Shèng Zhěn Zhōng Dān* (Sagacious Confucius' Pillow Elixir, 孔圣 枕中丹) with *Jiāo Tài Wán* (Peaceful Interaction Pill) modified. It is an empirial formula of the author. He has treated many people and all have had a certain treatment effectiveness. It has functions to calm and anchor the spirit, communicate the heart and the kidney harmoniously, strengthen the brain and benefit the intelligence. It can be used for the symptoms of neurasthenia, insomnia, forgetfulness and palpitations, etc.

Formula Anlysis

In the formula, the *Kǒng Shèng Zhěn Zhōng Dān* is also named as *Kǒng Zǐ Dà Shèng Zhī Zhěn Zhōng Fāng* (Confucius' Sagacious Wisdom Pillow Formula, 孔子大圣知枕中 方). In volume fourteen of the *Important Formulas Worth a Thousand Gold Pieces [for any Emergency]* (备急千金要方 , *Bèi Jí Qiān Jīn Yào Fāng*), it recorded that it is composed of *guī jiǎ* (Carapax et Plastrum Testudinis), *shēng lóng gǔ* (raw Os Draconis), *yuǎn zhì* (Radix Polygalae) and *shí chāng pú* (Rhizoma Acori Tatarinowii) in equal quantities. It functions to tonify the heart and the kidney, treat forgetfulness and insomnia. The characteristics of this formula use *guī jiǎ* (Carapax et Plastrum Testudinis), and *shēng lóng gǔ* (raw Os Draconis) to calm and anchor the spirit, nourish the yin and submerge the yang, and combine it with *shí chāng pú* (Rhizoma Acori Tatarinowii) and *yuǎn zhì* (Radix Polygalae) that are acrid and warm to open the orifices. Modern clinical practice has proven that if only using herbs to calm and anchor the spirit to treat insomnia, especially for the chronic stubborn insomnia, usually the results were not good. If some exciting herbs were added, then satisfactory results can be achieved. This formula is the compatibility of the inhibitatory and excitatory herbs. In addition, the compatibility of *chuān lián* (Rhizoma Coptidis) and *ròu guì* (Cortex Cinnamomi) in the *Jiāo Tài Wán* (Peaceful Interaction Pill) is also suitable for this situation. We can see that our ancestors' method of using this compatibility is based on long term clinical practice. It is consistant with the modern medicine knowledge. The author had treated a 38 year old woman, who had severe stress from worrying about her hepatitis, she was always worried about her illness and other things, that induced her insomnia and made it difficult to sleep. She had a couple of nights without closing her eyes, with anorexia, fatigue, lassitude and she could not get up from the bed to walk. Even after taking three tablets of diazepam she still could not sleep, she was extremely fearful and felt that her illness was severe without chance of recovery. Her aunt looked for the author and talked about her situation.

She was given a prescription and sent back to decoct it immediately.

Prescription

龟甲	*guī jiǎ*	10g	Carapax et Plastrum Testudinis
生龙骨	*shēng lóng gǔ*	15g	Os Draconis (raw)
远志	*yuǎn zhì*	12g	Radix Polygalae
石菖蒲	*shí chāng pú*	12g	Rhizoma Acori Tatarinowii
酸枣仁	*suān zǎo rén*	15g	Semen Ziziphi Spinosae
夜交藤	*yè jiāo téng*	30g	Caulis Polygoni Multiflori
川连	*chuān lián*	6g	Rhizoma Coptidis
肉桂	*ròu guì*	4g	Cortex Cinnamomi

This was decocted with water and divided to be taken twice, warm.

After taking one pack the patient could already sleep and after taking three packs she could sleep normally. It is evident that our ancestors' experiences, methods and formulae are worth inheriting and discovering.

2. *Huáng Lián* (Rhizoma Coptidis) —— *Dì Gǔ Pí* (地骨皮, Cortex Lycii)

Function

To clear heat and stop *xiao ke*.

Application

This is used for *xiao ke* syndrome, polydipsia and polyuria.

Compatibility Analysis

Huáng lián (Rhizoma Coptidis) is bitter and cold. It enters the stomach channel of the foot *yangming* meridian. For exuberant fire of the stomach, polyphagia and polydipsia of the middle *jiao xiao ke*. In the *Miscellaneous Records of Famous Physicians* it states that *huáng lián* (Rhizoma Coptidis) can "stop *xiao ke*". *Dì gǔ pí* (Cortex Lycii) is sweet, bland and cold. It enters the lung channel of the hand *taiyin* and the kidney channel of the foot *shaoyin* meridians. It not only has the power to clear lung heat but also has functions to nourish yin and tonify the kidney, and sedate the hidden fire in the kidney channel of the foot *shaoyin* meridian, therefore, it has the function to sedate the heat pathogen and stop severe thirst. In the *Pouch of Pearls* (珍珠囊 , *Zhēng Zhū Náng*) it states that *dì gǔ pí* (Cortex Lycii) treats "*xiao ke*". *Huáng lián* (Rhizoma Coptidis) and *dì gǔ pí* (Cortex Lycii) all can treat *xiao ke*. Meanwhile, *huáng lián* (Rhizoma Coptidis) enters the middle *jiao* and *dì gǔ pí* (Cortex Lycii) treats both the upper and lower *jiao*s. These two herbs combined together can clear heat in the upper, middle and lower *jiao*s. They also can nourish yin within their clearing heat function. They are really gentle herbs to treat diabetes (*xiao ke* syndrome). In clinical practice, regarding the dosage of the two herbs, there should be a large quantity of *dì gǔ pí* (Cortex Lycii). The best ratio is 5:1. *Huáng lián* (Rhizoma Coptidis) is extremely bitter, if using it in high doses it will injure the stomach, therefore, it is not appropriate to use it in high doses.

Case 1

Initial Visit

A female patient, 65 years old, July 20th, 1995.

The patient had a dry mouth, thirst and lassitude in her whole body and so she came to visit. On examining her empty stomach, the glucose in her blood was 11.56mmol/L and in her urine was ++++. The patient had good physical development. Her appetite was normal. Sometimes, she had a sense of hunger. There was significant lassitude in her lower limbs but without numbness. Her pulse was wiry and fast. Her tongue was red and her tongue coating was thin and white. She had a dry mouth, thirst with a like for drinking and excessive amounts of urination. This was kidney yin deficiency with deficiency fire flaring upward. The treatment plan was to use a formula to nourish yin and tonify the kidney, to clear the heat and stop thirst.

Prescription

地骨皮	*dì gǔ pí*	40g	Cortex Lycii
黄连	*huáng lián*	8g	Rhizoma Coptidis
生黄芪	*shēng huáng qí*	20g	Radix Astragali (raw)
花粉	*huā fěn*	20g	Radix Trichosanthis
丹参	*dān shēn*	24g	Radix et Rhizoma Salviae Miltiorrhizae
生地黄	*shēng dì huáng*	15g	Radix Rehmanniae
山茱萸	*shān zhū yú*	15g	Fructus Corni
茯苓	*fú líng*	12g	Poria
牡丹皮	*mǔ dān pí*	12g	Cortex Moutan
泽泻	*zé xiè*	10g	Rhizoma Alismatis
山药	*shān yào*	20g	Rhizoma Dioscoreae
葛根	*gé gēn*	24g	Radix Puerariae Lobatae

This was decocted with water and divided to be taken twice, one pack per day.

Second Visit

July 28th: After taking seven packs of the above formula, her thirst and lassitude were reduced. She continued to follow the original formula.

Third Visit

August 15th: After taking fourteen packs of the above formula, her blood glucose was 7.3 mmol/L on an empty stomach, the glucose in her urine was +. Her symptoms of a dry mouth, thirst and excessive urination, etc. were improved, but she still had the lassitude sensation sometimes. Based on the original formula, *tài zǐ shēn* (Radix Pseudostellariae)

15g, and *xiān hè cǎo* (Herba Agrimoniae) 24g, were added and she continued taking it.

Fourth Visit

September 1st: All of her self appraised symptoms disappeared. Her empty stomach blood glucose was 5.4mmol/L and the glucose in her urine was (–). She was instructed to take the formula every other day for seven more packs and then stop the formula and check in one month later.

Fifth Visit

October 10th: Her empty stomach blood glucose and glucose in her urine were all normal.

Case 2

Initial Visit

A female patient, 69 years old, November 20th, 1995.

The patient had a dry mouth, and thirst and liked to drink, for the previous month. She ate a lot but felt lassitude all over her body and heaviness in her lower limbs, etc. and so she came to our hospital for treatments. According to laboratory tests, her empty stomach blood glucose was 9.98mmol/L and a routine urine test showed: Glucose in urine +++, the results were (–). Her lower limbs had oedema with pitting on pressing. Her pulse was deep and wiry. Her tongue coating was thin and white. This was the syndrome of the elderly who had both spleen and kidney, qi and yin deficiency. With qi and yin deficiency, then lassitude, dry mouth and thirst manifest. With spleen and kidney deficiency then the water and fluid were not transformed and transported thus inducing her oedema. The treatment plan was to tonify qi, nourish yin, strengthen the spleen and tonify the kidney.

Prescription

地骨皮	*dì gǔ pí*	40g	Cortex Lycii
黄连	*huáng lián*	8g	Rhizoma Coptidis
山药	*shān yào*	24g	Rhizoma Dioscoreae
苍术	*cāng zhú*	15g	Rhizoma Atractylodis
防己	*fáng jǐ*	15g	Radix Stephaniae Tetrandrae
茯苓	*fú líng*	30g	Poria
生黄芪	*shēng huáng qí*	20g	Radix Astragali (raw)
泽泻	*zé xiè*	15g	Rhizoma Alismatis
益母草	*yì mǔ cǎo*	24g	Herba Leonuri
泽兰	*zé lán*	15g	Herba Lycopi
川牛膝	*chuān niú xī*	15g	Radix Cyathulae
木瓜	*mù guā*	12g	Fructus Chaenomelis
白术	*bái zhú*	12g	Rhizoma Atractylodis Macrocephalae
丹参	*dān shēn*	20g	Radix et Rhizoma Salviae Miltiorrhizae

This was decocted with water and divided to be taken twice, warm.

Second Visit

November 27th: After taking seven packs of the above formula, her empty stomach blood glucose was checked as 6.5mmol/L and glucose in urine as +. As it was effective, the prescription was not changed and she continued to take it again.

Third Visit

December 8th: After taking another seven packs, her lower limbs oedema was visibly reduced. Sometimes, her sleeping was not good. Based on the above formula, *suān zǎo rén* (Semen Ziziphi Spinosae) 15g, was added and she continued to take it again.

Fourth Visit

January 16th, 1996: After totally taking more then thirty packs of the above modifications, all of her symptoms disappeared. Her empty stomach blood glucose was 6.01mmol/L and glucose in urine was (−).

3. *Huáng Lián* (Rhizoma Coptidis) ── *Bǔ Gǔ Zhī* (补骨脂, Fructus Psoraleae)

Function

To mutually regulate cold and heat, warm the kidney, sustain the intestines and stop diarrhoea.

Application

This is used for chronic diarrhoea without recovery, complicated by cold and heat and chronic colitis.

Compatibility Analysis

Bǔ gǔ zhī (Fructus Psoraleae) is acrid and warm. It enters the spleen channel of the foot *taiyin* and the kidney channel of the foot *shaoyin* meridians. It functions to tonify the spleen and the kidney. It can arouse the yang to transform the yin, and can constrain to bind the collapse. It is an important herb to treat the kidney and spleen yang deficiency. *Huáng lián* (Rhizoma Coptidis) is bitter and cold. It can clear heat, dry dampness and sustain the intestines. It is an important herb to treat dysentery, however, its coldness and bitterness easily injure the middle *jiao* qi, but if combined with the acridity and warmth of *bǔ gǔ zhī* (Fructus Psoraleae), it can prevent the disadvantage from the bitterness and cold. These two herbs mutually regulate cold and heat and have functions to warm, clear and stop diarrhoea. They also have the characteristics of stopping diarrhoea without trapping pathogens inside, clearing heat and sustaining the intestines without injuring the middle *jiao*. It is appropriate to use it for cold and heat complicated colitis.

Case Study

Initial Visit

A female patient, 31 years old, September 23rd, 1997.

The patient had abdominal pain and diarrhoea for three years. She had been treated in many places but did not see any improvement. One month ago, she was treated at our hospital's Gastroenterology Department. However, her illness and pain were not reduced. Later, they invited me to do a consultation. The intestinal endoscope inspection showed: The mucous membrane of the appendix, and the ascending and transverse colon had mild hyperaemia. The descending colon's mucous membrane hyperaemia was severe. The sigmoid colon had moderate hyperaemia. The interstitial markings of the blood vessels were not clear, and there was little mucilage adhering. She was diagnosed with chronic colitis. Her stool examination did not show bloody pus cells. The patient had two to three bowel movements daily. There was diarrhoea without formed stools. These were accompanied by lower abdominal pain and a sinking sensation. Her pulse was deep and thready. Her tongue was dark and the coating was thin and white. This was yang deficiency with dampness and blood stasis obstructing internally. Chronic diarrhoea without stopping with the deep, thready pulse was due to yang deficiency. Diarrhoea, abdominal pain, a dark tongue with a white tongue coating were the symptoms of dampness and blood stasis. The treatment plan was to tonify the kidney, strengthen the spleen, regulate qi and transform the blood stasis, sustain the intestines and stop diarrhoea.

Prescription

焦白术	*jiāo bái zhú*	20g	Rhizoma Atractylodis Macrocephalae (stir-baked)
茯苓	*fú líng*	20g	Poria
补骨脂	*bǔ gǔ zhī*	10g	Fructus Psoraleae
黄连	*huáng lián*	6g	Rhizoma Coptidis
陈皮	*chén pí*	9g	Pericarpium Citri Reticulatae
防风	*fáng fēng*	12g	Radix Saposhnikoviae
赤芍	*chì sháo*	12g	Radix Paeoniae Rubra
红花	*hóng huā*	10g	Flos Carthami
生黄芪	*shēng huáng qí*	20g	Radix Astragali (raw)
吴茱萸	*wú zhū yú*	5g	Fructus Evodiae
木香	*mù xiāng*	9g	Radix Aucklandiae
甘草	*gān cǎo*	6g	Radix Glycyrrhizae

This was decocted with water and divided to be taken twice, warm, one pack per day.

Second Visit

September 29th: After taking six packs of the above formula, her abdominal pain and the sinking sensation symptoms were improving but her bowel movements were still diarrhoea and not moving smoothly. Based on the original formula, *jiāo bīng láng piàn* (sliced Semen Arecae Praepareta, 焦槟榔片) 9g, was added and she continued to take it.

Third Visit

October 6th: After taking another seven packs, her symptoms were basically eliminated. Her bowel movements were once a day and had shape. For strengthening the treatment effectiveness, the following formula was prescribed, and she was instructed to take it every other day, continuously for seven packs.

Prescription

焦白术	*jiāo bái zhú*	12g	Rhizoma Atractylodis Macrocephalae (stir-baked)
茯苓	*fú líng*	15g	Poria
补骨脂	*bǔ gǔ zhī*	8g	Fructus Psoraleae
黄连	*huáng lián*	5g	Rhizoma Coptidis
木香	*mù xiāng*	8g	Radix Aucklandiae
黄芪	*huáng qí*	20g	Radix Astragali
红花	*hóng huā*	9g	Flos Carthami
赤芍	*chì sháo*	9g	Radix Paeoniae Rubra
防风	*fáng fēng*	9g	Radix Saposhnikoviae
甘草	*gān cǎo*	6g	Radix Glycyrrhizae

This was decocted with water and divided to be taken twice.

One year later, at a follow up visit, the patient stated her abdominal pain and diarrhoea never reoccured.

Wěi ruí (萎蕤 , Polygonatum odoratum) is also called *yù zhú* (Rhizoma Polygonati Odorati). Its taste and property are sweet and neutral. It enters the heart channel of the hand *shaoyin*, the lung channel of the hand *taiyin* and the stomach channel of the foot *yangming* meridians. It functions to nourish yin, moisten the lung, generate fluid and stop thirst. It is an herb to tonify deficiency and clear heat. The *Grand Materia Medica* lists it as a first class herb. Miao Zhong-chun said: *"Its property is gentle and mild. It can tonify and nourish."* It is best to use it for any of the lung dryness induced cough or the stomach heat induced severe thirst. Recently, animal laboratory tests proved that *yù zhú* (Rhizoma Polygonati Odorati) has the function to strengthen the heart, as well as improve an abnormal EKG that is due to insufficient blood supply to the cardiac muscle, and can increase the heart beat.

1. *Wěi Ruí* (Polygonatum Odoratum) ——
Gé gēn(葛根, Radix Puerariae Lobatae)

Function

To nourish yin, generate fluid, strengthen the heart and expand the coronary arteries.

Application

This can be used for rheumatic heart disease and coronary heart disease. It has functions to alleviate the clinical symptoms and improve the electrocardiogram report.

Compatibility Analysis

Wěi ruí (Polygonatum odoratum) contains cardiac glycosides. It functions to strengthen the heart and raise the blood pressure. *Gé gēn* (Radix Puerariae Lobatae) is sweet, acrid and neutral. It contains flavonoids and has the effect to expand the blood vessels in the heart and brain. These two herbs combined together can strengthen the heart and expand the coronary arteries as well as improve insufficient blood supply to the cardiac muscle. Meanwhile, *gé gēn* (Radix Puerariae Lobatae) expands the blood vessels and can lower the blood pressure. *Wěi ruí* (Polygonatum odoratum) strengthens the heart and can raise the blood pressure. These two herbs combined together can stabilise the blood pressure.

Case Study

Initial Visit

A female patient, 47 years old, April 24th, 1987.

The patient had rheumatism for more than twenty years. Ten years ago, she had symptoms of palpitations, lassitude and pain anterior to the heart area. She was diagnosed with rheumatic heart disease. After various treatments, it was sometimes better and sometimes worse. Recently, as her condition became severe, she came for treatments. Her EKG: III, aVF, T waves were low and flat, and there was high voltage on her left ventricle. Her chest x-ray showed: The heart had expanded and was leaning towards the lower left. The middle of the heart was concave; the arch of the ascending aorta had expanded to a boot shape. Her blood pressure was 180/100 mmHg. Her heart auscultation: From the aorta area a rough systolic murmur and diastolic murmur could be heard. At the heart apex a Ⅲ period systolic murmur could be heard. Her liver was enlarged, it could be palpated under the hypochondria and it was soft. Aside from the pain in the front of her heart area, she had shortness of breath and palpitations, with accompanying symptoms of anorexia, insomnia and headache, etc. Her tongue tip was red and the coating was thin and white. The biomedical diagnosis was: 1. Rheumatic heart disease, the left mitral valve of the left atrium and ventricle could not close fully, the pulmonary valve for the aorta was narrow and also could not close fully. 2. Chronic heart failure. 3. The cardiac muscle had an insufficient blood supply. The Chinese medicine diagnosis was chest impediment. This is both qi and yin deficiency induced blood stasis and qi stagnation syndrome.

Prescription

玉竹	yù zhú	30g	Rhizoma Polygonati Odorati
葛根	gé gēn	24g	Radix Puerariae Lobatae
丹参	dān shēn	24g	Radix et Rhizoma Salviae Miltiorrhizae
山楂	shān zhā	24g	Fructus Crataegi
乌药	wū yào	10g	Radix Linderae
百合	bǎi hé	20g	Bulbus Lilii
砂仁	shā rén	3g	Fructus Amomi
枳壳	zhǐ qiào	10g	Fructus Aurantii
白术	bái zhú	10g	Rhizoma Atractylodis Macrocephalae
佛手	fó shǒu	9g	Fructus Citri Sarcodactylis
党参	dǎng shēn	15g	Radix Codonopsis
麦门冬	mài mén dōng	15g	Radix Ophiopogonis
桂枝	guì zhī	9g	Ramulus Cinnamomi
甘草	gān cǎo	6g	Radix Glycyrrhizae

This was decocted with water and divided to be taken twice.

Second Visit

July 7th, 1987: After taking eight packs of the above formula, her appetite increased, and her insomnia and headache were eliminated. Her pain in the front of the heart area, palpitations and lassitude were also reduced a lot. She also could handle some light house-hold chores. Due to limited living resources, she delayed visiting in time. Recently, due to overexertion, she had symptoms of palpitations, lassitude, pain in the front of her heart area, and her blood pressure was slightly high. Again, the original formula was used with *dài zhě shí* (Haematitum) 20g, added, and she continued to take it.

Xióng huáng's (雄黄 , Realgar) taste and property are bitter and warm. It is toxic and is first recorded in the *Divine Husbandman's Classic of the Materia Medica* and listed as a middle grade herb. It functions to clear the turbid, detoxify toxins, dry dampness and kill parasites. In the *Divine Husbandman's Classic of the Materia Medica* it states that it can "......*kill pathogens and all kinds of insect's toxins.*" The chief content of this herb is arsenic disulphide (As2S2). The arsenic content is around 75% and sulphur is 24.9%. Laboratory observations proved that the concentration of xióng huáng (Realgar) at 1/100 has an inhibiting function on the growth of Mycobacterium tuberculosis (or Koch's bacillus) and Mycobacterium bovis'. As this herb contains arsenic, the oral dosage should be small.

Xióng Huáng (Realgar) ——
Zhū Shā (朱砂, Cinnabaris), *Liú Huáng* (硫黄, Sulfur), *Gān Suì* (甘遂, Radix Kansui), *Dà Suàn* (大蒜, Bulbus Allii)

Function

To kill parasites.

Application

This is used for pulmonary tuberculosis.

Compatibility Analysis

Xióng huáng (Realgar) kills parasites and contains arsenic. It has an inhibiting function on the pulmonary tuberculosis bacillus. Zhū shā (朱砂 , Cinnabaris) is sweet and slightly cold. It also functions to detoxify toxins and kill parasites. Liú huáng (硫黄 , Sulphur) is sour, warm and toxic. It also can kill parasites. Gān suí (甘遂 , Radix Kansui) is bitter, cold, and toxic. It enters the lung channel of the hand *taiyin*, the kidney channel of the foot *shaoyin* and the large intestine channel of hand *yangming* meridians. It can sedate the lung and expel water. Dà suàn (大蒜 , Bulbus Allii) is acrid and warm. It can kill parasites. All of the above mentioned herbs have a parasite killing function, therefore, they are effective for treating pulmonary tuberculosis.

Case Study

Initial Visit

A male patient, 29 years old, July 28th, 1985.

In 1982, due to symptoms of cough, fever and night sweating, he was investigated at a local hospital and diagnosed with right upper lung lobe tuberculosis. He had been treated with anti-tuberculosis medicine but the result was not significant. He was investigated again, in June 1984, and it was discovered that there was a tuberculosis cavity of 2cm × 3cm on his right upper lung lobe. In the beginning of 1985, he was hospitalised at Shijiazhuang a tuberculosis hospital, for three months treatment. His symptoms were still not reduced. He still had cough with a lot of phlegm, anorexia, a stifling sensation in his chest, lassitude, etc. After an x-ray he was diagnosed with spreading pulmonary tuberculosis cavities. On examination, the patient's face revealed chronic illness. His tongue was pale and his tongue coating was thin and white. His pulse was wiry, big and forceless. It was qi and yin deficiency induced pulmonary tuberculosis. The treatment plan was to tonify qi, nourish yin, moisten the lung and kill parasites.

Prescription

Internal use *Gōng Láo Bǎi Bù Tāng* (Folium Ilicis and Stemona Root Decoction):

功劳叶	*gōng láo yè*	30g	Folium Ilex
百部	*bǎi bù*	10g	Radix Stemonae
白及	*bái jí*	12g	Rhizoma Bletillae
川贝母	*chuān bèi mǔ*	9g	Bulbus Fritillariae Cirrhosae
桔梗	*jié gěng*	9g	Radix Platycodonis
地榆	*dì yú*	12g	Radix Sanguisorbae
太子参	*tài zǐ shēn*	15g	Radix Pseudostellariae
陈皮	*chén pí*	9g	Pericarpium Citri Reticulatae
生黄芪	*shēng huáng qí*	20g	Radix Astragali (raw)
茯苓	*fú líng*	15g	Poria
甘草	*gān cǎo*	6g	Radix Glycyrrhizae

External use *Kàng Láo Wán* (Anti-Consumption Pill, 抗痨丸):

雄黄	*xióng huáng*	3.5g	Realgar
朱砂	*zhū shā*	3g	Cinnabaris
硫黄	*liú huáng*	2.5g	Sulphur
甘遂	*gān suì*	3.5g	Radix Kansui

The above herbs were all ground together into a fine powder, the purple skin of *dà suàn* (Bulbus Allii) was added (the single clove of garlic is the best). This was beaten together to be a sludge like mud. If beating does not produce mud, add some alcohol in, to make a ball-shaped herbal pill.

Application

The doctor used his right thumb, index and middle fingers to hold the herbal pill, on the patient's back slowly moving from DU1 *cháng qiáng* (长强) up to DU14 *dà zhuī* (大椎). The width was 9 ~ 12cm. The moving should be uniform and should be carried out not less than 200 times. Before this treatment, a warm and damp towel was used to clean the back in order to improve the herbs' penetration and absorption. After applying the herb, it should not be washed out immediately. Gauze should be used to cover it well. After 6 ~ 8 hours wash it out. Apply once a week, four times as a course. Rest in between treatments and then start the second treatment course again.

Second Visit

December 29th: The patient was treated for a total of five months. All of his symptoms were eliminated. His chest x-ray reported: The tuberculosis on his right upper lung lobe had already calcified and been absorbed, the hole had become a fibroid.

Note

For treating pulmonary tuberculosis, external use of *Kàng Láo Wán* (Anti-Consumption Pill) has better results for improving appetite and improving symptoms. If possible, add some *shè xiāng* (Moschus) to the formula, in order to increase the permeability, and the treatment effectiveness will be even better. Most herbs of this formula are toxic and should not be taken orally.

Chán tuì's (蝉蜕 , Periostracum Cicadae) taste and property are salty, sweet and cold. It enters the lung channel of the hand *taiyin* and the liver channel of the foot *jueyin* meridians. It functions to expel wind-heat, disperse the lung, vent out rashes, extinguish wind and calm convulsions. Its qi is light and clear. It enters and opens the lung, and expels wind-heat in order to vent exterior pathogens out. As with the *Chán Tuì Săn* (Cicada Moulting Powder, 蝉蜕散) (*chán tuì*, Periostracum Cicadae; *bò hé*, Herba Menthae) in the *Shen's Respect for Life* (沈氏尊生 , *Shĕn Shì Zūn Shēng*), it enters the liver channel of the foot *jueyin* meridian, it also can extinguish wind and calm the shock, treat tetanus and little children's fever and convulsiosn, and with the *Five Tigers Extinguishing Wind Powder* (*Wŭ Hŭ Zhuī Fēng Săn*, 五虎追风散) (*chán tuì*, Periostracum Cicadae; *quán xiē*, Scorpio; *dăn nán xīng*, Arisaema cum Bile; *tiān má*, Rhizoma Gastrodiae; *jiāng cán*, Bombyx Batryticatus) in the *Historical Prescription of the En Family* (史传恩家传方 , *Shĭ Chuán Ēn Jiā Chuán Fāng*). Moreover, *chán tuì* (Periostracum Cicadae) also can treat lung heat inducing hoarse voice, as in the *Hăi Chán Săn* (Sterculia and Cicada Moulting Powder, 海蝉散) (*chán tuì*, Periostracum Cicadae; *pàng dà hăi*, Semen Sterculiae Lychnophora). *Bò He Tāng* (Mint Decoction, 薄荷汤) mixed with *chán tuì* (Periostracum Cicadae) can treat small children's crying at night and is called *Chán Huā Săn* (Cordyceps Cicadae Powder, 蝉花散). They are all lesser herbs but effective formulae and compatibilities.

1. *Chán Tuì* (Periostracum Cicadae) ——
Dì Fū Zĭ (地肤子, Fructus Kochiae)

Function

To expel wind and clear heat, eliminate dampness and stop itching.

Application

This is used for German measles or eczema with itching skin. It is also used in every kind of allergic disease, and can reduce nephritis induced albuminuria.

Compatibility Analysis

Chán tuì (Periostracum Cicadae) is also called *chán yī* (Periostracum Cicadae). Its quality is light, its qi is thin, its power to clear and expel wind-heat, vent the exterior and disperse the lung is good. It is beneficial for treating skin itching due to wind. The taste and property of *dì fū zĭ* (地肤子 , Fructus Kochiae) are bitter and cold. It has functions to clear damp-heat, detoxify toxins and treat rashes. Its tastes are bitter and cold and its property is descending. It can enter the urinary bladder channel of the foot *taiyang* meridian to unblock and promote the urination. It also can detoxify toxins and eliminate dampness to treat itching around the body. The *Miscellaneous Records of Famous Physicians* states that it can *"eliminate the heat qi in the skin, make people lustrous, and expel bad ulcers ......" Chán tuì* (Periostracum Cicadae) and *dì fū zĭ* (Fructus Kochiae) combined together can enter the lung to disperse and expel wind-heat to let toxic pathogens be expelled out from the skin. They also can enter the urinary bladder to eliminate dampness and promote urination to leach out the dampness pathogen with the urine. Moreover, the urinary bladder is the *taiyang* meridian, it also governs the skin, therefore, *chán tuì* (Periostracum Cicadae) combined with *dì fū zĭ* (Fructus Kochiae) can eliminate the wind-heat and damp-pathogen from the skin or the urinary bladder.

It can be used for skin itchiness with any kind of wind-damp-heat toxin induced ulcers. The author has experienced that this compatibility has a desensitisation function. It is effectively used for sensitivity dermatitis and urticaria, moreover, it is also used for chronic nephritis induced albuminuria. Whether they have similar functions to hormones or not is pending further research.

Case 1

Skin pruritus.

Initial Visit

A male patient, 43 years old, July 27th, 1987.

The patient's lower limbs and lower abdomen were itching and hard to bear. If scratched, then granular red rashes appeared. The symptoms were already two years old. Every spring (around March 21st) his condition became gradually worse and every winter (around November 7th), his condition gradually reduced. This time, he had had the itchiness for a couple of months already. He was treated with medicines like chlorphenamine, diphenhydramine and cyprohetadine, etc. but none gave significant results. His skin was still itching especially in the night and made it difficult to sleep. Later, through his friend's referral, he came to our hospital for treatments. The patient had good physical development and his body was healthy. Aside from skin itchiness, he had no other illness. On his lower limbs and lower abdomen, there were scattered rashes with scratched scars. The rashes were less on the upper limbs. His skin was rough without being shiny. His tongue was red and his tongue coating was thin, white and slightly slippery. His pulse was deficient, fast and slightly wiry. This was wind-heat toxin pathogen stagnated in the muscles and skin. The prolonged case meant it injured the collaterals and caused the blood to overflow to the skin inducing the rashes. The wind-heat toxin pathogen stagnated in the exterior and injured the collaterals producing the itchiness. The pathogens stagnated and the collaterals were injured, when this was prolonged then there would be blood deficiency, therefore, his skin was lustreless. The treatment plan was to nourish blood, expel wind, invigorate blood and detoxify toxins.

Prescription

防风	*fáng fēng*	12g	Radix Saposhnikoviae
地肤子	*dì fū zǐ*	15g	Fructus Kochiae
蝉蜕	*chán tuì*	10g	Periostracum Cicadae
桃仁	*táo rén*	9g	Semen Persicae
红花	*hóng huā*	9g	Flos Carthami
赤芍	*chì sháo*	15g	Radix Paeoniae Rubra
双花	*shuāng huā*	24g	Flos Loniocerae Japonicae
何首乌	*hé shǒu wū*	24g	Radix Polygoni Multiflori
丹参	*dān shēn*	24g	Radix et Rhizoma Salviae Miltiorrhizae
甘草	*gān cǎo*	9g	Radix Glycyrrhizae

This was decocted with water and divided to be taken twice, one pack per day.

Second Visit

July 31st :After taking four packs of the above formula, his itchiness was reduced and he could sleep at night. Due to its effectiveness, the prescription was not changed and he continued to take it again.

Third Visit

August 4th: After taking another four packs, his itchiness was mild. However, if scratched, there were still red rashes. Using the above formula, *chuān huáng bǎi* (Cortex Phellodendri Chinensis) 9g, and *cāng zhú* (Rhizoma Atractylodis) 10g, were added in order to increase the power to expel damp-heat, and he continued taking it but reduced it to one pack, every other day.

Fourth Visit

August 19th: After taking seven packs of the above formula, his symptoms basically disappeared. Occasionally, there was mild itching. He was instructed to take the same formula for another seven packs.

Later, the next spring, at a follow up visit, there was no reoccurrence.

Case 2

Urticaria.

Initial Visit

A female patient, 41 years old, June 19th, 1995.

The patient had whole body itchiness for more than a month. After treatments, it was repeated without recovery and so she came to visit. On examination, the patient had red wheals. Their sizes and shapes were not identical, raised off the skin and flat-topped. Sometimes, the itchiness was hard to bear and was accompanied by nausea, abdominal pain, and stools with blood. Her tongue was red and tongue coating was white. Her pulse was floating, fast and wiry. Wind-heat stagnation induced it. The wind-heat pathogen overflowed to the exterior then came out in the rashes with itchiness. The lung and large intestine are internally-externally related. Wind-heat pushed internally; therefore, she had the abdominal pain and nausea. The heat pathogen pushed blood to move restlessly so she had the bloody stool. The treatment plan was to clear heat, expel wind, cool the blood and release the exterior.

Prescription

荆芥	*jīng jiè*	10g	Herba Schizonepetae
防风	*fáng fēng*	10g	Radix Saposhnikoviae
地肤子	*dì fū zǐ*	15g	Fructus Kochiae
蝉蜕	*chán tuì*	10g	Periostracum Cicadae
丹参	*dān shēn*	20g	Radix et Rhizoma Salviae Miltiorrhizae
葛根	*gé gēn*	24g	Radix Puerariae Lobatae
侧柏叶	*cè bǎi yè*	15g	Cacumen Platycladi

当归	*dāng guī*	15g	Radix Angelicae Sinensis
赤芍	*chì sháo*	12g	Radix Paeoniae Rubra
川芎	*chuān xiōng*	12g	Rhizoma Chuanxiong
陈皮	*chén pí*	9g	Pericarpium Citri Reticulatae
大黄	*dà huáng*	5g	Radix et Rhizoma Rhei
甘草	*gān cǎo*	6g	Radix Glycyrrhizae

This was decocted with water and divided to be taken twice, once each in the morning and evening, one pack per day.

Second Visit

June 21st: After taking three packs of the above formula, her urticaria did not rise up again. Her bloody stool stopped and her itchiness was eliminated but her stomach and abdominal distention and nausea still existed. Her pulse became slower and tranquil. The formula was changed to harmonise the stomach and regulate qi as well as expel wind and detoxify toxins.

Prescription

百合	*bǎi hé*	30g	Bulbus Lilii
乌药	*wū yào*	12g	Radix Linderae
香附	*xiāng fù*	12g	Rhizoma Cyperi
砂仁	*shā rén*	5g	Fructus Amomi
檀香	*tán xiāng*	9g	Lignum Santali Albi
丹参	*dān shēn*	20g	Radix et Rhizoma Salviae Miltiorrhizae
陈皮	*chén pí*	9g	Pericarpium Citri Reticulatae
半夏	*bàn xià*	6g	Rhizoma Pinelliae
蝉蜕	*chán tuì*	9g	Periostracum Cicadae
地肤子	*dì fū zǐ*	12g	Fructus Kochiae
荆芥	*jīng jiè*	10g	Herba Schizonepetae
柴胡	*chái hú*	15g	Radix Bupleuri
葛根	*gé gēn*	20g	Radix Puerariae Lobatae
甘草	*gān cǎo*	5g	Radix Glycyrrhizae

This was decocted with water to be taken as before.

June 25th: The patient's family member came and said that after taking three packs of the above formula, the patient's symptoms were all eliminated and she did not take the formula again.

2. *Chán Tuì* (Periostracum Cicadae) —— *Jiāng Cán* (僵蚕, Bombyx Batryticatus)

Function

To soothe and expel wind-heat, benefit the throat and transform phlegm, and to clear heat and resolve spasm.

Application

This is used for small children's convulsions or external wind-heat induced swollen and painful throat, hoarse voice, or tonsillitis. It is also used for chronic pharyngitis inducing a choking cough or after an upper respiratory tract infection sequellae dry cough, throat itchiness, etc.

Compatibility Analysis

Chán tuì (Periostracum Cicadae) is salty, sweet and cold. Its functions are beneficial to clear heat, disperse the lung, stabilise convulsions and benefit the throat. *Jiāng cán* (Bombyx Batryticatus) is salty, acrid and neutral. Its functions are specialised to expel wind, resolve spasm, transform phlegm and expel clumps. Su Song said that it can "*treat throat bi after an almost fatal stroke. Once it gets into the throat, it recovers immediately.*" In the *Grand Materia Medica* it states that it can "*expel wind, phlegm, clumps and goitre, wind in the head...*" The qi of *chán tuì* (Periostracum Cicadae) and *jiāng cán* (Bombyx Batryticatus) are both thin, light, and clear and can float upward. *Chán tuì* (Periostracum Cicadae) disperses the lung, benefits the throat and can help *jiāng cán* (Bombyx Batryticatus) to transform phlegm and expel clumps. If the phlegm is transformed and the clump is expelled, it also benefits the dispersing and descending of the lung qi to free the passage through to the throat. These two herbs combined together are mutually enhancing and their effects to benefit the throat and expel clumps are even better. These two herbs also both enter the liver channel of the foot *jueyin* meridian and are beneficial for expelling wind, clearing heat and resolving convulsions, therefore, they have good results for small children's fever and convulsions.

Case 1

Initial Visit

A female patient, 3 years old, September 15th, 1998.

The child had an exterior invasion and fever for five days accompanied by a sore throat and dry cough. She had been treated with ready-made medicines both Chinese medicine and biomedicine but they did not work, and so she came for treatment. On physical examination, it was discovered that the child had fever, a red throat, a body temperature 38.5℃ , her tonsils were slightly big and red, no dry or wet rales were heard from her lungs, but her breathing sounds were slightly rough. Her routine blood test results were all normal. Her pulse was thready and fast. Her finger vein was red and floating. Her tongue coating was thin and white. This was exterior wind-heat invasion with lung heat being unable to disperse syndrome. The treatment plan was to use a formula to soothe and disperse the wind-heat, benefit the throat and stop cough.

Prescription

| 蝉蜕 | *chán tuì* | 8g | Periostracum Cicadae |
| 僵蚕 | *jiāng cán* | 8g | Bombyx Batryticatus |

桑叶	*sāng yè*	10g	Folium Mori
菊花	*jú huā*	10g	Flos Chrysanthemi
杏仁	*xìng rén*	6g	Semen Armeniacae Dulcis
桔梗	*jié gěng*	6g	Radix Platycodonis
薄荷	*bò he*	6g	Herba Menthae
芦根	*lú gēn*	10g	Rhizoma Phragmitis
板蓝根	*bǎn lán gēn*	10g	Radix Isatidis
川贝母	*chuān bèi mǔ*	8g	Bulbus Fritillariae Cirrhosae
百部	*bǎi bù*	8g	Radix Stemonae
甘草	*gān cǎo*	6g	Radix Glycyrrhizae

This was decocted with water and divided, to be taken three times, one pack per day.

Second Visit

September 17th: After taking two packs of the above formula, her body temperature became normal. Her dry cough and sore throat were also reduced a lot. She continued taking the above formula.

Third Visit

September 20th: After taking another three packs of the above formula, all of her symptoms disappeared.

Case 2

Initial Visit

A female patient, 35 years old, September 17th, 1996.

The patient had a medical history of chronic pharyngitis for a couple of years. Recently, because of her job, she was tired and also had flu and a sore, dry throat with a persistent choking cough. After taking medicines of fast-acting flu capsules and sarcandra tablets, the exterior disease symptoms and sore throat had some relief, but her dry throat, itching throat and persistent choking cough were not reduced and so she came for treatment. On physical examination, the patient had a red throat, her pharynx and larynx wall was red and was accompanied by lymph follicle hypertrophy. Her pulse was wiry, thready and fast. Her tongue coating was thin and white and her tongue was red. This was induced by lung heat stagnated and tied without dispersing and prolonged fluid injury. The treatment plan was to clear and disperse the lung heat, nourish the yin and benefit the throat.

Prescription

蝉蜕	*chán tuì*	10g	Periostracum Cicadae
僵蚕	*jiāng cán*	10g	Bombyx Batryticatus
桔梗	*jié gěng*	10g	Radix Platycodonis
百部	*bǎi bù*	15g	Radix Stemonae
麦门冬	*mài mén dōng*	15g	Radix Ophiopogonis

荆芥	jīng jiè	10g	Herba Schizonepetae
杏仁	xìng rén	9g	Semen Armeniacae Dulcis
川贝母	chuān bèi mǔ	9g	Bulbus Fritillariae Cirrhosae
黄芩	huáng qín	9g	Radix Scutellariae
桑白皮	sāng bái pí	12g	Cortex Mori
紫菀	zǐ wǎn	9g	Radix et Rhizoma Asteris
甘草	gān cǎo	6g	Radix Glycyrrhizae

This was decocted with water and divided to be taken twice, one pack per day.

Second Visit

September 21st: After taking four packs of the above formula, her choking cough was significantly improved and her dry throat and itching throat were also reduced. Again, she continued to take the same formula.

Third Visit

September 26th: After taking another five packs, her choking cough stopped. Her dry and itching throat were also eliminated. She was instructed to take three more packs to clear it up.

Màn jīng zǐ's (蔓荆子 , Fructus Viticis) tastes and property are acrid, bitter and slightly cold. It enters the liver channel of the foot *jueyin*, the lung channel of the hand *taiyin* and the urinary bladder channel of the foot *taiyang* meridians. It functions to expel wind, clear heat and alleviate pain. As its taste is acrid and its body is light and floating, it can expel wind in the head, and benefit the nine orifices. Its property is slightly cold, therefore, it also can brighten the eyes, cool blood and clear wind-heat. The *Divine Husbandman's Classic of the Materia Medica* lists it as a top grade herb and states that it can "*brighten the eyes and benefit the nine orifices*". Zhang Yuan-su said that it can treat "*taiyang headache, heaviness of the head, dizziness and stuffy sensations...*" The classic formula *Jú Huā Sǎn* (Chrysanthemun Flower Powder, 菊花散) treats wind in the head and headache. The *Chán Huā Sǎn* (Cordyceps Cicadae Powder) treats dizziness, red and swollen eyes. They all use *màn jīng zǐ* (Fructus Viticis). Moreover, *màn jīng zǐ* (Fructus Viticis) also can treat pain from wind-damp bi, using its expelling wind and alleviating pain functions, but in clinical practice, it is less used.

1. *Màn Jīng Zǐ* (Fructus Viticis) —— *Chōng Wèi Zǐ* (茺蔚子, Fructus Leonuri)

Function

To clear the head and brighten the eyes, expel wind and alleviate pain.

Application

This is used for exterior wind-heat induced headache, dizziness, and red, swollen and painful eyes. It is also used for high blood pressure induced headache.

Compatibility Analysis

Màn jīng zǐ (Fructus Viticis) is acrid, bitter and slightly cold. Acridity can disperse, cold can clear. It specialises in soothing and dispersing the wind-heat. It is more effective for treating wind-heat in the head or on the face. *Chōng wèi zǐ* (茺蔚子 , Fructus Leonuri) is the seed of *yì mǔ cǎo* (Herba Leonuri). Its functions are the same as *yì mǔ cǎo* (Herba Leonuri). It functions to invigorate blood, transform blood stasis, eliminate blood stasis and generate new blood. Its taste is sweet, therefore, within its transforming blood stasis function; it also can tonify the heart, nourish the liver and brighten the eyes. It is really a gentle transforming blood stasis herb. All seeds can descend therefore, *chōng wèi zǐ* (Fructus Leonuri) also can descend qi and move blood. When these two herbs are combined together, *màn jīng zǐ* (Fructus Viticis) expels wind, and clears heat and *chōng wèi zǐ* (Fructus Leonuri) nourishes the liver, brightens the eyes, descends qi, and invigorates blood. If one wants to treat wind, one should treat blood first. *Màn jīng zǐ* (Fructus Viticis) helps *chōng wèi zǐ* (Fructus Leonuri) to transform blood stasis and invigorate blood, its power to expel wind, and clear heat pathogens in the blood level is even stronger. If wind-heat could not be expelled, then blood will flow rebelliously without descending therefore, *chōng wèi zǐ* (Fructus Leonuri) helps *màn jīng zǐ* (Fructus Viticis) to clear and disperse wind-heat, its strength to descend rebellions and transform blood stasis is even bigger. These two herbs combined together are mutually enhancing.

They are effective for treating wind-heat induced headache, dizziness, red eyes and high blood pressure induced headache. In clinical practice, they also can be used for hyperthyroidism patients with protruding eyes, to help improve their clinical symptoms.

Case Study

Initial Visit

A female patient, 46 years old, April 30th, 1987.

The patient had a medical history of high blood pressure inducing headache for a couple of years. Recently due to an exterior invasion, she had headache and head distention; her head felt like it might explode; she had dizziness, red eyes, irritability and heat in her chest, a dry mouth, and her body was hot with slight aversion to cold. Her pulse was floating, wiry and fast. Her tongue was red and the coating was thin and white. Her blood pressure was 160/95 mmHg. This was an exterior pathogen attacking exteriorly with the interior heat being unable to disperse syndrome. Due to the external wind-heat invasion, she had body heat and aversion to cold with a floating pulse. The wind-heat had stagnated inside so she had a dry mouth and irritability with heat inside her chest. Heat is a yang pathogen, its property is flaring upward, therefore, she had dizziness, red eyes, and head distention with explosive pain. The treatment plan was to soothe and disperse the wind-heat, to clear the head and brighten the eyes.

Prescription

蔓荆子	*màn jīng zǐ*	20g	Fructus Viticis
茺蔚子	*chōng wèi zǐ*	30g	Fructus Leonuri
菊花	*jú huā*	15g	Flos Chrysanthemi
薄荷	*bò he*	9g	Herba Menthae
金银花	*jīn yín huā*	20g	Flos Lonicerae Japonicae
川芎	*chuān xiōng*	12g	Rhizoma Chuanxiong
荷叶	*hé yè*	9g	Folium Nelumbinis
天麻	*tiān má*	12g	Rhizoma Gastrodiae
芦根	*lú gēn*	30g	Rhizoma Phragmitis
甘草	*gān cǎo*	6g	Radix Glycyrrhizae

This was decocted with water and divided to be taken twice.

Second Visit

May 4th: After taking four packs of the above formula, her headache reduced a lot. Her body heat was eliminated but she still had irritability and heat, with dizziness sometimes. Using the above formula, *huáng qín* (Radix Scutellariae) 9g, was added and she continued taking it.

Third Visit

May 7th: After taking another three packs, all of her symptoms basically disappeared. Her blood pressure was 130/85mm Hg. She was instructed to take blood reducing medicine often to reinforce the treatment.

Yi yǐ rén's (薏苡仁 , Semen Coicis) taste and property are sweet, bland and slightly cold. It enters the spleen channel of the foot *taiyin*, the stomach channel of the foot *yangming* and the lung channel of the hand *taiyin* meridians. It functions to clear heat, resolve dampness, strengthen the spleen, eliminate bi syndrome and drain pus. It can be used for oedema and athletes foot, spleen deficiency induced diarrhoea and wind-damp induced impediment syndrome. It is also an important herb to treat internal ulcers. If used raw, it can resolve dampness, clear heat and drain pus. If used fried, it can treat spleen deficiency induced diarrhoea. It is a commonly used herb in clinical practices, however its property is gentle and slow. If used in a decoction, its dosage needs to be large in order to get effect.

1. *Shēng Yì Rén* (raw Semen Coicis) ——
Táo Rén (桃仁, Semen Persicae)

Function

To clear heat, drain pus, reduce swelling and expel clumps.

Application

It can be used for internal ulcers and inflammatory clumps such as with appendicitis and woman's pelvic cavity inflammatory clumps, etc.

Compatibility Analysis

The property of *shēng yì rén* (raw Semen Coicis) is cold, it can clear heat. Its tastes are sweet and bland, it can leach out dampness and promote urination. It is beneficial for treating internal ulcers and functions to drain pus, eliminate abscesses and expel clumps. *Táo rén* (Semen Persicae) is bitter, sweet and neutral. It is a blood level herb for the liver channel of the foot *jueyin* meridian. It functions to invigorate blood and expel blood stasis. In the *Divine Husbandman's Classic of the Materia Medica* it states that it *"governs blood stasis and blocked blood flow, abdominal abscesses and pathogenic qi."* Li Dong-yuan said that: *"Táo rén* (Semen Persicae) *is bitter more than sweet. Its qi is thin and its taste is thick. It sinks and descends. Bitterness can sedate stagnated blood. Sweetness can generate new blood; therefore, it can be used to break stagnated blood."* It explains that *táo rén* (Semen Persicae) can generate new blood as well as breaking stagnation and expelling blood stasis. It has eliminating blood stasis without injuring the normal qi characteristics. Though *shēng yì rén* (raw Semen Coicis) is beneficial for reducing swellings and expelling clumps, its strength is slow and gentle. If it has *táo rén* (Semen Persicae) to help, its power to reduce and expel can be enhanced. Meanwhile, *shēng yì rén* (raw Semen Coicis) can strengthen the spleen and *táo rén* (Semen Persicae) can eliminate blood stasis and generate new blood. These two herbs combined together, have the characteristics to reduce abscesses, drain pus without injury to the blood, expel clumps, and eliminate blood stasis without injury to the normal qi.

Case 1

Initial Visit

A male patient, 15 years old, March 25th, 1995.

Ten days ago, the patient suddenly had lower right abdominal pain accompanied by stomach and abdominal distention, anorexia, nausea, fever, etc. The local hospital treated him for appendicitis and injected penicillin for a week. His fever and nausea, etc. were eliminated but the lower right abdominal pain was not reduced. The treatment was changed to cephalosporin intravenous injection for three days but his abdominal pain was the same as before and so he came to our hospital for treatments. On physical examination, the patient's right lower abdomen had significant rebound tenderness at the McBurney's point, and a tumour could be palpated there. His B-ultrasound examination showed: There was an inflammatory clump in his right lower abdomen around the appendix area. Its size was around 4.5cm × 4.0cm. A routine blood test showed: WBC 14.0×10^9/L, neutrophils 0.80. His pulse was wiry and fast. His tongue was red and his tongue coating was slightly turbid. The patient did not want to have surgery and wanted to use Chinese medicine for conservative treatment. The treatment plan was to invigorate the blood, transform the blood stasis, reduce swelling and expel the clumping.

Prescription

生薏米	shēng yì mǐ	30g	Semen Coicis (raw)
桃仁	táo rén	10g	Semen Persicae
牡丹皮	mǔ dān pí	10g	Cortex Moutan
延胡索	yán hú suǒ	15g	Rhizoma Corydalis
赤芍	chì sháo	30g	Radix Paeoniae Rubra
白芍	bái sháo	30g	Radix Paeoniae Alba
大黄	dà huáng	6g	Radix et Rhizoma Rhei
甘草	gān cǎo	6g	Radix Glycyrrhizae

This was decocted with water and divided to be taken twice, once each in the morning and evening on an empty stomach, one pack per day.

Second Visit

April 1st: After taking seven packs of the above formula, his pain reduced a lot. On palpation, there was still pressing tenderness and a clump could be felt. His appetite was normal and his routine blood test was also normal. He continued taking the original formula again.

Third Visit

April 7th: After taking another seven packs, his abdominal pain was eliminated. There was no tenderness on pressure on his right lower abdomen. The B-ultrasound examination showed that the tumour had basically disappeared. He was instructed to take five packs more in order to strengthen the treatment effects.

Case 2

Initial Visit

A female patient, 25 years old, January 5th, 1987.

Since last August, the patient had felt lower abdominal pain and it was aggravated on exertion. The pain had become more severe over the previous month. She had been examined in an hospital's Gynaecology Department and was diagnosed with adnexitis. After treatment with medicine had no effect she came to visit. On physical examination, the patient had good physical development. There was significant tenderness when pressing on her left lower abdomen and a rectangular tumour, similar to a thumb, could be palpated. Her routine blood test was normal. Accompanying symptoms included lots of leukorrhoea, low back soreness, with pain and heaviness sensations. Her pulse was wiry and thready. Her tongue was dark and her tongue coating was thin and yellow. This was the blood stasis and abdominal abscess syndrome. The treatment plan was to invigorate blood, transform the blood stasis, eliminate the abscess and expel the clumping.

Prescription

生薏米	*shēng yì mǐ*	30g	Semen Coicis (raw)
桃仁	*táo rén*	12g	Semen Persicae
丹参	*dān shēn*	24g	Radix et Rhizoma Salviae Miltiorrhizae
当归	*dāng guī*	12g	Radix Angelicae Sinensis
赤芍	*chì sháo*	10g	Radix Paeoniae Rubra
桂枝	*guì zhī*	9g	Ramulus Cinnamomi
茯苓	*fú líng*	15g	Poria
车前子	*chē qián zǐ*	12g	Semen Plantaginis (wrapped up)
香附	*xiāng fù*	15g	Rhizoma Cyperi
黄柏	*huáng bǎi*	9g	Cortex Phellodendri Chinensis
苍术	*cāng zhú*	10g	Rhizoma Atractylodis
甘草	*gān cǎo*	6g	Radix Glycyrrhizae

This was decocted with water and divided to be taken twice, warm, one pack per day.

Second Visit

January 9th: After taking four packs of the above formula, her abdominal pain was slightly reduced. Her leukorrhoea was also reduced. The rest of the symptoms were the same as before. Using the original formula, herbs to tonify the kidney and qi were added to treat her.

Prescription

生薏米	*shēng yì mǐ*	30g	Semen Coicis (raw)
桃仁	*táo rén*	12g	Semen Persicae
丹参	*dān shēn*	20g	Radix et Rhizoma Salviae Miltiorrhizae
当归	*dāng guī*	12g	Radix Angelicae Sinensis
土鳖虫	*tǔ biē chóng*	9g	Eupolyphaga seu Steleophaga

赤芍	*chì sháo*	10g	Radix Paeoniae Rubra
茯苓	*fú líng*	15g	Poria
桂枝	*guì zhī*	9g	Ramulus Cinnamomi
川断	*chuān duàn*	15g	Radix Dipsaci
杜仲	*dù zhòng*	20g	Cortex Eucommiae
延胡索	*yán hú suǒ*	12g	Rhizoma Corydalis
川牛膝	*chuān niú xī*	15g	Radix Cyathulae
甘草	*gān cǎo*	6g	Radix Glycyrrhizae

This was decocted with water, to be taken as before.

Third Visit

January 17th: After taking seven packs of the above formula, her lower back soreness and pain disappeared and her abdominal pain also reduced a lot. On palpation, her tumour was significantly smaller than before. Using the original formula, *yán hú suǒ* (Rhizoma Corydalis) was eliminated and *bái zhú* (Rhizoma Atractylodis Macrocephalae) 12g, was added. She continued taking it.

Fourth Visit

January 24th: After taking another seven packs, her abdominal pain disappeared and the tumour could not be clearly palpated. However, there was still tenderness on pressure over the tumour. She was instructed to take another seven packs.

Later the patient stated that she had recovered.

2. *Shēng Yì Rén* (raw Semen Coicis) —— *Shān Jiǎ* (山甲, Squama Manis)

Function

To invigorate blood, unblock the collaterals, eliminate swelling and expel clumps, and eliminate blood stasis and drain pus.

Application

This is used for prostatitis with frequent urination, urgent urination, pain at the beginning or at the end of urination or when the bowels move, or when there is white fluid (prostatic fluid) excreted out from the urinary tract or dysuria, or when the perineum area and lower abdomen have pain and a sinking sensation or accompanied with spermatorrhoea, impotence, premature ejaculation, bloody sperm or reduced libido, etc. It also can be used for adnexitis.

Compatibility Analysis

Shēng yì rén (raw Semen Coicis) is sweet, bland and slightly cold. It can clear heat and eliminate dampness, drain pus and expel clumps. *Chuān shān jiǎ* (Squama Manis) is salty and slightly cold. It enters the liver channel of the foot *jueyin* and the stomach channel of the foot *yangming* meridians. Its property is free moving. It functions to unblock the meridians, invigorate blood, eliminate swelling and drain pus. The *Essentials of the Materia Medica* states that it can:"*treat wind-damp-cold bi, unblock the collaterals and promote lactation, eliminate swelling and broken abscesses, stop pain and drain pus.*" These two herbs combined together can invigorate blood, unblock the collaterals, expel blood stasis and drain pus, therefore, it is effective to treat prostatitis.

Case Study

Initial Visit

A male patient, 41 years old, June 4th, 2000.

The patient had lower abdomen (above the pubis) distention and pain; and urinary tract dryness and coarseness; and pain for a year. He was treated at a local hospital without effect and so came to visit.

Last year, after he slept outside the house, due to hot weather, he had frequent and urgent urination. He thought that he had caught cold and used *shēng jiāng* (Rhizoma Zingiberis Recens) and sugar juice to treat it without effect. Later, he was treated with antibiotics by a local doctor for a urinary tract infection. The symptoms were slightly relieved. Last autumn, he went out to work and due to overexertion, his symptoms got worse. Although he had treatment, the result was not significant. He had a burning pain in his urinary tract as well as soreness, distention and pain around his perineum and groin. After urination, white urine leaked out and he also had low libido.

On physical examination, the patient had good physical development. A routine urine test showed: Occasional WBCs visible. The B-ultrasound report for his prostate stated: His prostate was slightly enlarged. His prostatic fluid: WBC 12 ~ 15 /HP, pus cell ++. His pulse was wiry, slippery and fast. His tongue was dark and the coating was thin and white. His biomedicical diagnosis was prostatitis. The Chinese medicine diagnosis was unctuous strangury. The syndrome differentiation was: Damp-heat flowing downward, invading the testes. The damp-heat could not disperse or be excreted so it injured the qi and blood. This induced blood stagnation and clumping internally. Sperm could not be bound so they overflowed externally. The treatment plan was to clear the damp-heat, unblock the meridians and transform the blood stasis.

Prescription

生薏仁	*Shēng yì rén*	30g	Semen Coicis (raw)
川牛膝	*chuān niú xī*	20g	Radix Cyathulae
穿山甲	*chuān shān jiǎ*	10g	Squama Manis (fried)
苍术	*cāng zhú*	15g	Rhizoma Atractylodis
黄柏	*huáng bǎi*	12g	Cortex Phellodendri Chinensis
金银花	*jīn yín huā*	20g	Flos Lonicerae Japonicae
丹参	*dān shēn*	20g	Radix et Rhizoma Salviae Miltiorrhizae
当归	*dāng guī*	15g	Radix Angelicae Sinensis
桃仁	*táo rén*	9g	Semen Persicae
牡丹皮	*mǔ dān pí*	10g	Cortex Moutan
甘草	*gān cǎo*	6g	Radix Glycyrrhizae

This was decocted with water, one pack per day and divided to be taken twice.

Second Visit

June 11th: After taking seven packs of the above formula, his urinary tract pain, frequent and urgent urination, etc. all were visibly reduced. Using the original formula, *pú gōng yīng* (Herba Taraxaci) 5g, and *fú líng* (Poria) 15g, were added and he continued to take it.

Third Visit

June 20th: After taking seven packs of the above formula, his frequent and urgent urination were eliminated and his urinary tract pain was not so obvious. His distention and pain sensations around his perineum area also disappeared. He kept taking the original formula again.

Fourth Visit

June 27th: His symptoms basically disappeared. His routine urine test was normal. His prostatic fluid test showed a few white blood cells still, but the pus cells disappeared. He was instructed to take the original formula every other day and continue for one month.

Later on, he stated that he had recovered.

3. *Shēng Yì Rén* (raw Semen Coicis) —— *Chuān Liàn Zǐ* (川楝子, Fructus Toosendan)

Function

To nourish the sinews, soften the liver, expel clumps and alleviate pain.

Application

This is used for lower abdominal pain and testes pain.

Compatibility Analysis

Shēng yì rén (raw Semen Coicis) is sweet and bland. It functions to resolve dampness, eliminate swelling and expel clumps. Meanwhile, it also softens the liver and nourishes sinews. The *Divine Husbandman's Classic of the Materia Medica* states that it *"governs tension and spasm of the sinews that cannot flex or extend."* *Chuān liàn zǐ* (Fructus Toosendan) is bitter and cold. It enters the liver channel of the foot *jueyin* meridian. It has the strongest power to resolve dampness and alleviate pain. The *Essentials of the Materia Medica* states that it *"can enter the liver channel of the foot jueyin meridian to soothe sinews......It is an important herb to treat hernias."* It, combined with *yán hú suǒ* (Rhizoma Corydalis), is called *Jīn Líng Zǐ Sǎn* (Toosendan Powder, 金铃子散). Zhang Shi-wan greatly praised this formula *"Its function is superior to that of Shī Xiào Sǎn* (Sudden Smile Powder) *but without injuring the middle jiao."* It is evident that *chuān liàn zǐ* (Fructus Toosendan) has good results to alleviate pain. The author in his clinical practice always uses *shēng yì rén* (raw Semen Coicis) with *chuān liàn zǐ* (Fructus Toosendan) to treat pain around the lower abdomen and testes areas, and the results are good. This is due to the fact that the liver channel of the foot *jueyin* meridian surrounds the genital area and enters the lower abdomen, therefore, if the sinews or the blood vessels of the liver channel of the foot *jueyin* meridian lose nourishment, or are obstructed by cold-damp, or due to the damp-heat flowing downward, or due to blood stasis and qi stagnation, they all can induce the pain in the lower abdomen and testis area. These two herbs combined together have functions to soften the liver, nourish the sinews, expel the clumps and alleviate pain; they are therefore, effective for use as treatment.

Case 1

Initial Visit

A male patient, 59 years old, May 30th, 1985.

The previous January the patient got an infection after a prostatectomy which induced left testis pain over the following months. The Department of Surgery diagnosed it as epididymitis. He was treated with antibiotics without effect and so came for Chinese medicine treatments.

On physical examination: His body had good physical development. His appetite, urination and bowel movements were all normal. His routine blood test was normal. On palpation, his left testis was swollen and big. It was around the size of a date seed. When pressed upon, he felt pain. The pain was significant on exertion. His pulse was deep and wiry. His tongue coating was thin and white. After surgery, due to blood stasis, the sinews and the blood vessels of the liver channel of the foot *jueyin* meridian were stagnated and obstructed which induced the pain. The treatment plan was to invigorate blood, transform blood stasis, expel clumps and alleviate pain.

Prescription

生薏仁	*shēng yì rén*	30g	Semen Coicis (raw)
川楝子	*chuān liàn zǐ*	10g	Fructus Toosendan
当归	*dāng guī*	12g	Radix Angelicae Sinensis
赤芍	*chì sháo*	10g	Radix Paeoniae Rubra
川牛膝	*chuān niú xī*	15g	Radix Cyathulae
桃仁	*táo rén*	9g	Semen Persicae
红花	*hóng huā*	9g	Flos Carthami
穿山甲	*chuān shān jiǎ*	10g	Squama Manis (fried)
沉香	*chén xiāng*	6g	Lignum Aquilariae Resinatum
皂角刺	*zào jiǎo cì*	12g	Spina Gleditsiae
柴胡	*chái hú*	10g	Radix Bupleuri
甘草	*gān cǎo*	9g	Radix Glycyrrhizae

This was decocted with water and divided to be taken twice warm, one pack per day.

Second Visit

June 18th: After taking a total of fifteen packs of the above formula, his swelling and pain had already reduced by more than half and his testis swelling was visibly reduced to the size of a soy bean, although it was still painful when pressed upon. His pulse was already slow and calm. The above formula was modified to treat him again.

Prescription

生薏仁	*shēng yì rén*	40g	Semen Coicis (raw)
川楝子	*chuān liàn zǐ*	10g	Fructus Toosendan
赤芍	*chì sháo*	15g	Radix Paeoniae Rubra
当归	*dāng guī*	12g	Radix Angelicae Sinensis
桃仁	*táo rén*	9g	Semen Persicae
红花	*hóng huā*	9g	Flos Carthami
皂角刺	*zào jiǎo cì*	10g	Spina Gleditsiae
沉香	*chén xiāng*	9g	Lignum Aquilariae Resinatum
生牡蛎	*shēng mǔ lì*	30g	Concha Ostreae (raw)
川牛膝	*chuān niú xī*	15g	Radix Cyathulae
甘草	*gān cǎo*	9g	Radix Glycyrrhizae

July 2nd: After taking ten packs of the above formula, his pain and swelling basically disappeared. He was instructed to take another five packs after which, he recovered.

Case 2

Initial Visit

A male patient, 38 years old, March 4th, 1986.

The patient had felt cold in his lower back for more than a year. Over the past ten days, his lower abdomen felt cold and painful. The pain expanded to his testes and was accompanied with impotence, and low libido. His pulse was wiry and deep. His tongue was pale and his tongue coating was thin and white. This was cold-damp in the lower *jiao*, with kidney yang deficiency, the sinews and the blood vessels in the liver channel of the foot *jueyin* meridian having lost warmth and nourishment. The treatment plan was to use a formula to warm the kidney to support the yang, nourish the blood and soften the liver.

Prescription

生薏仁	*shēng yì rén*	30g	Semen Coicis (raw)
川楝子	*chuān liàn zǐ*	10g	Fructus Toosendan
乌药	*wū yào*	12g	Radix Linderae
当归	*dāng guī*	12g	Radix Angelicae Sinensis
山茱萸	*shān zhū yú*	12g	Fructus Corni
菟丝子	*tù sī zǐ*	15g	Semen Cuscutae
覆盆子	*fù pén zǐ*	12g	Fructus Rubi
蜈蚣	*wú gōng*	1piece	Scolopendra
五味子	*wǔ wèi zǐ*	9g	Fructus Schisandrae Chinensis
枸杞子	*gǒu qǐ zǐ*	12g	Fructus Lycii
车前子	*chē qián zǐ*	12g	Semen Plantaginis
沉香	*chén xiāng*	9g	Lignum Aquilariae Resinatum

This was decocted with water and divided to be taken twice, one pack per day.

Second Visit

March 10th: After taking three packs of the above formula, his lower abdominal cold and pain were reduced a lot. His lower abdomen felt comfortable. After taking seven packs, his lower abdominal pain was already eliminated. His testis became warm and there was no testis coldness and pain. His impotence was also better. The formula was changed to tonify the essence, support the yang to treat his impotence.

4. *Shēng Yì Rén* (raw Semen Coicis) —— *Chuān Niú Xī* (川牛膝, Radix Cyathulae)

Function

To tonify the liver and the kidney, eliminate dampness and unblock the collaterals.

Application

This is used for wind-damp arthralgia below the waist and swollen and painful lower limbs joints.

Compatibility Analysis

Shēng yì rén (raw Semen Coicis) is sweet and bland. It can strengthen the spleen and transform dampness. It also can nourish the sinews and treat impediment syndrome. In the *Divine Husbandman's Classic of the Materia Medica* it states that it "*governs prolonged wind-damp impediment*". In the *Miscellaneous Records of Famous Physicians* it states that it can "*eliminate the pathogenic qi that induced numbness in the sinews and bone, benefit the intestines and reduce oedema.*" This herb is a gentle and effective herb to treat the impediment syndrome. (It, combined with *bái zhú* - Rhizoma Atractylodis Macrocephalae- treats impediment syndrome. Please refer to the compatibility of *bái zhú* - Rhizoma Atractylodis Macrocephalae). *Chuān niú xī* (Radix Cyathulae) tonifies the liver and the kidney, soothes sinews and benefits impediment syndrome, to invigorate blood and unblock the meridians. Its property is beneficial for moving downward. Zhang Xi-chuan treats lower limb pain always using *chuān niú xī* (Radix Cyathulae). If combined with *shēng yì rén* (raw Semen Coicis), it has better results for treating lower limbs' wind-damp arthralgia, as well as for swollen and painful lower limbs, and lower limbs' oedema.

Case 1

Initial Visit

A female patient, 45 years old, December 14th, 1985.

A couple of months previously, the patient did a heavy job and injured her lower back. Her low back pain expanded and induced left leg pain. After treatment, it recovered. Two days before, due to over use of strength and catching cold, her left leg had severe pain as if cut by a knife. It went from her lower back and continued to her lower left leg along the lateral side, extending to her left foot. She could not step on the ground and walk. An hospital's Orthopaedics Department examined her and diagnosed it as sciatica. The patient had no cold or heat sensations. Her left lower limb was not red or swollen. When walking, it was very painful and hard for her to bear. When she came for a visit, her family member used a tricycle to bring her. Her pulse was wiry and deep. Her tongue was slightly dark and her tongue coating was thin and white. This was caused by cold-damp stagnating and obstructing, and the blood vessels flow was not smooth. When cold is obstructing and there is blood stasis, then her pain is hard to bear. The treatment plan was to invigorate blood, transform blood stasis, warm the meridians and transform the dampness.

Prescription

生薏仁	*shēng yì rén*	50g	Semen Coicis (raw)
川牛膝	*chuān niú xī*	20g	Radix Cyathulae
丹参	*dān shēn*	30g	Radix et Rhizoma Salviae Miltiorrhizae
当归	*dāng guī*	20g	Radix Angelicae Sinensis
杭白芍	*háng bái sháo*	20g	Radix Paeoniae Lactiflorae
川芎	*chuān xiōng*	15g	Rhizoma Chuanxiong
木瓜	*mù guā*	15g	Fructus Chaenomelis
制川乌	*zhì chuān wū*	9g	Radix Aconiti Praeparata
鸡血藤	*jī xuè téng*	30g	Caulis Spatholobi
透骨草	*tòu gǔ cǎo*	15g	Caulis Impatientis
制乳香	*zhì rǔ xiāng*	6g	Olibanum (processed)
制没药	*zhì mò yào*	6g	Myrrha (processed)
甘草	*gān cǎo*	6g	Radix Glycyrrhizae

This was decocted with water and divided to be taken three times a day, warm, one pack per day.

Second Visit

December 20th: After taking six packs of the above formula, her pain was reduced a lot. She could already walk on the ground and so came to visit by herself. However, she still felt lower limb numbness and lassitude. The above formula was then combined with *Dāng Guī Sì Nì Tāng* (Chinese Angelica Counterflow Cold Decoction, 当归四逆汤), and modified to treat her.

Prescription

生薏仁	*shēng yì rén*	30g	Semen Coicis (raw)
川牛膝	*chuān niú xī*	15g	Radix Cyathulae
桂枝	*guì zhī*	10g	Ramulus Cinnamomi
杭白芍	*háng bái sháo*	15g	Radix Paeoniae Lactiflorae
细辛	*xì xīn*	6g	Radix et Rhizoma Asari
木通	*mù tōng*	6g	Caulis Akebiae
丹参	*dān shēn*	30g	Radix et Rhizoma Salviae Miltiorrhizae
当归	*dāng guī*	12g	Radix Angelicae Sinensis

木瓜	*mù guā*	15g	Fructus Chaenomelis
鸡血藤	*jī xuè téng*	30g	Caulis Spatholobi
透骨草	*tòu gǔ cǎo*	20g	Caulis Impatientis
甘草	*gān cǎo*	6g	Radix Glycyrrhizae

This was decocted with water and divided to be taken twice.

Third Visit

Follow up visit January 20th,1986: After taking ten packs of the above formula, her pain stopped and her numbness also became better. As it was the new year, she did not use the formula again, but having stopped the formula for half a month, there was no reoccurrence.

Case 2

Initial Visit

A male patient, 57 years old, May 8th, 1989.

The patient had pain and heaviness sensations in both lower limbs' joints for more than a month and it was accompanied by mild oedema. It was better when he got up in the morning and worse in the afternoon. He had been treated in a local hospital for rheumatism without effect and so came for treatment. His pulse was wiry and fast. His tongue was red and the coating was thin and white. His knee joints were red and swollen and both his lower limbs had pitting oedema, leaving finger indentations after being pressed upon. This was damp pathogens going downward obstructing and stagnating the sinews and the blood vessels, as it was prolonged it transformed into heat. The treatment plan was to transform dampness and unblock the collaterals, clear heat and expel stasis.

Prescription

生薏米	*shēng yì mǐ*	30g	Semen Coicis (raw)
川牛膝	*chuān niú xī*	15g	Radix Cyathulae
木瓜	*mù guā*	12g	Fructus Chaenomelis
防己	*fáng jǐ*	15g	Radix Stephaniae Tetrandrae
丹参	*dān shēn*	20g	Radix et Rhizoma Salviae Miltiorrhizae
当归	*dāng guī*	15g	Radix Angelicae Sinensis
泽泻	*zé xiè*	15g	Rhizoma Alismatis
茯苓	*fú líng*	15g	Poria
络石藤	*luò shí téng*	24g	Caulis Trachelospermi
陈皮	*chén pí*	9g	Pericarpium Citri Reticulatae
紫苏叶	*zǐ sū yè*	9g	Folium Perillae
鸡血藤	*jī xuè téng*	30g	Caulis Spatholobi
透骨草	*tòu gǔ cǎo*	15g	Caulis Impatientis
甘草	*gān cǎo*	6g	Radix Glycyrrhizae

This was decocted with water and divided, to be taken twice, warm, one pack per day.

Second Visit

May 12th: After taking three packs of the above formula, all of his symptoms were reduced. The pain, redness, and swelling all were reduced by more than half. As it was effective, the prescription was not changed and he continued to take the original formula again.

Third Visit

June 15th: The patient wrote to say that he followed the above formula, and took the decotion locally, for a total of more than twenty packs, and he recovered.

5. *Shēng Yì Rén* (raw Semen Coicis) ——
Hóng Huā (红花, Flos Carthami), *Bǎn Lán Gēn* (板蓝根, Radix Isatidis)

Function

To expel dampness and detoxify toxins, to transform stasis and expel clumps.

Application

This is used for flat warts.

Compatibility Analysis

Flat warts belongs to a viral skin disease. It is a common disease and is an infection by the human papillomavirus (HPV) family (HPV 3, 10, 28, and 49). It more often occurs in the young and can be seen more often during puberty in young girls. They appear more on the face and dorsal side of the hand. The skin is damaged. They are smooth, flat-topped papules around a rice or soy bean size, and are light red, yellow-brown or normal skin colour. Their shape is round or oval with a clear boundary and they are large in number. Their distribution is scattered or clustered. Occasionally, there is an itching sensation. This disease is more often due to damp-heat toxin pathogens wrestling on the skin that obstructs and stagnates the qi and blood; therefore, treatment should eliminate dampness and detoxify toxins, transform blood stasis and expel clumps. *Shēng yì rén* (raw Semen Coicis) is sweet and bland. Its power to eliminate dampness and expel clumps is superior. *Hóng huā* (Flos Carthami) is acrid and warm. Its strength to invigorate blood and expel blood stasis is strong. *Bǎn lán gēn* (Radix Isatidis) is bitter and cold. Its strength to clear heat and detoxify toxins is excellent. These three herbs combined together achieve the functions to clear heat, eliminate dampness and detoxify toxins. *Shēng yì rén* (raw Semen Coicis) combined with *bǎn lán gēn* (Radix Isatidis) can clear the damp-heat toxic pathogens. If the toxic pathogen is eliminated, then the qi and blood stagnation and obstruction can be eliminated. *Shēng yì rén* (raw Semen Coicis) combined with *hóng huā* (Flos Carthami) can transform blood stasis and expel clumps. If the blood stasis and clumps are expelled, then the damp-heat pathogen can be eliminated. These three herbs are mutual assistants and envoys. They are mutually enhanced and just match the aetiology of the flat warts. The author used them to treat many patients and all had satisfactory results.

Case Study

Initial Visit

A female patient, 18 years old, May 13th, 1987.

The patient had more than ten flat and rounded, rice size papules, brown in colour and with mild itching sensation on the front of her head, for more than half a month. There were less in the beginning but it increased more later on. Her appetite, urination and bowel movement were normal. Her physical development was good and she had no other illness. Her pulse was wiry, floating and fast. Her tongue coating was thin and white and her tongue was red. This was damp-heat toxic pathogen stagnating and obstructing the skin syndrome. The treatment plan was to clear heat, detoxify toxins, eliminate dampness and transform the blood stasis.

Prescription

生薏仁	*shēng yì rén*	30g	Semen Coicis (raw)
红花	*hóng huā*	10g	Flos Carthami
板蓝根	*bǎn lán gēn*	15g	Radix Isatidis

This was decocted with water and divided to be taken twice, one pack per day.

Second Visit

May 13th: After taking ten packs of the above formula, her facial papules disappeared.

Jú yè's (橘叶 , Folium Citri) taste and property are bitter and neutral. Its qi is slightly aromatic. It enters the liver channel of the foot *jueyin* meridian. It functions to move qi, expel clumps, soothe the liver and resolve stagnation. It is an important herb to treat breast abscess. The author thought that *jú yè* (Folium Citri) is the king herb to treat abscesses, its qi and taste are all thin and its strength to expel clumps and move qi is also weak. If only using this single herb to treat a breast abscess it would be difficult to get any effect. It should be combined with other herbs that can invigorate blood and expel blood stasis. However, *jú yè* (Folium Citri) enters the liver channel of the foot *jueyin* meridian. Its qi is slightly aromatic and it is specialised to treat breast diseases, therefore, using it for any breast conditions like breast abscesses, breast ulcers and lack of lactation, etc. If used as a guiding herb, most of time it can improve the treatment effect.

1. *Jú Yè* (Folium Citri) ——
Guā Lóu (瓜蒌, Fructus Trichosanthis)

Function

To clear heat, expel clumps, soothe the liver and treat breast abscesses.

Application

It is mainly used for female breast abscesses, swelling and pain.

Compatibility Analysis

Jú yè (Folium Citri) can soothe the liver, resolve stagnation, move qi and expel clumps. It is an important herb to treat breast abscesses. *Guā lóu* (Fructus Trichosanthis) is sweet and cold. It is effective for clearing heat, expelling clumps and transforming phlegm. It is also a good herb to treat breast abscesses. In the *Materia Medica of the Ming Dynasty* it states that *guā lóu* (Fructus Trichosanthis) can *"eliminate swollen toxins, treat breast abscesses and abscesses on the back"*. These two herbs combined together function to invigorate blood, expel clumps, soothe the liver and eliminate abscesses. Any qi, blood or phlegm stagnation induced mammary gland diseases can all use them for treatment.

Case 1

Mammary gland hypertrophy.

Initial Visit

A female patient, 40 years old, March 7th, 1986.

The patient had a tumour on the upper right side of her right breast, for more than half a month. She was diagnosed by an Oncology Department with mammary gland hypertrophy. Later, through her friend's referral she came to our hospital's clinic for treatment. On examination, the patient had good physical development. Her menstruation was normal; there was a 7cm × 15cm tumour on upper right side of her right breast. It was painful on palpation, and a mild hardness could be felt, it had a smooth surface, and no redness or swelling. Her pulse was wiry and thready. Her tongue coating was thin and white. This was induced by qi stagnation and blood stasis, with phlegm condensed and clumped.

Prescription

丹参	*dān shēn*	20g	Radix et Rhizoma Salviae Miltiorrhizae
当归	*dāng guī*	20g	Radix Angelicae Sinensis
瓜蒌	*guā lóu*	24g	Fructus Trichosanthis
橘叶	*jú yè*	12g	Folium Citri
制乳香	*zhì rǔ xiāng*	6g	Olibanum (processed)
制没药	*zhì mò yào*	6g	Myrrha (processed)
元参	*yuán shēn*	30g	Radix Scrophulariae
柴胡	*chái hú*	12g	Radix Bupleuri
连翘	*lián qiáo*	10g	Fructus Forsythiae
王不留行	*wáng bù liú xíng*	10g	Semen Vaccariae
生牡蛎	*shēng mǔ lì*	30g	Concha Ostreae (raw)
丝瓜络	*sī guā luò*	10g	Retinervus Luffae Fructus
生黄芪	*shēng huáng qí*	20g	Radix Astragali (raw)
甘草	*gān cǎo*	9g	Radix Glycyrrhizae

This was decocted with water, to be taken one pack per day.

Second Visit

March 15th: After taking six packs of the above formula, her tumour was visibly reduced. On palpation it was soft with not much pain. As it was effective, the prescription was not changed and she continued to take the same formula again.

Third Visit

March 21st: After taking the formula, the tumour basically disappeared. She was instructed to take three packs again to complete the treatment.

Case 2

Breast abscess.

Initial Visit

A female patient, 32 years old, June 30th, 1987.

The patient had a red swollen nodule on the upper side of her left breast, for a couple of months. It was painful and hot to the touch. It was sometimes better and sometimes worse. Her appetite was normal. Her pulse was wiry and her tongue coating was thin and white. This was blood and heat stagnating and clumping together syndrome. The treatment plan was to clear heat, expel clumps, invigorate blood and transform blood stasis.

Prescription

赤芍	*chì sháo*	12g	Radix Paeoniae Rubra
当归	*dāng guī*	15g	Radix Angelicae Sinensis
金银花	*jīn yín huā*	20g	Flos Lonicerae Japonicae

连翘	*lián qiáo*	15g	Fructus Forsythiae
瓜蒌	*guā lóu*	15g	Fructus Trichosanthis
橘叶	*jú yè*	12g	Folium Citri
元参	*yuán shēn*	20g	Radix Scrophulariae
柴胡	*chái hú*	15g	Radix Bupleuri
穿山甲	*chuān shān jiǎ*	10g	Squama Manis
生黄芪	*shēng huáng qí*	20g	Radix Astragali (raw)
王不留行	*wáng bù liú xíng*	10g	Semen Vaccariae
甘草	*gān cǎo*	6g	Radix Glycyrrhizae

This was decocted with water and divided to be taken twice, warm, one pack per day.

Second Visit

July 5th: After taking five packs of the above formula, her redness and swelling had disappeared and her pain was stopped. Using the above formula, *chì sháo* (Radix Paeoniae Rubra) was eliminated and *dān shēn* (Radix et Rhizoma Salviae Miltiorrhizae) 20g, was added. She took five more packs again. Later, after half a year's follow up visits, there was no reoccurrence.

A Brief Discussion on *Chái Hú Tāng* Syndrome and

the Compatibility Application of *Xiǎo Chái Hú Tāng*

Xiǎo Chái Hú Tāng (Minor Bupleurum Decoction, 小柴胡汤) is an important formula in Zhang Zhong-jing's *Discussion on Cold Damage*. The discussion of this formula occupied a lot space in the *Discussion on Cold Damage*. It has a broader modification changes and compatibilities application scope. It also has had a big influence on later generations. Therefore, to discuss, and research, the *Xiǎo Chái Hú Tāng* (Minor Bupleurum Decoction) syndrome and its herbal compatibilities is beneficial to help us to apply it even more effectively. Now, based on the author's experience he presents the knowledge for other doctors' reference.

WHAT ARE THE *CHÁI HÚ* SYNDROMES?

The 101st item in the original *Discussion on Cold Damage* states that *"when catching cold or invasion by wind, there is Chái Hú Syndrome. If any one syndrome presents, Xiǎo Chái Hú Tāng (Minor Bupleurum Decoction) can be used, all of the syndromes, do not need to be present."* Therefore, *Chái Hú* Syndrome should include those syndromes that can be treated by the *Xiǎo Chái Hú Tāng* (Minor Bupleurum Decoction). This is the key point for our discussion, otherwise, it is hard to grasp the applications of the *Xiǎo Chái Hú Tāng* (Minor Bupleurum Decoction). In the *Discussion on Cold Damage*, it describes using *Xiǎo Chái Hú Tāng* (Minor Bupleurum Decoction) to treat many diseases. Is it that all of these syndromes belong to the *Chái Hú* Syndrome? If this is the understanding, then it is incorrect, as there are representative syndromes and associated syndromes of the *Chái Hú* Syndrome that were discussed in the *Discussion on Cold Damage*. The 101st item in the *Discussion on Cold Damage* pointed out that the *Chái Hú* Syndrome includes the representative syndromes, not the associated syndromes.

The representative syndromes of the *Xiǎo Chái Hú Tāng* (Minor Bupleurum Decoction), according to the 96th item in the *Discussion on Cold Damage* states that *"the Xiǎo Chái Hú Tāng (Minor Bupleurum Decoction) governs those that have caught a cold for five or six days, attacked by wind, alternating chills and fever, chest and hypochondria suffering from fullness, no appetite, irritablity with nausea, or hypersensitivity in the abdomen but without vomiting; or thirst; or pain in the abdomen; or hard clumps in the hypochondria; or palpitations under the heart with dysuria; or no thirst with slight body heat; or cough."* The initial part of this item is the representative syndrome of the *Chái Hú* Syndrome. All symptoms after *"hypersensitivity in the abdomen but without vomiting"* above, are all secondary syndromes. They are also called probable syndromes. The initial part listed syndromes that can use *Xiǎo Chái Hú Tāng* (Minor Bupleurum Decoction) for treatment. The later following syndromes must come with the representative syndromes in order to use the *Chái Hú Tāng* (Minor Bupleurum Decoction) for treatment.

Zhang Zhong-jing thought that the aetiology and the mechanism of the *Chái Hú* Syndrome was *"blood deficiency and exhausted qi, with opened skin pores, and the pathogens taking advantage to invade and clumping in the hypochondria. The normal qi and pathogens were battling, manifesting the alternated chills and fever, with a fixed time for the outbreak or remission, and no appetite. Due to the fact that the zang and fu organs always influence each other, the pathogen invading the liver also influences the spleen, and therefore, cause vomiting. The Xiǎo Chái Hú Tāng (Minor Bupleurum Decoction) can treat it.*" (97th item in the original article)

It has already been clearly pointed out that the chief pathological change of the *Xiǎo Chái Hú Tāng* (Minor Bupleurum Decoction) syndrome "*is the pathogens taking the advantage and invading inward and clumping at the hypochondria.*" In other words the pathological change of the *Chái Hú* syndrome is at the hypochondria. Its clinical expressions are all related to hypochondriac pathological change. If analysed from Zhang Zhong-jing's items, there are three levels for the representative syndromes of the *Chái Hú* syndrome.

1. Chest and Hypochondria Syndrome: This is one of the representative syndromes of the *Chái Hú* syndrome. Chest and Hypochondria Syndrome can appear due to pathogenic qi clumped under the hypochondria inducing qi and blood stagnation and obstruction. In other words the chest and hypochondria suffer fullness, or there are hard clumps or pain under the hypochondria. The chest and hypochondria suffering fullness means that the patient feels fullness or stuffiness of the chest and hypochondria. Based on Zhang Zhong-jing's description for the Chest and Hypochondria Syndrome in the *Discussion on Cold Damage*, the syndromes includes: Chest fullness, or chest and hypochondriac fullness, or fullness in the chest with irritability, or fullness with hardness or stuffiness at hypochondria, or chest fullness and hypochondriac pain, or hypochondriac fullness and pain, or hypochondria and heart pain, etc.

There are three conditions related to the chest feeling stuffy: Only chest fullness, or only fullness beneath the hypochondria, or both chest and hypochondriac fullness.

In relation to the hardness and pain, they are only limited to the hypochondria and are not connected to the chest.

All the above syndromes are classified as the Chest and Hypochondria Syndrome of *Chái Hú* Syndrome.

2. Alternated Chills and Fever Syndrome: This manifests at a fixed time for the outbreak or remission, and is the main characteristic of the heat type of the *Chái Hú* Syndrome. Zhang Zhong-jing mentioned that the normal qi and pathogens are battling, clumped at the hypochondria, therefore, there is alternated chills and fever. You Zai-jing explained that: "*The hypochondria is the membrane of the shaoyang. Shaoyang is the intersection of yin and yang. If there are pathogens present, yin comes out to battle with the pathogens, and then there is cold. If yang enters in to battle with the pathogens then there is fever. Yin and yang coming out or going in, all have their own time, therefore, there are alternated chills and fever and their outbreak and remission have fixed times*".

3. Stomach and Intestines Syndrome: This manifests with no appetite, irritability and vomiting. Zhang Zhong-jing thought this was "*When the zang and fu organs are connected with each other, the pain must go down. The pathogen is higher and the pain is lower, therefore, vomiting occurs.*" It also means that the pathogens are clumped at the hypochondria and can connect to the other *zang* or *fu* organs. The liver and the gallbladder belong to the wood, if the qi of *shaoyang* is stagnated or clumped, then it can influence the spleen of earth in the middle *jiao*, which therefore, manifests in no appetite, irritability and vomiting, etc.

The above three syndromes are the representative syndromes of the *Chái Hú Tāng* (Minor Bupleurum Decoction), also called *Chái Hú* Syndrome. In clinical practice, if seeing any one of them the *Xiǎo Chái Hú Tāng* (Minor Bupleurum Decoction) can be used to treat it. Zhang Zhong-jing further explained: "*If a patient has a cold for four or five days, with fever, aversion to cold, neck stiffness, fullness of the hypochondria, warm hands and feet, with thirst, use the Xiǎo Chái Hú Tāng (Minor Bupleurum Decoction) for treatment.*" (99ᵗʰ item in the original article.)

This item uses hypochondriac fullness as the representative symptom of the *Chái Hú* Syndrome. When still experiencing symptoms such as fever, aversion to cold and neck stiffness of the *taiyang* exterior syndrome, there appears the representative syndrome "*hypochondriac fullness*" of the *Chái Hú* Syndrome of the *shaoyang* meridian, the *Xiǎo Chái Hú Tāng* (Minor Bupleurum Decoction) can be used to treat it.

"*A female had a cold for seven or eight days, continuous chills and fever, occurring at a fixed time, her menstruation had just stopped. This is heat entering the uterus, causing the blood to clump, therefore, its appearance is similar to malaria, with a fixed time for the outbreak. Xiǎo Chái Hú Tāng (Minor Bupleurum Decoction) governs their treatment.*"(144ᵗʰ item in the original article)

This item uses chills and fever, and a fixed time for the outbreak as the evidence to apply the *Xiǎo Chái Hú Tāng* (Minor Bupleurum Decoction).

"*If there is vomiting with fever, the Xiǎo Chái Hú Tāng (Minor Bupleurum Decoction) governs them*" (379ᵗʰ item in the original article).

This item uses Stomach and Intestines Syndrome as the representative syndrome to apply the *Xiǎo Chái Hú Tāng* (Minor Bupleurum Decoction).

The above three items all use one representative syndrome of the *Chái Hú* Syndrome, to apply the *Xiǎo Chái Hú Tāng* (Minor Bupleurum Decoction). For two or three representative syndromes, then one should apply the *Xiǎo Chái Hú Tāng* (Minor Bupleurum Decoction). For example:

"*Where there is yangming illness, hypochondriac hardness and fullness, constipation and vomiting, with a white tongue coating, patients can be given the Xiǎo Chái Hú Tāng (Minor Bupleurum Decoction) for treatment.*" (230ᵗʰ item in the original article).

This item has two representative syndromes of the *Chái Hú* syndrome: Hypochondriac hardness and fullness, and vomiting, therefore, use the *Xiǎo Chái Hú Tāng* (Minor Bupleurum Decoction).

"*If the taiyang illness is without relief, it enters the shaoyang and presents as hypochondriacal hardness with fullness, dry vomiting preventing eating, alternating chills and fever, that have not been treated with the vomiting or draining downward method, with a deep and tense pulse, give them Xiǎo Chái Hú Tāng (Minor Bupleurum Decoction) for treatment.*"(266ᵗʰ item in the original article.)

This item has three representative syndromes; therefore, the *Xiǎo Chái Hú Tāng* (Minor Bupleurum Decoction) should be used to treat it.

The above stated three kinds of syndromes are called the three representative syndromes of the *Xiǎo Chái Hú Tāng* (Minor Bupleurum Decoction). Aside from them, in the *Discussion on Cold Damage*, Zhang Zhong-jing also mentioned: "*If the shaoyang is sick, there is a bitter mouth, a dry throat and dizziness.*"(263th item in the original article.)

"*If the shaoyang meridian is attacked by wind, the ears cannot hear, there are red eyes, fullness and irritability inside the chest......*" (264th item in the original article.)

Many doctors use these two items as the compendium of the *shaoyang* illness. However, from the contents of the two items, it is far from complete in describing the representative syndromes of the *shaoyang* syndrome; therefore, some doctors suspect that they are not the original articles of Zhang Zhong-jing. For example Lu Yuan-lei said in the *Modern Annotation of Discussion on Cold Damage* (伤寒今释 , *Shāng Hán Jīn Shì*):"*The compendium of the shaoyang illness in this item listed the small things but left out the main ones of the representative syndrome.*" It is thought that the three symptoms of a bitter mouth, a dry throat and dizziness explained the exterior expression of the *shaoyang* illness. A bitter mouth is the gallbladder heat steaming upward. The pathogenic heat burns fluid and then there is the dry throat. The *shaoyang* also governs the wind and the fire. The wind and the fire flare upward therefore, there is dizziness. The *shaoyang* meridian and its collaterals coil around the head and eyes, travels in the chest, and it is the organ of wind and wood. If the pathogen resides in the *shaoyang*, then the wind moves and the fire flares, therefore, the ears cannot hear, and red eyes, fullness and irritability in the chest manifest. Those are the symptoms of the pathogen residing in the *shaoyang* with some exterior expressions. We can treat them as the associated syndromes but should not treat them as the representative syndromes of the *Chái Hú* Syndrome.

THE COMPOSITION AND MODIFICATIONS OF THE *XIǍO CHÁI HÚ TĀNG*

1. The Composition and Compatibility of the *Xiǎo Chái Hú Tāng* (Minor Bupleurum Decoction)

柴胡	*chái hú*	0.5 *jin*	Radix Bupleuri
黄芩	*huáng qín*	3 *liang*	Radix Scutellariae
半夏	*bàn xià*	0.5 *jin*	Rhizoma Pinelliae
生姜	*shēng jiāng*	3 *liang*	Rhizoma Zingiberis Recens
人参	*rén shēn*	3 *liang*	Radix et Rhizoma Ginseng
甘草	*gān cǎo*	2 *liang*	Radix Glycyrrhizae
大枣	*dà zǎo*	12 pcs	Fructus Jujubae

Decoct the above seven herbs, in 1 *dou* and 2 *sheng* of water and reduce to 6 *sheng*, filter and use the dregs to decoct again to make 3 *sheng*. Each time, take 1 *sheng*, warm, three times a day.

From the composition of the *Xiǎo Chái Hú Tāng* (Minor Bupleurum Decoction), based on the herbs' characteristics and their functions, they can be divided into the three following compatibilities:

• *Chái hú* (Radix Bupleuri) combined with *huáng qín* (Radix Scutellariae): Zhang

Zhong-jing was a *Han* Dynasty person. His herbal knowledge was mostly related to the *Divine Husbandman's Classic of the Materia Medica*. In the *Divine Husbandman's Classic of the Materia Medica* it states of *chái hú* (Radix Bupleuri): "*Its tastes are bitter and neutral. It governs the clumped qi in the heart, abdomen, intestines and stomach, food accumulation, cold and heat pathogens, expelling the old and generating the new.*" Of *huáng qín* (Radix Scutellariae) it states "*Its tastes are bitter and neutral. It governs all kinds of heat and jaundice, diarrhoea and dysentery of the intestines, expels water and drains stagnated blood downward.*"

These two are bitter herbs. They have functions to soothe the liver, resolve stagnation, reduce cold or heat, unblock the stomach and intestines and promote bowel movement. They are also a representative compatibility in the *Xiǎo Chái Hú Tāng* (Minor Bupleurum Decoction). Any fever that is induced by pathogenic heat in the *shaoyang* meridian, cannot be resolved without this compatability.

Chái hú (Radix Bupleuri) and *huáng qín* (Radix Scutellariae) in the *Chái Hú Tāng* (Minor Bupleurum Decoction) have functions to: 1. Harmonise the *shaoyang* to reduce cold and heat. They are the main herbs to relieve the cold-heat state in the *Chái Hú* Syndrome. 2. Resolve stagnation to treat the Chest and Hypochondria Syndrome which is, the pathogen residing in the *shaoyang* meridian and clumping at the hypochondria. Both *chái hú* (Radix Bupleuri) and *huáng qín* (Radix Scutellariae) enter the liver channel of the foot *jueyin* and the gallbladder channel of the foot *shaoyang* meridians. *Chái hú* (Radix Bupleuri) resolves stagnation and soothes the liver. *Huáng qín* (Radix Scutellariae) clears heat and benefits the gallbladder. The properties of *chái hú* (Radix Bupleuri) are raising and dispersing. The properties of *huáng qín* (Radix Scutellariae) are bitter and descending. These two herbs combined together can regulate the pivot mechanism of the liver and the gallbladder to benefit eliminating the Chest and Hypochondria Syndrome. 3. Help acrid herbs to eliminate Stomach and Intestines Syndrome. *Chái hú* (Radix Bupleuri) and *huáng qín* (Radix Scutellariae) have functions to soothe the liver and benefit the gallbladder; therefore, they can unblock the stomach and intestines in order to help eliminate the symptoms in the stomach and intestinal tract.

- *Bàn xià* (Rhizoma Pinelliae) combined with *shēng jiāng* (Rhizoma Zingiberis Recens). In the *Divine Husbandman's Classic of the Materia Medica* it states that: "*The tastes of bàn xià* (Rhizoma Pinelliae) *are acrid and neutral. It governs cold and heat, firmness under the heart, descending qi, swollen and painful throat, dizziness and chest distention, hiccough, borborygumus, and stops sweating.*" Of *shēng jiāng* (Rhizoma Zingiberis Recens) it states: "*Its tastes are acrid and warm. It governs chest fullness, cough and qi rebelling upward, warms the middle jiao, stops bleeding, sweating, expels wind-damp bi syndrome, and intestinal afflux with dysentery.*"

Both of these herbs are acrid. They are the chief herbs to descend rebellions and stop vomiting. Acridity can disperse. It also functions to resolve stagnation and eliminate fullness. These two herbs mainly have functions to: 1. Eliminate the representative syndrome of the Stomach and Intestines Syndrome. The property of *bàn xià* (Rhizoma Pinelliae) is slippery and beneficial for descending. It is an important herb to descend rebellions and stop vomiting. *Shēng jiāng* (Rhizoma Zingiberis Recens) is acrid and warm. It can strengthen the stomach and stop vomiting. It also can eliminate cold or heat pathogens. These two herbs combined together, have a better function for treating the stomach and intestinal tract induced vomiting without appetite. The representative syndrome of *chang wei zheng* in the *Chái Hú* Syndrome, cannot be eliminated without these two herbs. 2. Help *chái hú* (Radix Bupleuri) and *huáng qín* (Radix Scutellariae) to resolve cold and heat, to eliminate Chest and Hypochondria Syndrome. As the acrid taste has functions to resolve stagnation, and eliminate cold or heat, if the pathogen resides in the *shaoyang* meridian, with clumping at the hypochondria, using an acrid and dispersing formula can soothe and resolve the pathogen.

● For the compatibility of *rén shēn* (Radix et Rhizoma Ginseng), *gān cǎo* (Radix Glycyrrhizae) and *dà zǎo* (Fructus Jujubae), in the *Divine Husbandman's Classic of the Materia Medica* it states that: *"The taste of rén shēn (Radix et Rhizoma Ginseng) is sweet and slightly cold. It governs tonifying of the five zang organs, calms the spirit, anchors the ethereal soul and the corporeal soul, stops palpitations and eliminates pathogenic qi."* For *gān cǎo* (Radix Glycyrrhizae): *"Its tastes are sweet and neutral. It governs the cold or heat pathogenic qi in the five zang and six fu organs as well as detoxifies toxins."* For *dà zǎo* (Fructus Jujubae): *"Its tastes are sweet and neutral. It governs the pathogenic qi in the heart and abdomen. It can calm the middle jiao, nourish the spleen, help the twelve meridians, calm the stomach qi, open the nine orifices, and tonify qi deficiency, fluid deficiency, and poor constitution; calm severe fright, alleviate heaviness of the four limbs, and harmonise all kinds of herbs."*

All herbs in this compatibility taste sweet. They have functions to tonfy the qi of the middle *jiao*, calm the five *zang* organs, eliminate the cold and heat pathogenic qi from the five *zang* and six *fu* organs. They have the following functions in the *Chái Hú Tāng* (Minor Bupleurum Decoction): 1. To support the normal qi and secure the root. In the *Divine Husbandman's Classic of the Materia Medica* it states: *"If there are pathogens coming together, the normal qi must be deficient."* The aetiology of the *Chái Hú* Syndrome is *"blood deficiency and exhausted qi, opened skin pores, and the pathogens taking the advantage to invade"*. When the qi and blood are deficient, the defense function of the exterior is insufficient, therefore, there is pathogenic invasion, so, to support the normal qi and secure the root is a compatibility that cannot be missed in the *Chái Hú Tāng* (Minor Bupleurum Decoction). 2. They can help the above two compatibilities to eliminate Stomach and Intestines Syndrome and Alternated Chills and Fever Syndrome. 3. They can harmonise all herbs. Due to the sweet taste they function to rectify tastes, and can harmonise the taste of bitter, cold, acrid, and warm herbs to make them return to pure harmony. In addition, the property of the sweet, is tonifying, it can reduce the side effects from the bitter, cold and acrid, warm herbs.

Though the number of herbs in the above three compatibilities is few, however, for the *Chái Hú* Syndrome, it can be said that the herbs and the illness are mutually matched. With the right compatibility, among the herbs, they can mutually enhance and also can mutually inhibit. They not only can have good pharmacological effects, but also can prevent bad reactions. Therefore, in clinical practice, if one only accurately identifies the *Xiǎo Chái Hú Tāng* (Minor Bupleurum Decoction), its effect is very quick. It is one of the doctors' favourite formulae from previous generations.

2. The Combination of the Xiǎo Chái Hú Tāng (Minor Bupleurum Decoction)

Zhang Zhong-jing listed a couple of modification methods behind the formula of the *Xiǎo Chái Hú Tāng* (Minor Bupleurum Decoction). In clinical practice, it has very important guiding significance, due to the fact that the disease changes unceasingly instead of being fixed or invariable. Zhang Zhong-jing listed the modifications, telling us that the changes of the *Chái Hú* Syndrome are complicated. During the development of the changes of the *Chái Hú* Syndrome, some belong to *shaoyang*, some still are exterior syndromes, some have entered internally already, some still do not have all of the representative syndromes. Moreover, it has a huge variety of associated syndromes. If only one formula without modification is used, it it is difficult to get satisfactory results. Zhang Zhong-jing in discussing the application of the *Xiǎo Chái Hú Tāng* (Minor Bupleurum Decoction) was concerned it was difficult to understand, so he also listed some associated syndromes and their treatment methods. His intention was to focus attention on differentiation, grasp the disease changes and not to be conservative. To apply herbs one must be flexible to change. Below are listed some cases of Zhang Zhong-jing's modification changes, to observe his medication characteristics, for better application in clinical practice.

● For patients with irritability in the chest without vomiting, eliminate *bàn xià* (Rhizoma Pinelliae) and *rén shēn* (Radix et Rhizoma Ginseng), add *guā lóu* (Fructus Trichosanthis) one piece (96th item in the original article).

For the *Chái Hú* Syndrome with *"irritability in the chest"*, it explained that there is stagnated heat in the chest, therefore, add *guā lóu* (Fructus Trichosanthis) to clear the heat inside the chest, and clear the fire in the upper *jiao*, in order to resolve irritability and stuffiness in the chest.

Eliminate *bàn xià* (Rhizoma Pinelliae), as *bàn xià* (Rhizoma Pinelliae) is acrid and warm, it is not appropriate to use it for heat in the chest, and there is no vomiting, so, it is better to eliminate it.

Eliminate *rén shēn* (Radix et Rhizoma Ginseng), as *rén shēn* (Radix et Rhizoma Ginseng) tends towards being warm and tonifying. For stagnated heat induced irritability in the chest, it is not appropriate to use it, therefore, eliminate it also.

In the *Xiǎo Chái Hú Tāng* (Minor Bupleurum Decoction) eliminate *rén shēn* (Radix et Rhizoma Ginseng) and *bàn xià* (Rhizoma Pinelliae), and add *guā lóu* (Fructus Trichosanthis) to increase the cold herbs. This is better for the *Chái Hú* Syndrome with internal heat stagnation excess syndrome. If in clinical practice, it manifests with irritability in the chest with vomiting, then *bàn xià* (Rhizoma Pinelliae) need not be eliminated. The applications should be grasped flexibly.

● If there is thirst, eliminate *bàn xià* (Rhizoma Pinelliae), and add more *rén shēn* (Radix et Rhizoma Ginseng) to increase it to 4.5 *tael*, and add *guā lóu* gēn (栝楼根 , Radix Trichosanthis) 4 *tael*.

You Zai-jing said that: *"For the thirst, it is the wood fire irritably inside and fluid deficiency with qi dryness, therefore, eliminate the warmth and dryness of bàn xià (Rhizoma Pinelliae), add sweet and moist rén shēn (Radix et Rhizoma Ginseng), bitter and cold guā lóu gēn (Radix Trichosanthis) in order to eliminate heat and generate fluid."* This explanation of You's is very reasonable. Thirst is a syndrome of exuberant heat and injured fluid, therefore, eliminate the acridity and warmth of *bàn xià* (Rhizoma Pinelliae). If it is used, it will further injure the fluid. Add *guā lóu gēn* (Radix Trichosanthis) to generate fluid, stop thirst and clear heat. In the *Divine Husbandman's Classic of the Materia Medica* it states: *Guā lóu gēn* (Radix Trichosanthis) *"governs xiao ke, fever, irritability and fullness, and exuberant heat."* Increasing the dosage of *rén shēn* (Radix et Rhizoma Ginseng) is due to the heat injuring qi and yin. If qi and yin are injured, then there is thirst, and deficiency, therefore, use the sweetness and moistness of *rén shēn* (Radix et Rhizoma Ginseng) to tonify qi and generate fluid. In clinical practice, if the qi and yin deficiency syndrome is not so significant, using the original dosage of *rén shēn* (Radix et Rhizoma Ginseng) is adequate. If there is exuberant excess heat with severe thirst and the pulse is forceful, *shēng shí gāo* (raw Gypsum Fibrosum) also can be added as, Zhang Zhong-jing said in the *Discussion on Cold Damage* *"Observe the pulse......follow the syndrome to treat."* It precisely explains the truth that herbs should follow the syndrome for modification.

● If there is abdominal pain, eliminate *huáng qín* (Radix Scutellariae), and add *sháo yào* (Radix Paeoniae Rubra) 3 *tael*.

For *sháo yào* (Radix Paeoniae Rubra), the *Divine Husbandman's Classic of the Materia Medica* states: *"Its tastes are bitter and neutral. It governs pathogenic qi and abdominal pain, eliminates blood bi syndrome, breaks firm accumulations, cold, heat hernia or intestinal obstruction, alleviates pain, promotes urination, and tonifies qi."* *Sháo yào* (Radix Paeoniae Rubra) is beneficial for resolving spasm and treating abdominal pain, therefore, it should be added in.

Huáng qín (Radix Scutellariae) is bitter and cold. The *Miscellaneous Records of Famous Physicians* also mentions that *huáng qín* (Radix Scutellariae) can *"treat phlegm heat, heat in the stomach, and lower abdominal cramping pain."* Why eliminate it? This is due to the fact that the abdominal pain that *huáng qín* (Radix Scutellariae) treats belongs to excess heat abdominal pain, however, the abdominal pain of the *Chái Hú* Syndrome is more for deficiency abdominal pain. For example, the 100th item in the *Discussion on Cold Damage* states: *"If one has a cold, the yang pulse is choppy and the yin pulse is wiry, there should be urgent pain in the abdomen"* and *"blood deficiency, qi exhausted, opened pores"* etc. that belong to the *Chái Hú* syndromes, are all explained by this point, therefore, it is better to eliminate *huáng qín* (Radix Scutellariae).

You Zai-jing said that: *"Pain in the abdomen is induced by the wood pathogen injuring the earth induced. Huáng qín (Radix Scutellariae) is bitter and cold, it is not good for the spleen yang. Sháo yào (Radix Paeoniae Rubra) is sour and cold. It can sedate wood within the earth, eliminate the pathogenic qi, and alleviate the abdominal pain."*

In addition, if in clinical practice, there is abdominal pain leaning to excess, *sháo yào* (Radix Paeoniae Rubra) should be added. *Huáng qín* (Radix Scutellariae) is not necessary to eliminate it. For example in *Dà Chái Hú Tāng* (Major Bupleurum Decoction) *huáng qín* (Radix Scutellariae) and *sháo yào* (Radix Paeoniae Rubra) are used together.

● If there is firm stuffiness in the hypochondria, eliminate *dà zǎo* (Fructus Jujubae), add *mǔ lì* (牡蛎 , Concha Ostreae) 4 *tael*.

For *mǔ lì* (Concha Ostreae), the *Divine Husbandman's Classic of the Materia Medica* states: "*It governs chills and fever of cold damage…scrofula.*" In the *Miscellaneous Records of Famous Physicians*, it states that *mǔ lì* (Concha Ostreae) can "*treat hypochondriac stuffiness induced heat.*" Wang Hao-gu said that in the *Miscellaneous Records of Famous Physicians*: "*Using chái hú* (Radix Bupleuri) *to guide it, it eliminates hardness in the hypochondria. Using tea to guide it, it eliminates neck nodules. Using dà huáng* (Radix et Rhizoma Rhei) *to guide it, it eliminates swelling between the thighs. Using bèi mǔ* (Bulbus Fritillariae) *as envoy, they eliminate accumulations and clumps.*" The above explained that *mǔ lì* (Concha Ostreae) is salty and cold. It has functions to soften hardness and expel nodules. It is the same as *chái hú* (Radix Bupleuri), which is beneficial for eliminating hardness of stuffiness in the hypochondria. Moreover, in the *Divine Husbandman's Classic of the Materia Medica* it states that *mǔ lì* (Concha Ostreae) also can govern chills and fever when getting cold, it is also more suitable for the representative *Chái Hú* Syndrome.

Dà zǎo (Fructus Jujubae) tastes sweet, it can increase fullness and is not appropriate for the hard *bi* in the hypochondria. Moreover, there are an excess of sweet herbs in the formula, which could affect the salty taste of *mǔ lì* (Concha Ostreae) and reduce its softening hardness and expelling stuffiness functions, therefore, it should be eliminated.

● If there are palpitations under the heart and dysurination, eliminate *huáng qín* (Radix Scutellariae) and add *fú líng* (Poria) 4 *tael*.

For *fú líng* (Poria), the *Divine Husbandman's Classic of the Materia Medica* states that "*its tastes are sweet and neutral. It governs rebellious qi in the chest and hypochondria inducing worry, fright, palpitations, clumping pain under the heart, chills and fever, irritability, fullness, cough and hiccough, severe dry mouth, dry tongue, promotes urination……*" *Fú líng* (Poria) has functions to promote urination and calm palpitations, meanwhile it also treats chills, fever, irritablity and fullness, and rebellious qi in the chest and hypochondria. This is a perfect match for the *Chái Hú* Syndrome; therefore, adding *fú líng* (Poria) is needed and appropriate.

In relation to the question of eliminating *huáng qín* (Radix Scutellariae), in the *Divine Husbandman's Classic of the Materia Medica*, it states that *huáng qín* (Radix Scutellariae) can "*expel water*". Since *huáng qín* (Radix Scutellariae) also can promote urination, why eliminate it? There are two reasons: Firstly if it is the *Chái Hú* Syndrome with palpitations under the heart and dysuria, it is due to spleen deficiency induced water and dampness stagnating internally. If water and dampness are stagnated internally, there will be palpitations under the heart and dysuria. If using *fú líng* (Poria) to strengthen the spleen and promote urination, when water is eliminated, then the symptoms will be eliminated. Secondly *huáng qín* (Radix Scutellariae) is bitter and cold. It can damage the spleen yang. If the spleen yang is deficient then water and dampness cannot be transformed and transported. Therefore, using *huáng qín* (Radix Scutellariae) is not appropriate. *Huáng qín* (Radix Scutellariae) promoting urination is to treat heat stagnation and qi obstruction induced dysuria. It belongs to the excess syndromes, therefore, eliminate it.

● If there is no thirst with slight fever on the exterior, eliminate *rén shēn* (Radix et Rhizoma Ginseng) and add *guì zhī* (Ramulus Cinnamomi) 3 *tael*, cover the patient so they get warm and sweat slightly, then the patient can recover.

Chái Hú Syndrome has an associated syndrome that has no thirst and has a slight fever on the exterior. It explained that though the illness has already developed into the *shaoyang* meridian, the *taiyang* syndrome is still not fully eliminated, showing no thirst and slight fever on the exterior. This is the pathogen not having totally entered the interior. It is a mix of *taiyang* and *shaoyang* complicated syndromes, therefore, add *guì zhī* (Ramulus Cinnamomi) to release the exterior.

Eliminating *rén shēn* (Radix et Rhizoma Ginseng) is because the property of *guì zhī* (Ramulus Cinnamomi) is hot. If *rén shēn* (Radix et Rhizoma Ginseng) is not eliminated, it may increase the heat property of the whole formula. It is not good for the *shaoyang* illness. Therefore, if adding *guì zhī* (Ramulus Cinnamomi), *rén shēn* (Radix et Rhizoma Ginseng) needs to be eliminated, to balance the property of the formula, but it's not all due to *rén shēn's* (Radix et Rhizoma Ginseng) sweet flavour, hot property and tonifying action trapping pathogens to eliminate it within the formula. For example the *Chái Hú Guì Zhī Tāng* (Bupleurum and Cinnamon Twig Decoction, 柴胡桂枝汤) the *Discussion on Cold Damage* also treats *taiyang* and *shaoyang* complicated syndrome. *Taiyang* syndrome is not finished but the pathogen has already invaded *shaoyang*. The formula is the *Xiǎo Chái Hú Tāng* (Minor Bupleurum Decoction) with *guì zhī* (Ramulus Cinnamomi) and *sháo yào* (Radix Paeoniae Rubra), it is the *Guì Zhī Tāng* (Cinnamon Twig Decoction) plus the *Xiǎo Chái Hú Tāng* (Minor Bupleurum Decoction). This formula does not eliminate *rén shēn* (Radix et Rhizoma Ginseng), this is due to the fact that *guì zhī* (Ramulus Cinnamomi) and *sháo yào* (Radix Paeoniae Rubra) are herbs where one is hot and the other is cold and are added into *Xiǎo Chái Hú Tāng* (Minor Bupleurum Decoction) at the same time. It will not cause the property of the whole formula to lose balance and appear as a cold or heat syndrome, therefore, *rén shēn* (Radix et Rhizoma Ginseng) does not need to be eliminated. Zhang Zhong-jing's herbal application of adding or eliminating an herb, from this angle, is very subtle. Both the representative syndromes and the associated syndromes all need to be considered. To add or eliminate an herb is not done at will, but also, it is not a stereotype or absolutely invariable.

- If there is cough, eliminate *rén shēn* (Radix et Rhizoma Ginseng), *dà zǎo* (Fructus Jujubae) and *shēng jiāng* (Rhizoma Zingiberis Recens), and add *wǔ wèi zǐ* (Fructus Schisandrae Chinensis) 0.5 *catty* and *gān jiāng* (Rhizoma Zingiberis) 2 *tael*.

Here the cough belongs to the cold qi rebellion induced cough. It belongs to the excess cough; therefore, it is not appropriate to use the tonifying herbs of *rén shēn* (Radix et Rhizoma Ginseng) or *dà zǎo* (Fructus Jujubae) to prevent the sweet taste inducing stagnation and becoming a disadvantage to the rebellious qi induced cough.

The *Divine Husbandman's Classic of the Materia Medica* states of *wǔ wèi zǐ* (Fructus Schisandrae Chinensis): "*It governs tonifying qi, cough of qi rebellion, over-exertion injury, emaciation and strengthening the yin.*" From this, it can be seen that *wǔ wèi zǐ* (Fructus Schisandrae Chinensis) is a good herb to calm the cough. Its taste is sour and its property is warm. It can restrain the lung qi. It is the best one for the lung coldness and qi rebellion inducing cough.

In the *Divine Husbandman's Classic of the Materia Medica* it states of *gān jiāng* (Rhizoma Zingiberis): "*It governs chest fullness and rebellious qi inducing cough.*" It can warm the lung and stop cough. It is also an herb to treat the lung coldness induced cough.

In the *Discussion on Cold Damage*, Zhang Zhong-jing treated the lung coldness induced cough more often using *wǔ wèi zǐ* (Fructus Schisandrae Chinensis) and *gān jiāng* (Rhizoma Zingiberis) together, as in *Xiǎo Qīng Lóng Tāng* (Minor Green-Blue Dragon Decoction). It uses sourness and acridity together and has a mutually regulating function. The sourness of *wǔ wèi zǐ* (Fructus Schisandrae Chinensis) can retain the lung, strengthen the yin, generate fluid, stop cough and prevent the acridity and dispersing nature of *gān jiāng* (Rhizoma Zingiberis) from consuming the yin. The acridity and dispersing nature of *gān jiāng* (Rhizoma Zingiberis) also can prevent *wǔ wèi zǐ* (Fructus Schisandrae Chinensis) from over retaining due to its sour taste.

Why did this item eliminate *shēng jiāng* (Rhizoma Zingiberis Recens) and add *gān jiāng* (Rhizoma Zingiberis)? We say that basically the properties of *gān jiāng* (Rhizoma Zingiberis) and *shēng jiāng* (Rhizoma Zingiberis Recens) are the same, but *shēng jiāng* (Rhizoma Zingiberis Recens) leans towards dispersing and *gān jiāng* (Rhizoma Zingiberis) leans towards warmth and heat. As cough belongs to the lung coldness induced qi rebellion, therefore, the warmth of *gān jiāng* (Rhizoma Zingiberis) to warm the lung and stop cough should be used, instead of the dispersing and releasing the exterior of *shēng jiāng* (Rhizoma Zingiberis Recens). *Shēng jiāng* (Rhizoma Zingiberis Recens) and *gān jiāng* (Rhizoma Zingiberis) all belong to one product, what is their difference? *Shēng jiāng* (Rhizoma Zingiberis Recens) is dried under the sun, has lost most of its dispersing property but still retains its non-volatile ingredients, therefore, relatively, *shēng jiāng* (Rhizoma Zingiberis Recens) has more volatile ingredients and *gān jiāng* (Rhizoma Zingiberis) has more non-volatile ingredients, so, *shēng jiāng* (Rhizoma Zingiberis Recens) leans towards dispersing and *gān jiāng* (Rhizoma Zingiberis) leans towards warmth and heat.

The modification of this item, is versatile. It is not an absolute method. In clinical practice, *gān jiāng* (Rhizoma Zingiberis) and *wǔ wèi zǐ* (Fructus Schisandrae Chinensis) are not added simply because *Chái Hú* Syndrome is combined with cough. If it belongs to lung coldness and rebellious qi, this modification is correct and adequate. If it belongs to the excess heat with an excess pulse, yellow and sticky phlegm, irritability and stuffy sensation in the chest, then, this modification is not appropriate. If further adding *gān jiāng* (Rhizoma Zingiberis) and *wǔ wèi zǐ* (Fructus Schisandrae Chinensis), it will cause a bad result. The prescription should be changed with the change of the syndrome without being conservative.

3. The Principles of Modificaiton of *Chái Hú* Syndrome

Through Zhang Zhong-jing's analysis in the 96th item of the *Discussion on Cold Damage*, draws the following conclusions for the formula modification rule of the *Chái Hú* Syndrome:

• First, the essence of the syndrome should be observed. In other words, to apply a modification method for the *Chái Hú* Syndrome, first differentiate if the syndrome belongs to the *shaoyang* syndrome or not. If it belongs to the *shaoyang* syndrome, then use this method.

• Secondly observe with dedication. As illness is always changing, the chills, fever, deficiency or excess conditions of the representative syndrome and its associated syndromes needs to be observed. In doing so, correct modifications can be made, without mistake.

● Thirdly, it should be modified according to the disease changes and the associated syndromes' expressions, but firstly research if there are any herbs that do not fit the syndromes, what should be omitted and what should be added?

● If the syndrome has *taiyang* symptoms, add herbs that can release the exterior. If the syndrome has interior excess, add herbs that can clear the interior. It follows that if it has both exterior and interior signs, the exterior releasing and interior clearing herbs can be added in at the same time.

● *Chái hú* (Radix Bupleuri) and *huáng qín* (Radix Scutellariae) are bitter and cold herbs. *Rén shēn* (Radix et Rhizoma Ginseng) and *dà zǎo* (Fructus Jujubae) are sweet and warm herbs. These two categories of herbs have the function to control the cold or heat property of the *Xiǎo Chái Hú Tāng* (Minor Bupleurum Decoction), especially, *huáng qín* (Radix Scutellariae) and *rén shēn* (Radix et Rhizoma Ginseng), as one is cold and one is tonifying. If the syndrome belongs to excess, do not use *rén shēn* (Radix et Rhizoma Ginseng). If the syndrome belongs to deficiency, do not use *huáng qín* (Radix Scutellariae).

● If add in salty herbs, it is better to reduce the sweet tasting herbs.

● *Chái Hú* Syndrome is a formula composed of three categories of herbs: sweet herbs, bitter herbs and acrid herbs. The acrid, warm *shēng jiāng* (Rhizoma Zingiberis Recens) and *bàn xià* (Rhizoma Pinelliae) are beneficial for descending rebellions and stopping vomiting, however, their property is warm and dry which will injure the yin. The sweetness of *rén shēn* (Radix et Rhizoma Ginseng), *dà zǎo* (Fructus Jujubae) and *gān cǎo* (Radix Glycyrrhizae) can prevent its disadvantage in consuming the fluid. *Chái hú* (Radix Bupleuri) and *huáng qín* (Radix Scutellariae) can prevent it leaning towards dryness and heat. Therefore, in clinical practice, depending on the condition and according to the syndromes, the herbs and dosages in the formula can be adjusted.

● To add any herb into the formula, it is not only required to fit the associated syndromes, it should also not affect the representative syndromes. It should take care of both the associated syndrome and the representative syndrome.

THE CLINICAL APPLICATIONS OF THE *XIǍO CHÁI HÚ TĀNG*

The above already explained the compatibilities and modifications of the *Chái Hú* Syndrome and the *Chái Hú Tāng* (Minor Bupleurum Decoction). Its purpose is to raise and guide clinical applications and expand the clinical application scope of the *Xiǎo Chái Hú Tāng* (Minor Bupleurum Decoction). The author on the applications of *Xiǎo Chái Hú Tāng* (Minor Bupleurum Decoction), mainly based his applications on the three representative syndromes of the *Chái Hú* Syndrome. These are the previously mentioned: Chest and Hypochondria Syndrome, Alternated Chills and Fever Syndrome and Stomach and Intestines Syndrome. Meanwhile, using Zhang Zhong-jing's modification rule to adjust herbs, some case studies are outlined below to clarify his rules.

1. Chest and Hypochondria Syndrome Case Studies: The Chest and Hypochondria Syndrome of the *Xiǎo Chái Hú Tāng* (Minor Bupleurum Decoction) is the chest and hypochondria suffering from fullness, or fullness and irritability in the chest, or fullness in hypochondria, or hardness and fullness in the hypochondria, or fullness and pain in the hypochondria, etc. These symptoms can appear in many diseases in clinical practice, such as hepatitis, liver cirrhosis, cholecystitis, gallstones and intercostal neuralgia, etc. Using Zhang Zhong-jing's rule of *"When seeing only one syndrome it can be applied, it does not need to have all of them"* to apply the *Xiǎo Chái Hú Tāng* (Minor Bupleurum Decoction) for treatment can produce satisfactory results.

Case 1

Hepatitis B.

Initial Visit

A female patient, 35 years old, April 29th, 1997.

The patient had a medical history as a hepatitis B carrier. Recently, due to overexertion, she had lassitude, anorexia, fullness and stuffy sensations in her chest and diaphragm, fullness and pain discomfort sensations on the liver area in the hypochondria. Her liver function test at a local hospital was abnormal: TTT 6.7U, ALT 86U/L, and three big positives for her hepatitis B five indicators test. Her pulse was deep and slow. Her tongue coating was white but slightly turbid. The treatment plan was to use a formula to soothe the liver, harmonise the stomach, strengthen the spleen, expel dampness, transform blood stasis and detoxify toxins.

Prescription

白花蛇舌草	*bái huā shé shé cǎo*	30g	Herba Hedyotis Diffusae
蒲公英	*pú gōng yīng*	15g	Herba Taraxaci
柴胡	*chái hú*	15g	Radix Bupleuri
黄芩	*huáng qín*	9g	Radix Scutellariae
虎杖	*hǔ zhàng*	15g	Rhizoma et Radix Polygoni Cuspidati
丹参	*dān shēn*	20g	Radix et Rhizoma Salviae Miltiorrhizae
生薏仁	*shēng yì rén*	24g	Semen Coicis (raw)
枳壳	*zhǐ qiào*	12g	Fructus Aurantii
白术	*bái zhú*	12g	Rhizoma Atractylodis Macrocephalae
香附	*xiāng fù*	12g	Rhizoma Cyperi
砂仁	*shā rén*	8g	Fructus Amomi
甘草	*gān cǎo*	6g	Radix Glycyrrhizae

This was decocted with water and divided to be taken twice, one pack per day.

Second Visit

July 2nd: After taking a total of more than sixty packs of the above formula, her liver function test was normal, all of her symptoms disappeared and her hepatitis B five indicators became three small positives.

Note

The patient in this case had fullness and pain discomfort sensations in her hypochondria, as well as the fullness and stuffy sensation in her chest of the Chest and

Hypochondria Syndrome. She also had the anorexia (no appetite) of the Stomach and Intestines Syndrome. She had two representative syndromes of the *Chái Hú* syndrome; therefore, *Xiǎo Chái Hú Tāng* (Minor Bupleurum Decoction) was used as the chief. As the patient had a white turbid tongue coating, a fullness and stuffy sensation in her chest and diaphragm, and a slow pulse, these indicated a dampness pathogen obstructing the interior syndrome. Therefore, the sweet and moist herbs of *rén shēn* (Radix et Rhizoma Ginseng) and *dà zǎo* (Fructus Jujubae) in the *Chái Hú Tāng* (Minor Bupleurum Decoction) were omitted, to prevent them trapping the pathogen and increasing the dampness. As the patient did not vomit, therefore, he did not use *shēng jiāng* (Rhizoma Zingiberis Recens) and *bàn xià* (Rhizoma Pinelliae), instead, he added in the herbs that can soothe the liver, harmonise the stomach, transform dampness and detoxify toxins and the treatment was effective.

Case 2

Gallstone syndrome.

Initial Visit

A female patient, 36 years old, March 19th, 1985.

The patient's chief complaint: Ever since the patient was young, she had a medical history of abdominal pain. For years, it did not get better. When she had severe pain, she was treated for epigastric pain and it was sometimes better and sometimes worse. At the beginning of the year, she had a B-ultrasound test in an hospital and was diagnosed with cholecystitis with a gallstone. Her gallbladder wall was slightly thicker at 0.4cm, not smooth and the stone size was 1.3cm. The patient had good physical development. Her right upper abdominal pain extended to her right hypochondria gallbladder area and was accompanied by anorexia, lassitude and abdominal distention. Her pulse was wiry. Her tongue was thin and white. The treatment plan was to soothe the liver, benefit the gallbladder, sterilise and expel the stone.

Prescription

柴胡	*chái hú*	20g	Radix Bupleuri
黄芩	*huáng qín*	9g	Radix Scutellariae
半夏	*bàn xià*	10g	Rhizoma Pinelliae
杭白芍	*háng bái sháo*	30g	Radix Paeoniae Lactiflorae
大黄	*dà huáng*	12g	Radix et Rhizoma Rhei
鸡内金	*jī nèi jīn*	20g	Endothelium Corneum Gigeriae Galli
郁金	*yù jīn*	20g	Radix Curcumae
枳实	*zhǐ shí*	12g	Fructus Aurantii Immaturus
金钱草	*jīn qián cǎo*	40g	Herba Lysimachiae
蒲公英	*pú gōng yīng*	20g	Herba Taraxaci
藿香	*huò xiāng*	9g	Herba Agastachis
甘草	*gān cǎo*	6g	Radix Glycyrrhizae

This was decocted with water and divided to be taken twice, once each in the morning and evening, one pack per day.

Second Visit

March 21st: After taking three packs of the above formula, her pain was reduced and her appetite increased. She had two to three bowel movements a day. She was instructed to continue with the same formula.

Third Visit

April 12th: After taking more than twenty packs of modifications of the above formula, her symptoms basically disappeared except that there was a heavy sensation in her upper abdomen. This indicated that the patient's cholecystitis was basically improving but the gallstone was still there. The treatment plan was to use a formula to soothe the liver, benefit the gallbladder and expel the gallstone to attack it.

Prescription

柴胡	*chái hú*	20g	Radix Bupleuri
黄芩	*huáng qín*	10g	Radix Scutellariae
金钱草	*jīn qián cǎo*	60g	Herba Lysimachiae
郁金	*yù jīn*	20g	Radix Curcumae
大黄	*dà huáng*	10g	Radix et Rhizoma Rhei
鸡内金	*jī nèi jīn*	30g	Endothelium Corneum Gigeriae Galli
杭白芍	*háng bái sháo*	20g	Radix Paeoniae Lactiflorae
丹参	*dān shēn*	20g	Radix et Rhizoma Salviae Miltiorrhizae
半夏	*bàn xià*	9g	Rhizoma Pinelliae
钩藤	*gōu téng*	30g	Ramulus Uncariae Cum Uncis
川楝子	*chuān liàn zǐ*	10g	Fructus Toosendan
甘草	*gān cǎo*	10g	Radix Glycyrrhizae

She was instructed that 10 minutes after taking the decoction, she was to use animal oil to fry two eggs and eat them.

Fourth Visit

April 15th: After taking one pack of the formula, she had right upper abdominal pain but it was not severe. One hour later, her pain stopped. There was no pain after taking decoction. The B-ultrasound test showed there was no gallstone.

Note

This case used hypochondriac pain, anorexia, and abdominal distention as the main symptoms to apply the *Chái Hú Tāng* (Minor Bupleurum Decoction). This case also had both the Chest and Hypochondria Syndrome and Stomach and Intestines Syndrome but was mainly the Chest and Hypochondria Syndrome. Due to the gallstone treatment plan it was better to soothe the liver, benefit the gallbladder and expel the stone, therefore, in the formula, *rén shēn* (Radix et Rhizoma Ginseng), *dà zǎo* (Fructus Jujubae) and *gān jiāng* (Rhizoma Zingiberis) that are sweet, warm, stagnating and tonifying were eliminated due to concern that they could not benefit expelling the stone. Instead, herbs were added in that could soothe the liver, benefit the gallbladder, clear heat and expel the stone.

Case 3

Cholecystitis.

Initial Visit

A male patient, 16 years old, January 13th, 1988.

The patient had right hypochondriac pain, accompanied by epigastric and abdominal distention and fullness, worse after eating, for the previous couple of months. He took medicine to eliminate food and strengthen the stomach but they did not work and so he came for treatment. On examination: His pulse was wiry and forceful. His tongue coating was white and his tongue was red. His right hypochondria in the gallbladder area had tenderness on pressing and the pain became worse when percussing. The B-ultrasound test showed: The gallbladder was enlarged, the gallbladder wall thickness was 0.5cm and not smooth. He was diagnosed with cholecystitis. The treatment plan was to soothe the liver, benefit the gallbladder, clear inflammation and alleviate pain.

Prescription

柴胡	chái hú	15g	Radix Bupleuri
黄芩	huáng qín	9g	Radix Scutellariae
半夏	bàn xià	9g	Rhizoma Pinelliae
白芍	bái sháo	15g	Radix Paeoniae Alba
枳实	zhǐ shí	10g	Fructus Aurantii Immaturus
大黄	dà huáng	5g	Radix et Rhizoma Rhei
蒲公英	pú gōng yīng	15g	Herba Taraxaci
延胡索	yán hú suǒ	10g	Rhizoma Corydalis
郁金	yù jīn	9g	Radix Curcumae
金钱草	jīn qián cǎo	24g	Herba Lysimachiae
甘草	gān cǎo	6g	Radix Glycyrrhizae

This was decocted with water and divided to be taken twice, warm, once each in the morning and evening, one pack per day.

Second Visit

January 19th: After taking six packs of the above formula, his hypochondriac pain was eliminated and his abdominal distention and fullness were also significantly reduced. On pressing on his gallbladder area, there was still discomfort. Again, he was instructed to take the formula as before.

Third Visit

January 28th: His pain disappeared. His appetite increased and after eating, there were no distention or fullness sensations. As it was inconvenient for a student to take decoctions, his prescription was changed to the patent remedy *Xiāo Yán Lì Dǎn* tablets (Anti-Inflammatory and Gallbladder-Disinhibiting Tablet) to clear it up.

Note

This cholecystitis case had hypochondriac pain and abdominal fullness as the representative syndromes of the *Chái Hú Tāng* (Minor Bupleurum Decoction), therefore, a modified *Chái Hú Tāng* (Minor Bupleurum Decoction) was used for treatment and he recovered. As the inflammatory symptom is related to heat, therefore, the following herbs were not used: *rén shēn* (Radix et Rhizoma Ginseng), *shēng jiāng* (Rhizoma Zingiberis Recens) and *dà zǎo* (Fructus Jujubae), all the sweet, acrid, warm and tonifying herbs in the *Xiǎo Chái Hú Tāng* (Minor Bupleurum Decoction). The concern was that they could add heat and trap the pathogen. Instead, *bái sháo* (Radix Paeoniae Alba), *yán hú suǒ* (Rhizoma Corydalis), *yù jīn* (Radix Curcumae) and *zhǐ shí* (Fructus Aurantii Immaturus) were added to soothe the liver, resolve stagnation and alleviate pain as well as adding *dà huáng* (Radix et Rhizoma Rhei), *pú gōng yīng* (Herba Taraxaci) and *jīn qián cǎo* (Herba Lysimachiae) that can clear heat, eliminate inflammation and benefit the gallbladder.

Case 4

Intercostal neuralgia.

Initial Visit

A male patient, 34 years old, February 28th, 1986.

The patient suddenly had pain on both sides of the rib cage. It was significant especially around three to four o'clock in the morning. Most of the time, the pain woke him up from sleeping. After getting up with some exertion, the pain was slightly relieved. Every time, after he awoke, he could not fall back to sleep. His appetite, urination and bowel movements were all normal. His pulse was wiry. His tongue coating was thin and white. This was liver stagnation and qi obstruction syndrome. The morning is the time for the yang to start growing and the time between three to five o'clock in the morning governs the liver. Pain at a fixed time is more due to blood stasis, therefore, the treatment plan was to soothe the liver and resolve stagnation.

Prescription

柴胡	*chái hú*	15g	Radix Bupleuri
杭白芍	*háng bái sháo*	30g	Radix Paeoniae Lactiflorae
延胡索	*yán hú suǒ*	15g	Rhizoma Corydalis
木香	*mù xiāng*	9g	Radix Aucklandiae
陈皮	*chén pí*	10g	Pericarpium Citri Reticulatae
川楝子	*chuān liàn zǐ*	10g	Fructus Toosendan
甘草	*gān cǎo*	9g	Radix Glycyrrhizae

This was decocted with water and divided to be taken twice, one pack per day.

Second Visit

March 4th: After taking three packs of the above formula, his pain was stopped and there was no feeling of discomfort. The patient was concerned about reoccurrence and so he was instructed to take another three packs to make a full recovery.

Note

There was only one syndrome in this case, as there was no chills, fever or Stomach and Intestines Syndrome. Therefore, *huáng qín* (Radix Scutellariae), *rén shēn* (Radix et Rhizoma Ginsen), *shēng jiāng* (Rhizoma Zingiberis Recens), *dà zǎo* (Fructus Jujubae) and *bàn xià* (Rhizoma Pinelliae) in the *Xiǎo Chái Hú Tāng* (Minor Bupleurum Decoction) were all unnecessary and only *chái hú* (Radix Bupleuri) and *gān cǎo* (Radix Glycyrrhizae), that can soothe the liver, relax the urgency and alleviate pain, were kept. For enhancing the function of resolving stagnation and alleviating pain, *yán hú suǒ* (Rhizoma Corydalis), *chuān liàn zǐ* (Fructus Toosendan), *háng bái sháo* (Radix Paeoniae Lactiflorae), *mù xiāng* (Radix Aucklandiae) and *chén pí* (Pericarpium Citri Reticulatae), etc, were added in in order to enhance *chái hú* (Radix Bupleuri) to soothe the liver and resolve stagnation, and *gān cǎo* (Radix Glycyrrhizae) to relax urgency and alleviate pain.

2. Alternated Chills and Fever Syndrome Case Studies: Alternating chills and fever, that stops and manifests at fixed times. This is the *Chái Hú* Syndrome specialised heat pattern. *Shaoyang* is the pivot of the three yangs. Once pathogens have invaded the *shaoyang* and are switching back and forth between half exterior and half interior, the pathogen and the yang qi battling manifests as the heat, and battling with the yin as the cold. With yin or yang entering or leaving, each has its time, therefore, chills and fever are alternated, and its outbreak and remission has a fixed time. If in clinical practice, seeing any alternating chills and fever that belong to the heat patten, giving *Xiǎo Chái Hú Tāng* (Minor Bupleurum Decoction) to treat it will be highly effective. In clinical practice, aside from an exterior invasion syndrome with a chills and fever pattern, syndromes such as malaria and tuberculosis will also have this kind of heat pattern. If treated with the *Xiǎo Chái Hú Tāng* (Minor Bupleurum Decoction) it has the same effectiveness.

Case 1

Exterior invasion.

Initial Visit

A male patient, 61 years old, September 20th, 1997.

The patient had had a cold and fever for six or seven days. He was treated by Chinese and biomedicine without effect. He also used antibiotics for three days. His high fever had dropped but he still had lassitude, chest fullness, anorexia, no appetite, a bitter mouth, a dry throat, chills and fever discomfort symptoms without relief. On examination, his pulse was wiry and fast. His tongue coating was thin and white. This was the pathogen invading the *shaoyang* syndrome. His illness was at the half-exterior and half-interior phase. All of the exterior relieving and interior attacking herbs could not expel the pathogen out of the body. *Xiǎo Chái Hú Tāng* (Minor Bupleurum Decoction) should be used to harmonise for treatment. He was given the original formula of the *Xiǎo Chái Hú Tāng* (Minor Bupleurum Decoction):

柴胡	*chái hú*	15g	Radix Bupleuri
黄芩	*huáng qín*	10g	Radix Scutellariae
半夏	*bàn xià*	9g	Rhizoma Pinelliae

党参	*dǎng shēn*	12g	Radix Codonopsis
炙甘草	*zhì gān cǎo*	6g	Radix et Rhizoma Glycyrrhizae Praeparata cum Melle
生姜	*shēng jiāng*	3slices	Rhizoma Zingiberis Recens
大枣	*dà zǎo*	5pcs	Fructus Jujubae

This was decocted with water and divided to be taken twice, once each in the morning and evening, one pack per day.

Second Visit

September 23rd: After taking one pack of the above formula, his chills and fever were eliminated. After taking three packs, all of his symptoms disappeared.

Note

This case is the typical *Chái Hú* Syndrome, therefore, the original formula was used and the patient recovered. The alternating chills and fever in this case were significant, therefore, *chái hú* (Radix Bupleuri) and *huáng qín* (Radix Scutellariae) must be used. As it was accompanied by chest fullness, anorexia, and no appetite, etc., he had both the Chest and Hypochondria Syndrome and Stomach and Intestines Syndrome, therefore, *bàn xià* (Rhizoma Pinelliae) and *shēng jiāng* (Rhizoma Zingiberis Recens) could not be deleted. The exterior invasion was already six or seven days old, and was accompanied by lassitude and dry mouth, etc. This indicated that his normal qi was already deficient and his qi and yin were also injured, therefore, *dǎng shēn* (Radix Codonopsis), *dà zǎo* (Fructus Jujubae) and *zhì gān cǎo* (Radix et Rhizoma Glycyrrhizae Praeparata cum Melle) also could not be reduced. There was no need for modification as there were no other syndromes associated with this one. The formula and the syndrome were exactly matched, and the effects were also quick.

Case 2

Malaria.

Initial Visit

A male patient, 42 years old, July 3rd, 1972.

Since the previous week, the patient had afternoon shivers followed by high fever. It occurred once every other day, three times so far. His Town Health Centre diagnosed him with malaria. He had been treated with chloroquine tablets but his condition was still not controlled, therefore, he came for Chinese medicine treatment. On examination, his physique was strong. His pulse was wiry. His tongue was white and slightly turbid. The patient's chief complaint was that in the afternoons around 1:00 PM, he had shivers, and aversion to cold. Although he wrapped up with blankets, he still could not keep out the cold, then, around half an hour later, there was fever, accompanied by a headache, and a red face. About two to three hours later, his whole body was sweating profusely and his body temperature gradually dropped to normal. His whole body felt lethargic and fatigued. He needed to rest in bed for half a day to recover. This occurred every other day. The diagnosis was tertian malaria. Modified *Xiǎo Chái Hú Tāng* (Minor Bupleurum

Decoction) was used to treat him because of the alternating chills and fever, and its fixed times of manifesting and remitting.

Prescription

柴胡	chái hú	20g	Radix Bupleuri
黄芩	huáng qín	15g	Radix Scutellariae
常山	cháng shān	20g	Radix Dichroae
半夏	bàn xià	9g	Rhizoma Pinelliae
党参	dǎng shēn	15g	Radix Codonopsis
槟榔片	bīng láng piàn	10g	Semen Arecae (slice)
青蒿	qīng hāo	15g	Herba Artemisiae Annuae
甘草	gān cǎo	9g	Radix Glycyrrhizae

This was decocted with water and divided to be taken twice, one pack per day.

Second Visit

July 7th: After taking one pack, his symptoms were reduced the next time it occurred. The degree of shivers and high fever, and the duration was also reduced. After taking four packs, his Alternated Chills and Fever Syndrome basically disappeared. He was instructed to continuously take three more packs of the same formula and he recovered.

Note

Malaria is an infection of plasmodium. It is a disease that after the plasmodium has entered the human's body, it multiplies in the liver cells and red blood cells causing the red blood cells to be damaged periodically. As its outbreak is of the heat type, it belongs to the typical heat type of the *Chái Hú* Syndrome, therefore, previous generations of doctors all used the *Chái Hú Tāng* to treat it. As malaria is not an exterior invasion heat disease, if the *Xiǎo Chái Hú Tāng* (Minor Bupleurum Decoction) is only used, though it can alleviate its symptoms, it cannot treat its root problem. Therefore, into the *Xiǎo Chái Hú Tāng* (Minor Bupleurum Decoction), *cháng shān* (Radix Dichroae) and *bīng láng piàn* (sliced Semen Arecae) were added to cut off the malaria and kill the plasmodium in order to increase the treatment effect. *Qīng hāo* (Herba Artemisiae Annuae) was added to help *chái hú* (Radix Bupleuri), and *huáng qín* (Radix Scutellariae) was added to clear heat and pierce the pathogen. Modern pharmacological experiments have shown that the *Xiǎo Chái Hú Tāng* (Minor Bupleurum Decoction) has functions to protect the liver, relieve heat, calm and increase immunity, therefore, it is effective to treat malaria.

Case 3

Pulmonary tuberculosis with fever.

Initial Visit

A male patient, 29 years old, October 7th, 1998.

The patient had had right chest and hypochondriac pain, cough, lassitude, and afternoon hot flushes for more than half a month. He was investigated in a local hospital and diagnosed with infiltration of the right upper lung by tuberculosis. After he was treated with anti-tuberculosis medicine of isoniazid and streptomycin for a week, tinnitus and dizziness manifested. Then, he stopped the streptomycin and came to our hospital to ask for treatment.

The patient's physical development was satisfactory. His body leaned towards being emaciated and weak. He complained of right chest pain, cough, a little phlegm, poor appetite, chills and fever with a low grade fever in the afternoon, his body temperature was around 37.5℃, accompanied by lassitude, fatigue, and night sweating, etc. His pulse was wiry, thready and fast. His tongue was red and his tongue coating was thin and white. This was both qi and yin deficiency induced tuberculosis. He worked hard everyday, so this first caused his normal qi to be deficient. Mycobacterium tuberculosis seized the opportunity of deficiency to sneak in, and injured the lung qi and consumed the lung yin. Both his qi and yin were then deficient; therefore lassitude, body tiredness, cough, chills and fever, night sweating, etc. appeared. The treatment plan was to tonify qi, nourish yin, relieve the heat and clear the consumption.

Prescription

柴胡	*chái hú*	15g	Radix Bupleuri
黄芩	*huáng qín*	10g	Radix Scutellariae
百部	*bǎi bù*	15g	Radix Stemonae
功劳叶	*gōng láo yè*	30g	Folium Ilex
青蒿	*qīng hāo*	12g	Herba Artemisiae Annuae
大贝母	*dà bèi mǔ*	10g	Bulbus Fritillariae Thunbergii
杏仁	*xìng rén*	9g	Semen Armeniacae Dulcis
桔梗	*jié gěng*	10g	Radix Platycodonis
百合	*bǎi hé*	30g	Bulbus Lilii
地榆	*dì yú*	10g	Radix Sanguisorbae
半夏	*bàn xià*	9g	Rhizoma Pinelliae
太子参	*tài zǐ shēn*	15g	Radix Pseudostellariae
甘草	*gān cǎo*	6g	Radix Glycyrrhizae

This was decocted with water and divided to be taken twice, one pack per day. He was instructed to stay away from tobacco, alcohol, acrid and spicy food.

Externally an anti-tuberculosis pill was used to rub on his back-shu points, once a week. (For its usage and compatibility, please refer to *xióng huáng* – Realgar).

Second Visit

October 14th: After taking seven packs of the above formula, his afternoon chills and fever symptom was reduced. His appetite increased. His cough, lassitude and night sweating were also improved. The original formula was used to let him continue taking it.

Third Visit

October 21st: All of his symptoms were visibly improved. Again, the above formula was used and modified to treat him.

Fourth Visit

October 28th: His afternoon chills and fever symptoms were eliminated. His cough, chest and hypochondriac pain were already insignificant. His chest x-ray reported: The right upper lung tuberculosis infiltration had improved. His prescription was changed to *Gōng Láo Bǎi Bù Tāng* (Folium Ilicis and Stemona Root Decoction) for treatment (for its usage and composition please refer to *gōng láo yè* - Folium Ilex), for three months and his illness recovered.

Note

The pulmonary tuberculosis patient in this case had chills and fever, occurring at fixed times in the afternoon. This belonged to the heat type of the *Chái Hú* Syndrome. Meanwhile, there were chest and hypochondriac pain, poor appetite, etc. hence, the *Xiǎo Chái Hú Tāng* (Minor Bupleurum Decoction) was used as the chief to treat it. Due to the fact that the heat was induced by the lung yin deficiency, not the pathogen invading the exterior, therefore, the regulating, nutritive and defensive qi herbs of *shēng jiāng* (Rhizoma Zingiberis Recens) and *dà zǎo* (Fructus Jujubae) were eliminated, but anti-tuberculosis herbs of *bǎi bù* (Radix Stemonae), *gōng láo yè* (Folium Ilex) and *dì yú* (Radix Sanguisorbae) were added in, as well as nourishing yin and moistening lung herbs of *bǎi bù* (Radix Stemonae), *dà bèi mǔ* (Bulbus Frittilariae Thunbergii) and *xìng rén* (Semen Armeniacae Dulcis), etc. The formula was close to the aetiology, therefore, it was effective to use it.

Case 4

Heat invades uterus.

During the female menstrual period when invaded by wind-cold, the exterior pathogen takes advantage of the deficiency and transforms into heat, enters internally and finally becomes heat entering the uterus syndrome. The fever symptoms sometimes manifest as the chills and fever episode with fixed times, or similar to malaria, or only fever. In clinical practice, if a female patient presents during her menstruation period and has exterior invasion induced fever, the *Xiǎo Chái Hú Tāng* (Minor Bupleurum Decoction) can be used for treatment.

Initial Visit

A female patient, 29 years old, April 18th, 1985.

Since the previous January, after the patient got cold during her menstrual period, every time for ten days before her menstruation, she had cough, rapid asthma, with a phlegm noise and she could not lie down flat. After menstruation, all of her symptoms reduced. She had been treated in a hospital with antibiotics and medicines for stopping cough, and calming asthma without effect. After she was discharged from the hospital, she came to our clinic for treatment. The patient breathed rapidly with her mouth open and shoulders raised; there was a phlegm noise sounding from her throat, and was very painful and was accompanied by afternoon aversion to cold, and fever. Her pulse was thready and fast. Her tongue was red with a thin and white coating. This was exterior invasion during the menstruation period, with the pathogen taking the advantage of her deficiency, entering her uterus inducing obstruction in her penetrating vessel and qi to rebel upward. The qi rebelling upward pressed on the lung qi that could not disperse and descend, therefore, she had cough. The treatment plan was to regulate the penetrating vessel and descend rebellion.

Prescription

麦门冬	*mài mén dōng*	30g	Radix Ophiopogonis
半夏	*bàn xià*	15g	Rhizoma Pinelliae
代赭石	*dài zhě shí*	20g	Haematitum
党参	*dǎng shēn*	15g	Radix Codonopsis
川牛膝	*chuān niú xī*	15g	Radix Cyathulae
甘草	*gān cǎo*	10g	Radix Glycyrrhizae
粳米	*jīng mǐ*	1pinch	Oryza Sativa L.(as a guide)

This was decocted with water and divided to be taken twice, warm, one pack per day.

Second Visit

April 20th: After taking two packs of the above formula, her cough and phlegm noise had stopped and she could lie down flat, and all except her afternoon chills and fever were eliminated. At this stage it was suddenly realised that heat had entered the uterus, and a harmonising method should be used to treat it.

Prescription

麦门冬	*mài mén dōng*	24g	Radix Ophiopogonis
半夏	*bàn xià*	9g	Rhizoma Pinelliae
柴胡	*chái hú*	12g	Radix Bupleuri
黄芩	*huáng qín*	9g	Radix Scutellariae
党参	*dǎng shēn*	12g	Radix Codonopsis
桔梗	*jié gěng*	12g	Radix Platycodonis
百合	*bǎi hé*	20g	Bulbus Lilii
代赭石	*dài zhě shí*	15g	Haematitum
川牛膝	*chuān niú xī*	15g	Radix Cyathulae
甘草	*gān cǎo*	6g	Radix Glycyrrhizae

This was decocted with water and taken.

Third Visit

April 23rd: After taking three packs of the above formula, all of her symptoms disappeared. Later, through a follow up visit, when her menstruation came, there was no occurrence of cough, asthma and chills, or fever.

Note

This case is heat entering the uterus with penetrating vessel qi rebellion inducing cough and asthma syndrome. In the beginning the aim was to treat the penetrating vessel qi rebellion, and although her cough and asthma were reduced, the chills and fever were not eliminated. Later the *Xiǎo Chái Hú Tāng* (Minor Bupleurum Decoction) was used as the chief to treat her and she recovered. It is evident that Zhang Zhong-jing's theory is not erroneous

3. **Stomach and Intestines Syndrome:** This is also a representative syndrome of the *Chái Hú* Syndrome. *Shaoyang* belongs to the gallbladder. If the pathogen is knotted at the hypochondria, the qi of the gallbladder will be stagnated and bound to rebel horizontally. Then, it must affect the stomach of the middle *jiao* and intestines (wood overacting on earth). Therefore, Zhong-jing said that: *"Zang and fu are connected together, if there is pain it must go down. If the pathogen is higher and the pain is lower this will induce vomiting."* In clinical practice, the Stomach and Intestines Syndrome of the *Chái Hú* Syndrome are more often expressed as vomiting, hiccough, no appetite, abdominal pain, or irritability and a want to vomit, etc. These symptoms sometimes appear with Chest and Hypochondria Syndrome, sometimes with Alterllated Chills and Fever Syndrome, but sometimes, it appears alone. When two representative syndromes appear simultaneously, it is easier to differentiate for using the *Chái Hú Tāng* (Minor Bupleurum Decoction). When only Stomach and Intestines Syndrome appears alone, it is more difficult to know how to apply the *Xiǎo Chái Hú Tāng* (Minor Bupleurum Decoction), therefore, we must differentiate from the overall picture and carefully observe and analyse in order to be accurate, without error.

Case 1

Pancreatitis

Initial Visit

A male patient, 45 years old, April 12th, 1988.

The patient had abdominal pain and vomiting for a couple of days. He was investigated in a hospital and diagnosed with pancreatitis. After treatment, his pain had still not stopped and so, he came to our clinic for treatment. On physical examination, his upper abdominal pain refused pressure. When in pain, it expanded to his lower back, and was accompanied by nausea and vomiting that became worse after eating, with epigastric and abdominal distention, and fullness and discomfort. His pulse was wiry and forceful. His tongue coating was white and slightly turbid and his tongue was red. This was the liver and stomach disharmony and pathogenic heat stagnated internally syndrome. The treatment plan was to soothe the liver, harmonise the stomach, descend rebellions and alleviate pain.

Prescription

柴胡	*chái hú*	15g	Radix Bupleuri
半夏	*bàn xià*	9g	Rhizoma Pinelliae
黄芩	*huáng qín*	9g	Radix Scutellariae
大黄	*dà huáng*	6g	Radix et Rhizoma Rhei
金银花	*jīn yín huā*	24g	Flos Lonicerae Japonicae
杭白芍	*háng bái sháo*	20g	Radix Paeoniae Lactiflorae
郁金	*yù jīn*	12g	Radix Curcumae

枳实	*zhǐ shí*	10g	Fructus Aurantii Immaturus
延胡索	*yán hú suǒ*	15g	Rhizoma Corydalis
甘草	*gān cǎo*	8g	Radix Glycyrrhizae

This was decocted with water and divided to be taken twice, one pack per day.

Second Visit

April 17th: After taking five packs of the above formula, his abdominal pain had reduced a lot. There was no nausea and vomiting after eating, but he still had the abdominal discomfort. Again, the original formula was used to treat him.

Third Visit

April 23rd: After taking another five packs, all of his symptoms disappeared.

Note

This case used abdominal pain and vomiting of the gastrointestinal tract as the representative syndrome of the *Chái Hú* Syndrome and the *Xiǎo Chái Hú Tāng* (Minor Bupleurum Decoction) was applied to treat it. As it belonged to a heat pathogen stagnating internally, therefore, the warm and tonifying herbs of *rén shēn* (Radix et Rhizoma Ginseng), *dà zǎo* (Fructus Jujubae) and *shēng jiāng* (Rhizoma Zingiberis Recens) were not appropriate and were eliminated and *jīn yín huā* (Flos Lonicerae Japonicae) and *dà huáng* (Radix et Rhizoma Rhei) were added to clear and sedate the pathogenic heat and *sháo yào* (Radix Paeoniae Rubra), *yù jīn* (Radix Curcumae) and *yán hú suǒ* (Rhizoma Corydalis) were added in to relieve stagnation, soothe the liver and alleviate pain. The formula and the syndrome were exactly matched; therefore, the treatments were effective.

Case 2

Vomiting during pregnancy.

Initial Visit

A female patient, 25 years old, June 4th, 1991.

The patient had amennorrhoea for three months, and nausea, vomiting and anorexia for a week and so came to visit. She complained that after being married for one year, the previous menstruations were normal. Now, she had amennorrhoea for three months already. Recently, she had nausea, vomiting, epigastric and abdominal distention and fullness, etc. She wanted to vomit at the sight of food. A urine test indicated she was pregnant. Her tongue was red and her tongue coating was white. Her pulse was wiry and fast. This was pregnancy severely obstructing the gallbladder, and liver heat stagnation with the foetal qi rebelling upward syndrome. The treatment plan was to harmonise the stomach, descend rebellions, clear heat and calm the foetus.

Prescription

柴胡	*chái hú*	12g	Radix Bupleuri
黄芩	*huáng qín*	12g	Radix Scutellariae
半夏	*bàn xià*	9g	Rhizoma Pinelliae

竹茹	zhú rú	10g	Caulis Bambusae in Taenia
白术	bái zhú	12g	Rhizoma Atractylodis Macrocephalae
太子参	tài zǐ shēn	10g	Radix Pseudostellariae
藿香	huò xiāng	9g	Herba Agastachis
生姜	shēng jiāng	3slices	Rhizoma Zingiberis Recens

This was decocted with water and divided to be taken often. She was instructed to slowly finish drinking it within half a day. If vomiting occurred after taking the decoction, clean water should be used to wash the mouth out, and to wait for a little bit before drinking again.

Second Visit

June 7th: After taking one pack of the above formula, her nausea and vomiting stopped. After taking three packs, her symptoms disappeared and she could eat. Again, three more packs of the same formula were used to complete the treatment and she recovered.

Note

This pregnancy severe obstruction case used nausea, vomiting and anorexia as the representative syndrome. It belonged to the Stomach and Intestines Syndrome of the *Chái Hú* Syndrome, therefore, the *Xiǎo Chái Hú Tāng* (Minor Bupleurum Decoction) was used to treat her and she recovered. As the vomiting patient does not like sweet flavours *gān cǎo* (Radix Glycyrrhizae) and *dà zǎo* (Fructus Jujubae) were eliminated, instead, *zhú rú* (Caulis Bambusae in Taenia), *bái zhú* (Rhizoma Atractylodis Macrocephalae) and *huò xiāng* (Herba Agastachis), etc. were added in to clear heat, calm the foetus, transform the turbid and descend the rebellion as herbs to increase the effectiveness of the descending rebellion, calming the foetus functions of the the *Xiǎo Chái Hú Tāng* (Minor Bupleurum Decoction).

Case 3

Meniere's disease.

Initial Visit

A male patient, 50 years old, March 12th, 1993.

The patient had a dizziness history for two years, there was an outbreak every time when he was tired or caught a cold. When he had an outbreak, his head was dizzy and he could not open his eyes. He felt the house was spinning, and it was accompanied by vomiting, nausea, and tinnitus, etc. This time due to an exterior invasion and fever, his dizziness happened again and so, he came to visit. He complained that he felt dizzy, he had ringing in his ears, nausea and vomiting, hearing loss, and could not open his eyes and move them. His pulse was wiry. His tongue coating was thin and white. Biomedicine diagnosed it as meniere's disease. The syndrome differentiation was heat stagnation in the liver channel of the foot *jueyin* and the gallbladder channel of the foot *shaoyang* meridians. *Shaoyang* governs wind and fire. If wind and fire flare upward, then he felt dizzy, with ringing in his ears. The pathogenic heat attacked his stomach, so he had vomiting. The treatment plan was to soothe the liver, clear the gallbladder, descend the rebellions and stop vomiting. A modified *Xiǎo Chái Hú Tāng* (Minor Bupleurum Decoction) was used to treat him.

Prescription

柴胡	*chái hú*	15g	Radix Bupleuri
黄芩	*huáng qín*	9g	Radix Scutellariae
半夏	*bàn xià*	12g	Rhizoma Pinelliae
茯苓	*fú líng*	30g	Poria
泽泻	*zé xiè*	15g	Rhizoma Alismatis
丹参	*dān shēn*	30g	Radix et Rhizoma Salviae Miltiorrhizae
板蓝根	*bǎn lán gēn*	24g	Radix Isatidis
甘草	*gān cǎo*	6g	Radix Glycyrrhizae
生姜	*shēng jiāng*	3slices	Rhizoma Zingiberis Recens

This was decocted with water and divided to be taken a couple of times, to drink it often, one pack per day.

Second Visit

March 15th: After taking three packs of the above formula, his dizziness was reduced and he had no vomiting. As it was effective, the prescription was not changed and he continued to take it.

Third Visit

March 20th: After taking another five packs, all of his symptoms disappeared. He was given three packs of the same formula to clear it up and he recovered later.

Note

This case used dizziness, vomiting, nausea, tinnitus, and hearing loss as the main symptoms. The 263rd item in the *Discussion on Cold Damage* states that: "*If the shaoyang is sick, there are bitter mouth, dry throat and dizziness.*" In the 264th item it states that: "*If shaoyang is attacked by wind, the two ears cannot hear......*" From analysis of the items in the *Discussion on Cold Damage*, aside from vomiting and nausea belonging to the gastrointestinal tract as the representative syndrome of the *Chái Hú* Syndrome, dizziness and tinnitus are also an expression of the *shaoyang* syndrome, therefore, this case adopted the *Xiǎo Chái Hú Tāng* (Minor Bupleurum Decoction). This syndrome is the gallbladder heat flaring upward inducing dizziness and tinnitus. It belongs to excess syndromes, hence, *rén shēn* (Radix et Rhizoma Ginseng) and *dà zǎo* (Fructus Jujubae) were eliminated, as they are warm and tonifying herbs, to prevent it helping the heat to raise the fire up. Modern medical science believes that Meniere's disease is related to the endolymph collecting in the inner ear; therefore, *fú líng* (Poria) and *zé xiè* (Rhizoma Alismatis) were added in to strengthen the spleen, leach out the water and promote urination in order to benefit the elimination of the inner ear's oedema. *Dān shēn* (Radix et Rhizoma

Salviae Miltiorrhizae) was added in to invigorate the blood and transform blood stasis, improve micro-circulation, and eliminate spasm of the small arteries in the labyrinth. *Bǎn lán gēn* (Radix Isatidis) was added as it is bitter, cold and can clear heat, and is anti-viral to reduce viral damage to the inner ear. In addition, the *Xiǎo Chái Hú Tāng* (Minor Bupleurum Decoction) also has an anti-viral and regulating function on the autonomic nerves, therefore, they are effective to treat meniere's disease.

Case 4

Bile reflux gastritis.

Initial Visit

A male patient, 40 years old, April 15th, 1987.

The patient had epigastric distention, fullness and pain for more than two months. When he had pain, he took medicines like aluminum hydroxide, etc. without effect and so he came to visit. The patient's chief complaint: His epigastric pain, distention and fullness, accompanied by a burning sensation, a dry and bitter mouth with a want to vomit, anorexia, and his epigastric area had pressing tenderness. His GI tract endoscope report showed: His gastric mucosa had a severe mottled pattern, hyperaemia, the pylorus was opening continually and yellowish fluid was visible. He was diagnosed with bile reflux gastritis. His pulse was wiry. His tongue was dark with a white and turbid coating. This was the syndrome of liver stagnation and qi obstruction, with the stomach having lost its harmony and its descending function. The treatment plan was to soothe the liver, benefit the gallbladder, harmonise the stomach and descend the rebellion.

Prescription

柴胡	*chái hú*	15g	Radix Bupleuri
黄芩	*huáng qín*	10g	Radix Scutellariae
半夏	*bàn xià*	9g	Rhizoma Pinelliae
丹参	*dān shēn*	20g	Radix et Rhizoma Salviae Miltiorrhizae
百合	*bǎi hé*	20g	Bulbus Lilii
枳壳	*zhǐ qiào*	12g	Fructus Aurantii
白术	*bái zhú*	12g	Rhizoma Atractylodis Macrocephalae
砂仁	*shā rén*	8g	Fructus Amomi
蒲公英	*pú gōng yīng*	12g	Herba Taraxaci
乌药	*wū yào*	12g	Radix Linderae
藿香	*huò xiāng*	9g	Herba Agastachis
紫苏梗	*zǐ sū gěng*	9g	Caulis Perillae

This was decocted with water and divided to be taken twice, one pack per day.

Second Visit

April 22nd: After taking seven packs of the above formula, all of his symptoms were reduced but he still had anorexia and a turbid tongue coating. Based on the above formula, *jī nèi jīn* (Endothelium Corneum Gigeriae Galli) 12g, was added and he continued to take it.

Third Visit

April 29th: After taking another seven packs of the above formula, all of his symptoms were reduced a lot. His epigastric pain and burning sensation basically disappeared. His appetite increased. The same formula was used to treat him again. This patient took a total of more than thirty packs of the formula and he recovered.

Note

This case focused on the stomachache, anorexia and nausea, etc., that were mainly the Stomach and Intestines Syndrome, to apply the *Chái Hú Tāng*. It is effective to use the *Chái Hú Tāng* as it has functions to soothe the liver, resolve stagnation, benefit the gallbladder, harmonise the stomach and descend rebellions. The patient had stomach and abdominal fullness and distention, nausea, accompanied by a burning sensation, and stomachache. It was considered improper to use *rén shēn* (Radix et Rhizoma Ginseng), *shēng jiāng* (Rhizoma Zingiberis Recens) and *dà zǎo* (Fructus Jujubae) that are sweet and warm herbs in the *Chái Hú Tāng* (Minor Bupleurum Decoction) as they could add more heat and increase the fullness. Instead, *dān shēn* (Radix et Rhizoma Salviae Miltiorrhizae), *pú gōng yīng* (Herba Taraxaci) were added to transform blood stasis, benefit the gallbladder and alleviate pain as well as *bǎi hé* (Bulbus Lilii), *wū yào* (Radix Linderae), *shā rén* (Fructus Amomi), *bái zhú* (Rhizoma Atractylodis Macrocephalae) and *zhǐ qiào* (Fructus Aurantii), etc. that can harmonise the stomach, regulate qi and resolve stagnation in order to reinforce the power of the *Chái Hú Tāng* (Minor Bupleurum Decoction) in soothing the liver, benefitting the gallbladder, harmonising the stomach and descending rebellions.

Closing Words

The *Xiǎo Chái Hú Tāng* (Minor Bupleurum Decoction) is a commonly used formula in clinical practice. Before learning the application of the *Xiǎo Chái Hú Tāng* (Minor Bupleurum Decoction), one first needs to study the *Discussion on Cold Damage* about the *Chái Hú* Syndrome discussions and the representative syndromes, following syndromes and its modifications. In doing so, one can be better and more accurate in applying the *Chái Hú Tāng* (Minor Bupleurum Decoction) and can expand its application scope. The author, in his applications of *Xiǎo Chái Hú Tāng* (Minor Bupleurum Decoction), based on Zhang Zhong-jing's: *"Not all syndromes are needed one is enough"* rule, and following the three representative syndromes' differentiation method, listed case studies on the *Chái Hú Tāng's* (Minor Bupleurum Decoction) application experience for reference.

Index by Chinese Medicinals and Formulas

Notes

Notes

Notes

图书在版编目（CIP）数据

中药配伍应用心得（英文）/ 何秀川编著 . —北京：人民
卫生出版社，2008.4
ISBN 978-7-117-09208-1

Ⅰ.中…　Ⅱ.何…　Ⅲ.中药配伍—英文　Ⅳ.R289.1

中国版本图书馆 CIP 数据核字（2007）第 137438 号

中药配伍应用心得（英文）

_____

编　　著：何秀川
出版发行：人民卫生出版社（中继线 +8610-6761-6688）
地　　址：中国北京市丰台区方庄芳群园三区 3 号楼
邮　　编：100078
网　　址：http://www.pmph.com
E - mail：pmph @ pmph.com
发　　行：zzg@pmph.com.cn
购书热线：+8610-6769-1034（电话及传真）
开　　本：787×1092　1/16
版　　次：2008 年 4 月第 1 版　　2008 年 4 月第 1 版第 1 次印刷
标准书号：ISBN 978-7-117-09208-1/R · 9209